MICROELECTRONICS

A **SCIENTIFIC AMERICAN** *Book*

MICROELECTRONICS

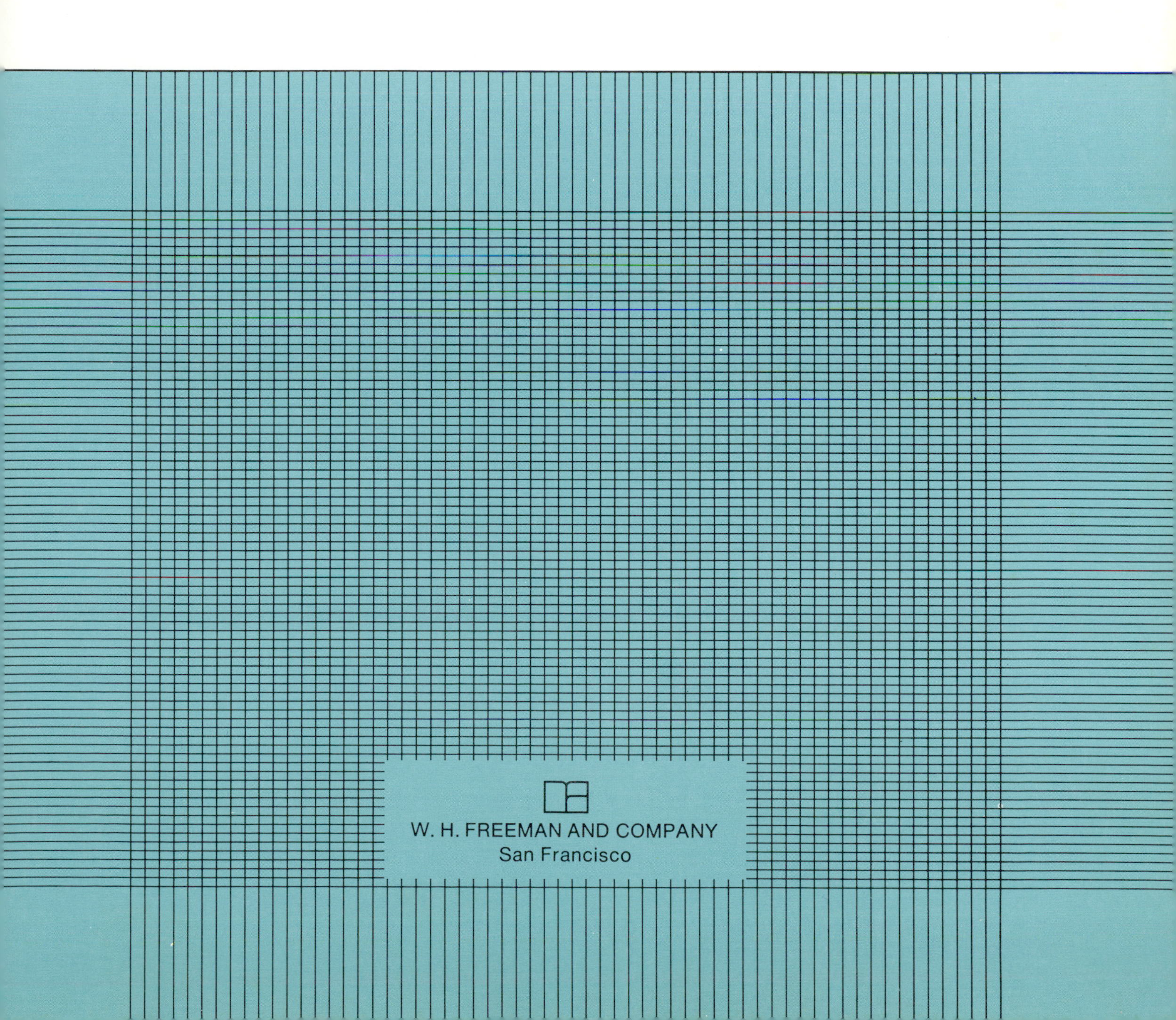

The Cover

The photograph on the cover symbolizes the theme of this issue of SCIENTIFIC AMERICAN: microelectronics, the art of putting complex electronic circuits on "chips" of silicon roughly a quarter of an inch square. The blue area in the center of the photograph is a single chip, a high-speed current-mode logic integrated circuit made by Texas Instruments Incorporated for a Honeywell Information Systems computer. Arrayed around the chip are 40 leads that connect it with its external environment. The leads are made of tin-plated copper. They are connected by an automatic process in which they are bonded by heat and compression to gold bumps plated at edge of the chip. The photograph was made by Fritz Goro in the Laboratory for Applied Microscopy of E. Leitz, Inc.

Library of Congress Cataloging in Publication Data

Main entry under title:

Microelectronics.

"A Scientific American book."
"The chapters in this book were first published in the September 1977 issue of Scientific American."
Bibliography: p.
Includes index.
1. Microelectronics—Addresses, essays, lectures.
2. Microcomputers—Addresses, essays, lectures.
I. Scientific American.
TK7874.M479   621.381'7   77-13955
ISBN 0-7167-0067-0
ISBN 0-7167-0066-2 pbk.

Copyright© 1977 by Scientific American, Inc.
All rights reserved. No part of this book may be reproduced by any mechanical, photographic, or electronic process, or in the form of a phonographic recording, nor may it be stored in a retrieval system, transmitted, or otherwise copied for public or private use, without written permission from the publisher.

The eleven chapters in this book originally appeared as articles in the September 1977 issue of *Scientific American*.

Printed in the United States of America

9 8 7 6 5 4 3 2 1

CONTENTS

FOREWORD

This book tells of the virtual disappearance of the computer. Within 25 years, this glamorous piece of hardware has come and gone! *Computer* was, in any case, a poverty-stricken name for the universal machine that was first described in the 1930s by the young British mathematician A. M. Turing and given its definitive mathematical statement by John von Neumann in the 1940s. These men showed that it was possible for us to build a machine that would do anything we could think of telling it to do. The realization of this possibility in practice has, until now, been encumbered by hardware. Now, microelectronics is eliminating the hardware, reducing to the ultimate minimum the physical barriers between the thinking and the doing.

Present microelectronics technology can pack 10,000 computer circuit elements (the capacity of a big central processor) into a chip of semiconductor material measuring a centimeter or so on a side. The memory circuits, formerly contained in a separate magnetic-core unit, are being inscribed on the same chip alongside the processor or incorporated into its circuitry. With electron beams and x-rays to write and print the circuits, computer technologists are learning to pack them even tighter—more than 20,000 circuits per square centimeter. The computer has, in effect, imploded into the crystal lattice of a chip of semiconductor.

The writing of the computer program that lays out the circuit elements on the chip is the most costly process in the production of one of these tiny computers. It is cheaper thereafter to reproduce the program on a chip than on paper by high-speed printer. Thus, along with the distinction between central processor and memory, the distinction between software and hardware is disappearing.

Finally, the computer itself is disappearing into the devices and machines it operates or that operate it. Thus, the typewriter that once typed messages into and out of the computer now incorporates the computer and becomes the intelligent terminal, proliferating into office jobs throughout the economy. The scientific instrument that formerly reported its readings to the computer now becomes an intelligent oscilloscope or spectrograph, or whatever, and interprets its own readings. The on-line process controller now makes the decision as well as the reading, and raises the temperature, squeezes down the press, or adjusts the cutting blade to a ten-thousandth.

The hugest single impact of microelectronics will be felt in the storage, retrieval, processing, communication and display of information. At the storage end, the cost per bit of computer memory has been declining on a steep slope of 30 percent per year. A writer may conjure, for example, with the prospect of a dictionary incorporated into the intelligent typewriter—a typewriter that talks back. At the display end, it is well known that a picture is worth a ton of computer print-out. Graphic display is the only way to make sense of much of the product of the enormous information-crunching power of the computer. The generation of such displays has been found to require ten times the circuitry of the computer that drives it. Now the cathode-ray-tube display unit will contain not only its own circuitry but the computer as well. The designer, engineer and architect will visualize their ideas as they think them. With the country wired with high-capacity, two-way communication channels, also made possible by microelectronics (plus the laser and fiber optics), the scholar (or perhaps even the high-school student doing homework) will one day have the Library of Congress on line, with all its wealth of graphic as well as textual information.

The chapters in this book were first published in the September 1977 issue of SCIENTIFIC AMERICAN, which was the twenty-eighth in the series of single-topic issues published annually by the magazine. The editors herewith express appreciation to their colleagues at W. H. Freeman and Company, the book-publishing affiliate of SCIENTIFIC AMERICAN, for the enterprise that has made the contents of this issue so speedily available in book form.

THE EDITORS*

September, 1977

*BOARD OF EDITORS: Gerard Piel (Publisher), Dennis Flanagan (Editor), Francis Bello (Associate Editor), Philip Morrison (Book Editor), Trudy E. Bell, Brian P. Hayes, Jonathan B. Piel, John Purcell, James T. Rogers, Armand Schwab, Jr., Jonathan B. Tucker, Joseph Wisnovsky

I

MICROELECTRONICS

Microelectronics

by ROBERT N. NOYCE

Introducing a volume on the microelectronic revolution, in which putting large numbers of electronic elements on silicon "chips" has profoundly increased the capabilities of electronic devices

The evolution of electronic technology over the past decade has been so rapid that it is sometimes called a revolution. Is this large claim justified? I believe the answer is yes. It is true that what we have seen has been to some extent a steady quantitative evolution: smaller and smaller electronic components performing increasingly complex electronic functions at ever higher speeds and at ever lower cost. And yet there has also been a true revolution: a qualitative change in technology, the integrated microelectronic circuit, has given rise to a qualitative change in human capabilities.

It is not an exaggeration to say that most of the technological achievements of the past decade have depended on microelectronics. Small and reliable sensing and control devices are the essential elements in the complex systems that have landed men on the moon and explored Mars, not to speak of their similar role in the intercontinental weapons that dominate world politics. Microelectronic devices are also the essence of new products ranging from communications satellites to hand-held calculators and digital watches. Somewhat subtler, but perhaps eventually more significant, is the effect of microelectronics on the computer. The capacity of the computer for storing, processing and displaying information has been greatly enhanced. Moreover, for many purposes the computer is being dispersed to the sites where it is operated or where its output is applied: to the "smart" typewriter or instrument or industrial control device.

The microelectronics revolution is far from having run its course. We are still learning how to exploit the potential of the integrated circuit by developing new theories and designing new circuits whose performance may yet be improved by another order of magnitude. And we are only slowly perceiving the intellectual and social implications of the personal computer, which will give the individual access to vast stores of information and the ability to learn from it, add to it and communicate with others concerning it.

This *Scientific American* book is devoted to microelectronics and explores the nature of microelectronic circuit elements, the design and fabrication of large-scale integrated circuits, a broad range of their applications and some of their implications for the future. Here I want primarily to show how the evolution of microelectronics illustrates the constant interaction of technology and economics. The small size of microelectronic devices has been important in many applications, but the major impact of this new technology has been to make electronic functions more reproducible, more reliable and much less expensive. With each technical development costs have decreased, and the ever lower costs have promoted a widening range of applications; the quest for technical advances has been required by economic competition and compensated by economic reward.

It all began with the development 30 years ago of the transistor: a small, low-power amplifier that replaced the large, power-hungry vacuum tube. The advent almost simultaneously of the stored-program digital computer provided a large potential market for the transistor. The synergy between a new component and a new application generated an explosive growth of both. The computer was the ideal market for the transistor and for the solid-state integrated circuits the transistor spawned, a much larger market than could have been provided by the traditional applications of electronics in communications. The reason is that digital systems require very large numbers of active circuits compared with systems having analogue amplification, such as radios. In digital electronics a given element is either on or off, depending on the input. Even when a large number of elements are connected, their output will still be simply on or off; the gain of the individual stage is unity, so that even cascading several stages leaves the gain still unity. Analogue circuits, on the other hand, typically require amplification of the input. Since the gain of each amplifier may typically be 10, only a few stages can be cascaded before the practical limit of voltage levels for microelectronic elements is reached. An analogue system therefore cannot handle large numbers of microcircuits, whereas a digital system requires them; a pocket calculator contains 100 times as many

MICROELECTRONIC DEVICES arrayed on the opposite page are microprocessors, each the equivalent of the central processing unit of a small computer. The photograph shows a portion of a "wafer," a thin slice of silicon on which the devices have been fabricated, enlarged 10 diameters. An individual device, or chip, is shown at its actual size (.164 inch by .222 inch) above this caption. After the devices have been fabricated on the wafer and tested they are separated, fitted with leads and packaged. This microprocessor is the Intel Corporation's 8085, a general-purpose device that has 6,200 transistors and can execute 770,000 instructions per second. Smaller chip at the far right is used for testing wafer at various stages of its manufacture.

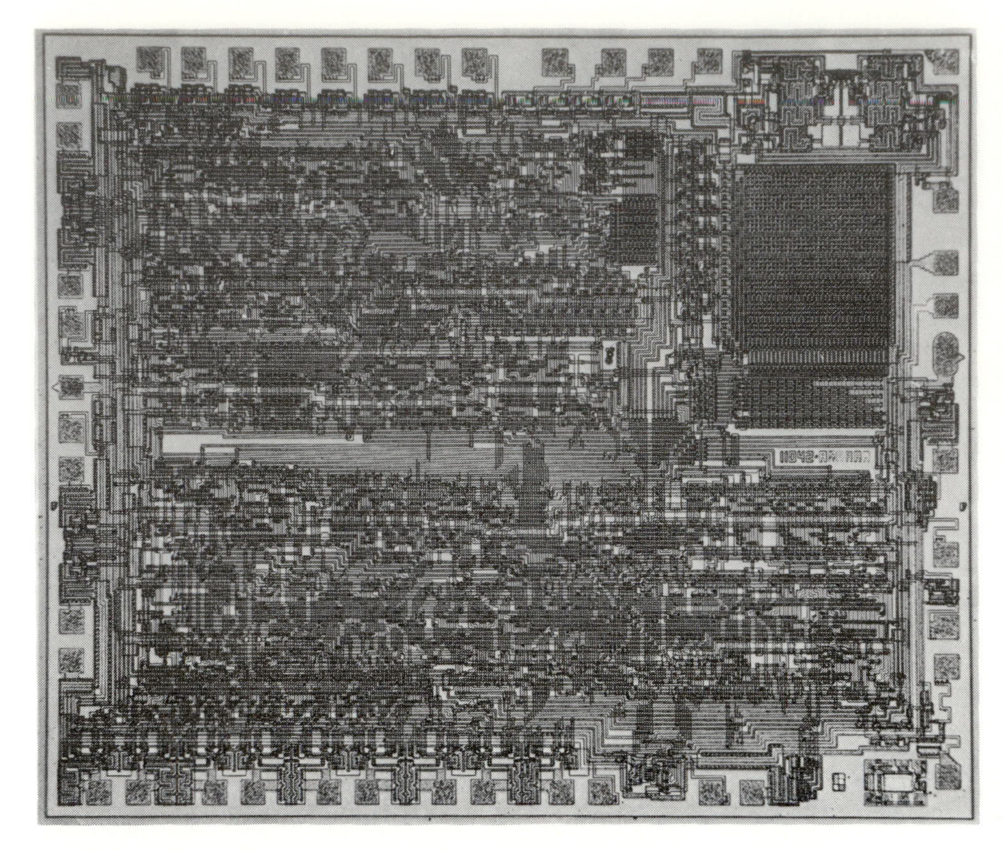

FOUR DIVERSE CHIPS suggest the wide range of purposes served by microelectronic devices and the patterns characteristic of those purposes. At the left is a large-scale integrated circuit: a microprocessor made by the Rockwell International Corporation. In addition to the logic elements of a central processing unit it incorporates input-output circuitry and a read-only memory (*regular area at upper right*). Second from the left is another large-scale integrated circuit: a semiconductor memory made by Fairchild Semiconductor. Its reg-

transistors as a radio or a television receiver.

In spite of the inherent compatibility of microelectronics and the computer, the historical fact is that early efforts to miniaturize electronic components were not motivated by computer engineers. Indeed, the tremendous potential of the digital computer was not quickly appreciated; even the developers of the first computer felt that four computers, more or less, would satisfy the world's computation needs! Various missile and satellite programs, however, called for complex electronic systems to be installed in equipment in which size, weight and power requirements were severely constrained, and so the effort to miniaturize was promoted by military and space agencies.

The initial approach was an attempt to miniaturize conventional components. One program was "Project Tinkertoy" of the National Bureau of Standards, whose object was to package the various electronic components in a standard shape: a rectangular form that could be closely packed rather than the traditional cylindrical form. Another approach was "molecular engineering." The example of the transistor as a substitute for the vacuum tube suggested that similar substitutes could be devised: that new materials could be discovered or developed that would by their solid-state nature allow electronic functions other than amplification to be performed within a monolithic solid. These attempts were largely unsuccessful, but they publicized the demand for miniaturization and the potential rewards for the successful development of some form of microelectronics. A large segment of the technical community was on the lookout for a solution of the problem because it was clear that a ready market awaited the successful inventor.

What ultimately provided the solution was the semiconductor integrated circuit, the concept of which had begun to take shape only a few years after the invention of the transistor. Several investigators saw that one might further exploit the characteristics of semiconductors such as germanium and silicon that had been exploited to make the transistor. The body resistance of the semiconductor itself and the capacitance of the junctions between the positive (*p*) and negative (*n*) regions that could be created in it could be combined with transistors in the same material to realize a complete circuit of resistors, capacitors and amplifiers [see "Microelectronic Circuit Elements," by James D. Meindl, page 12]. In 1953 Harwick Johnson of the Radio Corporation of America applied for a patent on a phase-shift oscillator fashioned in a single piece of germanium by such a technique. The concept was extended by G. W. A. Dummer of the Royal Radar Establishment in England. Jack S. Kilby of Texas Instruments Incorporated and Jay W. Lathrop of the Diamond Ordnance Fuze Laboratories.

Several key developments were required, however, before the exciting potential of integrated circuits could be realized. In the mid-1950's engineers learned how to define the surface configuration of transistors by means of photolithography and developed the method of solid-state diffusion for introducing the impurities that create *p* and *n* regions. Batch processing of many transistors on a thin "wafer" sliced from a large crystal of germanium or silicon began to displace the earlier technique of processing individual transistors. The hundreds or thousands of precisely registered transistors that could be fabricated on a single wafer still had to be separated physically, assembled individually with tiny wires inside a protective housing and subsequently assembled into electronic circuits.

The integrated circuit, as we conceived and developed it at Fairchild Semiconductor in 1959, accomplishes

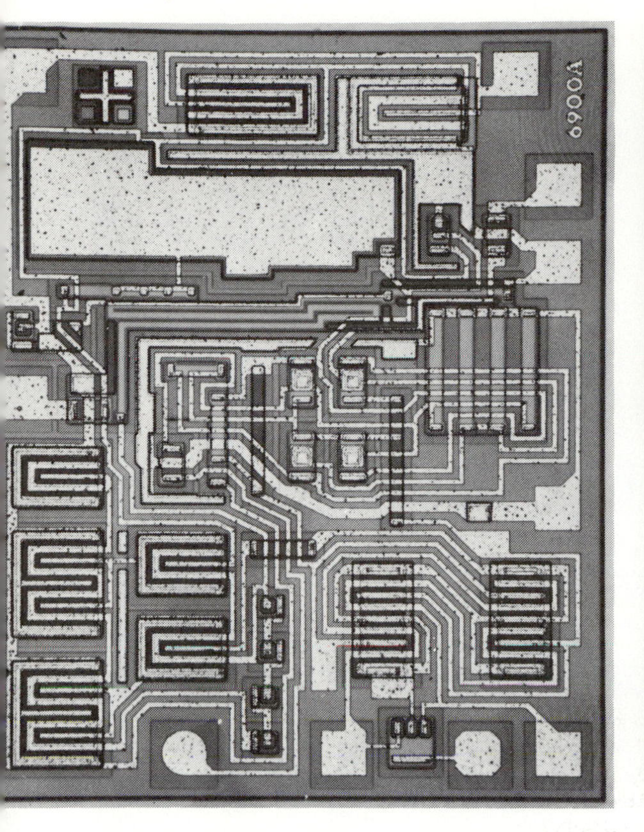

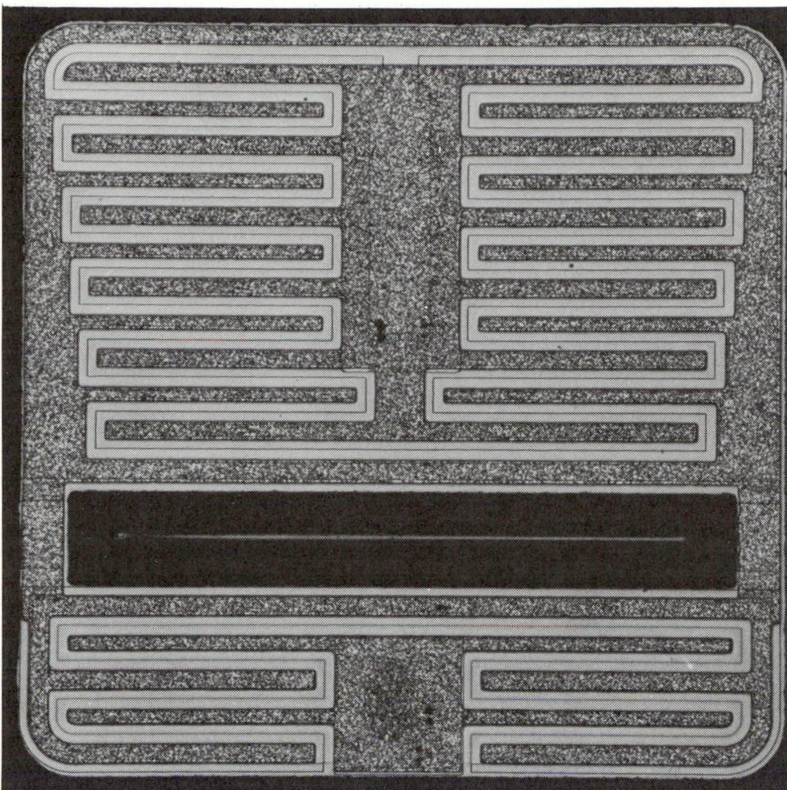

ular pattern is characteristic of memory devices. Third from the left is a small-scale integrated circuit: an "op amp," or operational amplifier, made by the RCA Corporation. It is a variable amplifier whose output is linked to its input so that it can serve as an adaptable building block in various feedback applications. At the far right is a power device: a high-gain "Darlington transistor" made by Motorola Semiconductor Products Inc. In it two transistors provide enough power to drive the horizontal-deflection circuitry of a television receiver.

the separation and interconnection of transistors and other circuit elements electrically rather than physically. The separation is accomplished by introducing *pn* diodes, or rectifiers, which allow current to flow in only one direction. The technique was patented by Kurt Lehovec at the Sprague Electric Company. The circuit elements are interconnected by a conducting film of evaporated metal that is photoengraved to leave the appropriate pattern of connections. An insulating layer is required to separate the underlying semiconductor from the metal film except where contact is desired. The process that accomplishes this insulation had been developed by Jean Hoerni at Fairchild in 1958, when he invented the planar transistor: a thin layer of silicon dioxide, one of the best insulators known, is formed on the surface of the wafer after the wafer has been processed and before the conducting metal is evaporated onto it.

Since then additional techniques have been devised that give the designer of integrated circuits more flexibility, but the basic methods were available by 1960, and the era of the integrated circuit was inaugurated. Progress since then has been astonishing, even to those of us who have been intimately engaged in the evolving technology. An individual integrated circuit on a chip perhaps a quarter of an inch square now can embrace more electronic elements than the most complex piece of electronic equipment that could be built in 1950. Today's microcomputer, at a cost of perhaps $300, has more computing capacity than the first large electronic computer, ENIAC. It is 20 times faster, has a larger memory, is thousands of times more reliable, consumes the power of a light bulb rather than that of a locomotive, occupies 1/30,000 the volume and costs 1/10,000 as much. It is available by mail order or at your local hobby shop.

In 1964, noting that since the production of the planar transistor in 1959 the number of elements in advanced integrated circuits had been doubling every year, Gordon E. Moore, who was then director of research at Fairchild, was the first to predict the future progress of the integrated circuit. He suggested that its complexity would continue to double every year. Today, with circuits containing 2^{18} (262,144) elements available, we have not yet seen any significant departure from Moore's law. Nor are there any signs that the process is slowing down, although a deviation from exponential growth is ultimately inevitable. The technology is still far from the fundamental limits imposed by the laws of physics; further miniaturization is less likely to be limited by the laws of physics than by the laws of economics.

The growth of the microelectronics industry illustrates the extent to which investment in research can create entrepreneurial opportunity, jobs and a major export market for the U.S. After the introduction of the integrated circuit in the early 1960's the total world consumption of integrated circuits rose rapidly, reaching a value of nearly $1 billion in 1970. By 1976 world consumption had more than tripled, to $3.5 billion. Of this total U.S.-based companies produced more than $2.5 billion, or some 70 percent, about $1 billion of which was exported to foreign customers. The impact on the electronics industry is far greater than is implied by these figures. In electronic equipment less than 10 percent of the value is in the integrated circuits themselves; a $10,000 minicomputer contains less than $1,000 worth of integrated circuits, and a $300 television set contains less than $30 worth. Today most of the world's $80 billion electronics industry depends in some way on integrated circuits.

The substitution of microelectronic devices for discrete components reduces costs not only because the devices themselves are cheaper but for a variety of

MICROCOMPUTER, THE IMSAI 8048, is made by the IMSAI Manufacturing Corporation, which assembles integrated circuits with other devices on an eight-by-10-inch board. The microprocessor is the square chip in the center of the light gray package about a third of the way from the bottom. The 16 packages at bottom right constitute a 2K-byte (16,384-bit, or binary-digit) random-access memory; the light-colored package next to them provides programmable 2K-byte read-only memory in addition to the 1K-byte program memory in the microprocessor. Nine light-emitting diodes (*to left of the keyboard*) provide an alphanumeric display. This is a control computer that can, for example, monitor an operation and throw switches in order to control it by means of high-current relays (*large black boxes*).

other reasons. First, the integrated circuit contains many of the interconnections that were previously required, and that saves labor and materials. The interconnections of the integrated circuit are much more reliable than solder joints or connectors, which makes for savings in maintenance. Since integrated circuits are much smaller and consume much less power than the components they have displaced, they make savings possible in such support structures as cabinets and racks as well as in power transformers and cooling fans. Less intermediate testing is needed in the course of production because the correct functioning of the complex integrated circuits has already been ensured. Finally, the end user needs to provide less floor space, less operating power and less air conditioning for the equipment. All of this is by way of saying that even if integrated circuits were only equivalent in cost to the components they have displaced, other savings would motivate the use of fewer, more complex integrated circuits as they became available.

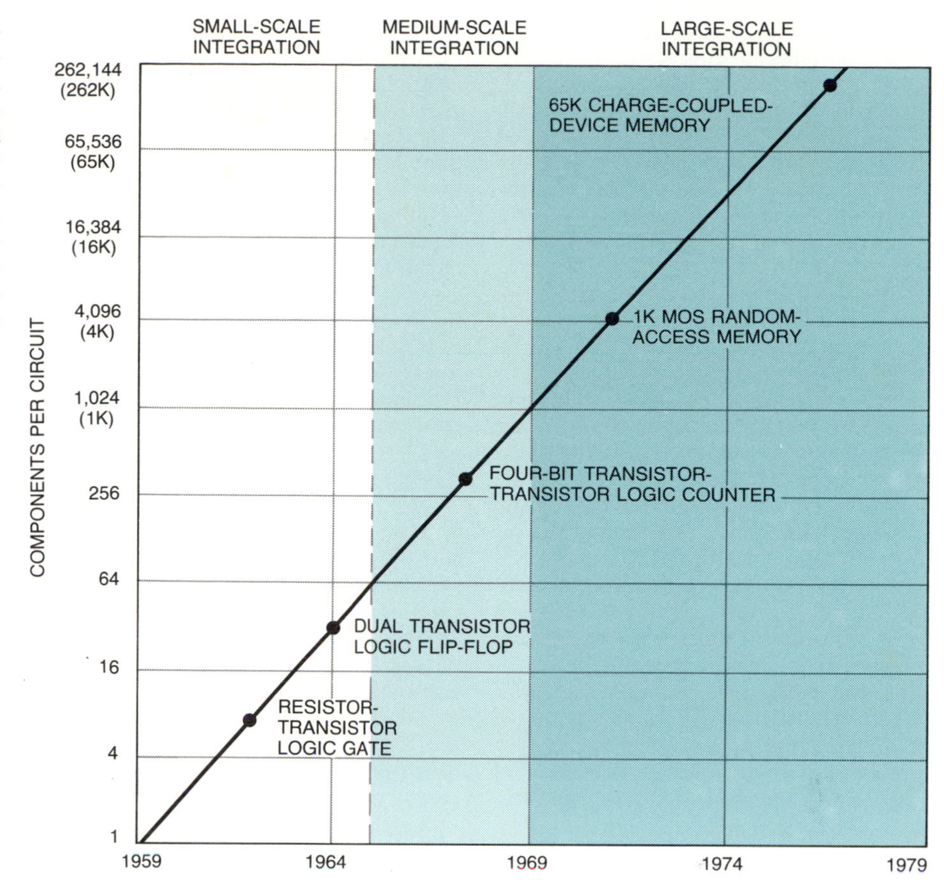

NUMBER OF COMPONENTS per circuit in the most advanced integrated circuits has doubled every year since 1959, when the planar transistor was developed. Gordon E. Moore, then at Fairchild Semiconductor, noted the trend in 1964 and predicted that it would continue.

The most striking characteristic of the microelectronics industry has been a persistent and rapid decline in the cost of a given electronic function. The hand-held calculator provides a dramatic example. Its cost has declined by a factor of 100 in the past decade. A portion of the rapid decline in cost can be accounted for in terms of a "learning curve": the more experience an industry has, the more efficient it becomes. Most industries reduce their costs (in constant dollars) by 20 to 30 percent each time their cumulative output doubles. Examining data for the semiconductor industry, we find that integrated-circuit costs have declined 28 percent with each doubling of the industry's experience. Because of the rapid growth of this young industry these cost reductions have come at a much more rapid pace than in mature industries; the electronics industry's experience has been doubling nearly every year. The cost of a given electronic function has been declining even more rapidly than the cost of integrated circuits, since the complexity of the circuits has been increasing as their price has decreased. For example, the cost per bit (binary digit) of random-access memory has declined an average of 35 percent per year since 1970, when the major growth in the adoption of semiconductor memory elements got under way. These cost declines were accomplished not only by the traditional learning process but also by the integration of more bits into each integrated circuit: in 1970 a change was made from 256 bits to 1,024 bits per circuit and now the number of bits is in the process of jumping from 4,096 per circuit to 16,384.

The hundredfold decline in prices for electronic components since the development of the integrated circuit is unique because, although other industries have shown similar experience curves, the integrated-circuit industry has been unique in its annual doubling of output over an extended number of years. Rather than serving a market that grows only in pace with the gross national product or the population, the industry has served a proliferating market of ever broadening applications. As each new application consumes more microelectronic devices more experience has been gained, leading to further cost reductions, which in turn have opened up even wider markets for the devices. In 1960, before any production of integrated circuits, about 500 million transistors were made. Assuming that each transistor represents one circuit function, which can be equated to a logic "gate" or to one bit of memory in an integrated circuit, annual usage has increased by 2,000 times, or has doubled 11 times, in the past 17 years. This stunning increase promotes continual cost reductions.

The primary means of cost reduction has been the development of increasingly complex circuits that lower the cost per function for both the circuit producer and the equipment manufacturer. The main technical barrier to achieving more functions per circuit is production yield. More complex circuits result in larger devices and a growing probability of defects, so that a higher percentage of the total number of devices must be scrapped. When the cost of scrapping exceeds the cost saving in subsequent assembly and test operations, the cost per function increases rather than decreases. The most cost-efficient design is a compromise between high assembly costs (which are incurred at low levels of integration) and high scrapping costs (which are incurred at high levels of integration).

Technological developments have concentrated primarily on increasing the production "yield," either by reducing the density of defects or by reducing dimensions. Meticulous attention to process control and cleanliness has been necessary to reduce defect density. A dust particle in any critical process is enough to make a device worthless, so that most operations must be carried out in "clean rooms." Reduction of the dimensions of the basic circuit elements, which enables one to crowd more complex circuits within a given area, has been accomplished by improving the resolution of the photoengraving processes. Now optical limits are being reached as dimensions in the circuit patterns enter the range of only a few wavelengths of light, and methods in which electron beams or X rays are substituted for visible light are being developed in order to reduce the dimensions even further [see "The Fabrication of Microelectronic Circuits," by William G. Oldham, page 40].

The reduction in size of the circuit elements not only reduces the cost but also improves the basic performance of the device. Delay times are directly proportional to the dimensions of circuit elements, so that the circuit becomes faster as it becomes smaller. Similarly, the power is reduced with the area of the circuits. The linear dimensions of the circuit elements can probably be reduced to about a fifth of the current size before any fundamental limits are encountered.

In an industry whose product declines in price by 25 percent a year the motivation for doing research and development is clearly high. A year's advantage in introducing a new product or new process can give a company a 25 percent cost advantage over competing companies; conversely, a year's lag puts a company at a significant disadvantage with respect to its competitors. Product development is a critical part of company strategy and product obsolescence is a fact of life. The return on successful investment in research and development is great, and so is the penalty for failure. The leading producers of integrated circuits spend approximately 10 percent of their sales income on research and development. In a constant-price environment one could say that investment for research and development buys an annuity paying $2.50 per year for each dollar invested! Clearly most of this annuity is either paid out to the purchasers of integrated circuits or reflected in price reductions that are necessary to develop new markets.

In this environment of rapid growth in market, rapid technological change and high returns on the successful development of a new product or process, a great number of entrepreneurial opportunities have been created and exploited. It is interesting that whereas the U.S. has led in both the development and the commercialization of the new technology, it was not the companies that were in the forefront of the vacuum-tube business that proceeded to develop its successor, the transistor. Of the 10 leading U.S. producers of vacuum tubes in 1955, only two are among today's top 10 U.S. semiconductor producers; four of the top 10 semiconductor companies were formed after 1955, and those four represent only a small fraction of the successful new ventures in the field. Time and time again the rapid growth of the market has found existing companies too busy expanding markets or product lines to which they were already committed to explore some of the more speculative new markets or technologies. And so the door was left open for new ventures, typically headed (originally at least) by an entrepreneur with a research or marketing background who had enough faith in the new market and technology to gamble. Fortunately for the U.S. economy capital was readily available in the late 1950's and the 1960's to finance these ventures, and approximately 100 new companies were formed to produce semiconductor devices; many of them made significant contributions to the development of microelectronics. Two such contributions in which I was directly involved were the development of the planar transistor and planar integrated circuit at Fairchild when that organization was only two years old, and the development of the microprocessor at the Intel Corporation only two years after that company was founded. There are many more examples. The environment for entrepreneurial innovation in the U.S. is not matched in other industrialized nations,

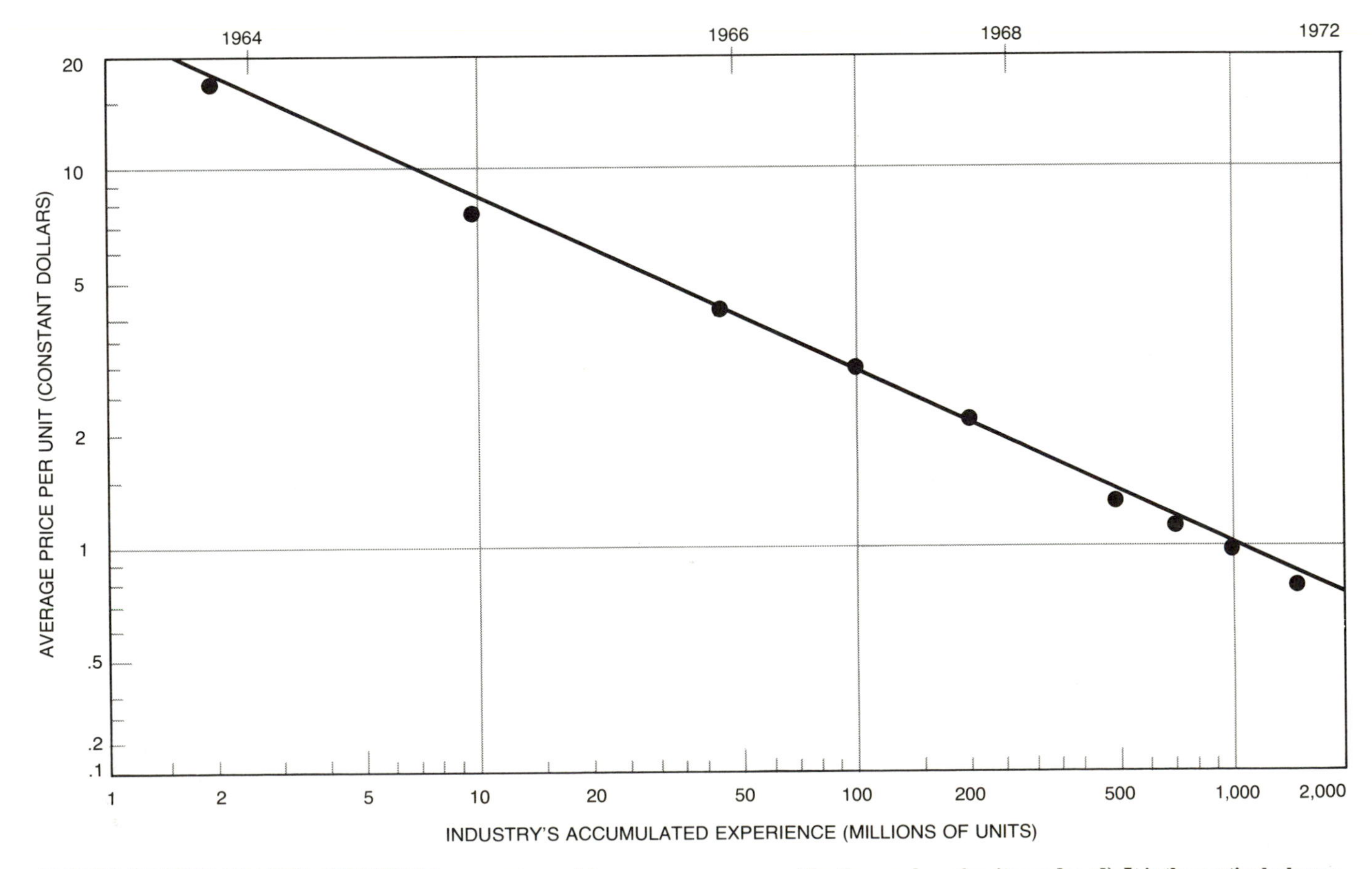

PRICES OF INTEGRATED CIRCUITS have conformed to an experience curve common to many industries, declining about 28 percent with each doubling of the industry's cumulative experience (as measured by the number of units produced). It is the particularly rapid growth of the microelectronics industry that has made the rate of decline in prices appear to be higher than the rate in other industries.

and it has been a major contributor to America's leadership in the field.

The growth of microelectronics has in turn created other opportunities. A host of companies have been established to serve the needs of the integrated-circuit producers. These companies supply everything from single-crystal silicon to computer-controlled design aids to automatic test equipment and special tooling. Often the novel consumer products that are spawned by developments in microelectronics have been manufactured and marketed initially by new companies. The digital watch and the television game are familiar examples.

When the integrated circuit was still an infant, Patrick E. Haggerty of Texas Instruments called attention to the increasing pervasiveness of electronics and predicted that electronic techniques would continue to displace other modes of control, reaching into nearly all aspects of our lives. Just such a displacement has been taking place, primarily because the microelectronics industry has been able to make ever more sophisticated functional elements at ever decreasing costs. Mechanical elements of the calculator and the watch have been displaced by integrated circuits that are less expensive and also offer more flexibility. Now the electromechanical functions of vending machines, pinball machines and traffic signals are being displaced. In the near future the automobile engine will be controlled by a computer, with a consequent improvement in efficiency and reduction of pollutants. All these applications are simply extensions of the traditional applications of electronics to the task of handling information in measurement, communication and data manipulation. It has often been said that just as the Industrial Revolution enabled man to apply and control greater physical power than his own muscle could provide, so electronics has extended his intellectual power. Microelectronics extends that power still further.

By 1986 the number of electronic functions incorporated into a wide range of products each year can be expected to be 100 times greater than it is today. The experience curve predicts that the cost per function will have declined by then to a twentieth of the 1976 cost, a reduction of 25 percent per year. At such prices electronic devices will be exploited even more widely, augmenting mail service, expanding the library and making its contents more accessible, providing entertainment, disseminating knowledge for educational purposes and performing many more of the routine tasks in the home and office. It is in the exponential proliferation of products and services dependent on microelectronics that the real microelectronic revolution will be manifested.

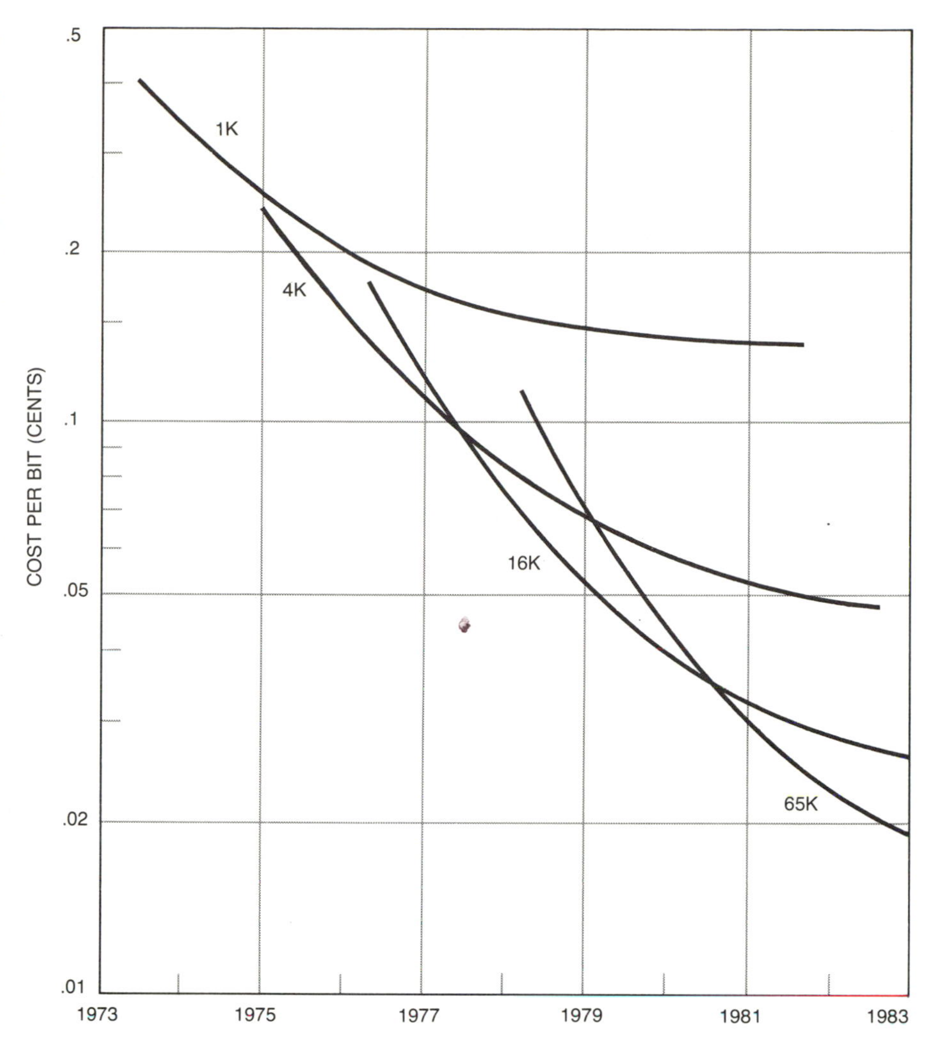

COST PER BIT of computer memory has declined and should continue to decline as is shown here for successive generations of random-access memory circuits capable of handling from 1,024 (1K) to 65,536 (65K) bits of memory. Increasing complexity of successive circuits is primarily responsible for cost reduction, but less complex circuits also continue to decline in cost.

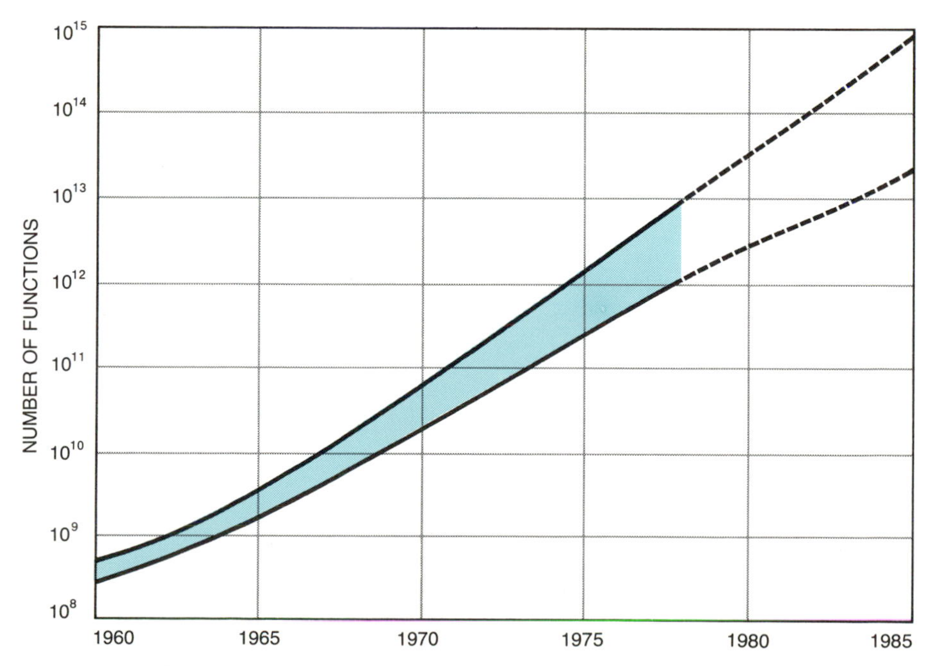

ANNUAL UTILIZATION of electronic functions (transistors, logic gates and bits of memory) worldwide has increased some 2,000 times since the integrated circuit was developed in 1960. Utilization can be expected to increase by another factor of 100 in the next 10 years.

2

MICROELECTRONIC CIRCUIT ELEMENTS

Microelectronic Circuit Elements

by JAMES D. MEINDL

The basic functional element of a modern electronic circuit is the transistor. Microelectronic technology has made it possible to employ large numbers of them in a single circuit

An electronic device is made up of "active" circuit elements, such as transistors, in combination with other components such as resistors, capacitors and inductors. It was once the universal practice to manufacture each of the components separately and then assemble the complete device by wiring the components together with metallic conductors. The advent of microelectronic circuits has not, for the most part, changed the nature of the basic functional units: microelectronic devices are also made up of transistors, resistors, capacitors and similar components. The major difference is that all these elements and their interconnections are now fabricated on a single substrate in a single series of operations.

The primary material of microelectronic circuits is silicon, and the development of microelectronics has therefore depended in large measure on the invention of techniques for making the various functional units on or in a crystal of this semiconducting material. Indeed, methods have been developed for fabricating most of the standard devices, although inductors cannot conveniently be employed. The designs of the earlier and bulkier technology, however, have not merely been adopted and miniaturized. With a change in scale has come a change in the resources available to the designer, followed by a change in the way those resources are deployed. In particular, a growing number of functions have been given over to the circuit elements that are most easily fabricated in silicon and that perform best: transistors. Several kinds of microelectronic transistor have been developed, and for each of them families of associated circuit elements and circuit patterns have evolved.

An important distinction can be made between active circuit elements, exemplified by transistors, and all other electronic devices, which by contrast can be called passive. An active element can change its state in response to an external signal; resistors, capacitors and inductors have no such capability.

Any material that carries an electric current—even a straight wire—exhibits all the characteristics of the passive circuit elements: resistance, capacitance and inductance. Useful devices simply possess one of these attributes in larger measure. The electrical resistance of a metal wire can be interpreted as the disruption of the orderly movement of electrons by interactions with the atomic structure of the material. Resistors constructed as discrete components generally employ a medium with a relatively high resistance per unit length, such as carbon or the alloy nichrome.

Capacitance and inductance are effects that can be attributed to the electromagnetic field generated by an electric charge or current. Capacitance is a measure of the electric field surrounding a conductor. The largest values of capacitance are attained when extensive areas of conductors bearing opposite electric charges are brought close together, and so discrete capacitors are often made of metal plates or foils separated by a thin layer of insulator. Inductance represents the energy stored in the magnetic field set up by an electric current. In order to concentrate the magnetic field, inductors are made by winding a conductor into a coil, sometimes with a core of a ferromagnetic material.

In a microelectronic device the carbon or nichrome of a resistor, the interleaved conductors of a capacitor and the windings of an inductor simply are not available. All the components of the circuit must be fabricated in a crystal of silicon or on the surface of the crystal. Silicon is far from being an ideal material for these functions, and only modest values of resistance and capacitance can be achieved. Practical microelectronic inductors cannot be formed at all. On the other hand, silicon is a material without equal for the fabrication of transistors, and the abundance of these active components in microelectronic devices more than compensates for the shortcomings of the passive elements.

The one property of transistors that makes them indispensable in microelectronics is the capacity for gain, or amplification. This property can be understood by considering both passive and active circuit elements as "black boxes" whose internal workings are immaterial and whose behavior can be examined only at their input and output terminals. A signal applied to the input terminals of a black box containing a resistor, a capacitor or an inductor can be transformed in a number of ways, but invariably the signal is reduced in power. A black box that contains a transistor, on the other hand, can transform a low-power signal into a high-power one.

In order to understand how transistors and other circuit elements can be made from silicon, it is necessary to consider the physical nature of semiconductor materials. In a conductor, such as a metal, current is carried by electrons that are free to wander throughout the

PAIR OF LOGIC CIRCUITS in the photomicrograph on the opposite page were fabricated as a unit in the surface of a silicon chip. The circuits consist of several interconnected transistors, which are by far the most important circuit elements in microelectronic devices. Of the four aluminum conductors that encircle the device and approach it from the right, the top one makes contact with one region of a transistor in which current flows parallel to the surface of the chip, emerging through the conductor that terminates on the bright blue horizontal bar. Under this conductor are three more transistors of a different type; in these, current flows up from the substrate to the row of contacts below the blue bar. The circuit is an example of the semiconductor technology called integrated-injection logic, or I^2L. A distinguishing feature of I^2L circuits is that some regions of the chip function as elements of more than one transistor. In the blue bar the surface of the chip can be seen; the pebbly texture elsewhere is polycrystalline silicon. Variations in color are not intrinsic to the silicon but are caused by interference in the layers of silicon dioxide that cover the surface. The device was made in the author's laboratory at Stanford University by Roderick D. Davies and was photographed by Fritz Goro.

atomic lattice of the substance. In an insulator all the electrons are tightly bound to atoms or molecules and hence none are available to serve as carriers of electric charge. The situation in a semiconductor is intermediate between the two: free charge carriers are not ordinarily present, but they can be generated with a modest expenditure of energy.

An atom of silicon has four electrons in its valence, or outermost, shell of electrons; in solid silicon, pairs of these electrons, shared by neighboring atoms, are arranged symmetrically so that each atom is surrounded by eight shared electrons. Since all the electrons are committed to the bonds between atoms, a crystal of pure silicon is a poor conductor of electricity.

Semiconductor devices are made by introducing controlled numbers of impurity atoms into the crystal, the process called doping. For example, part of a silicon crystal might be doped with phosphorus, an element whose atoms have five electrons in the valence shell. A phosphorus atom can displace a silicon atom without disrupting the crystal structure, but the extra electron it brings has no place in the interatomic bonds. In the absence of an external stimulus the extra electron remains in the vicinity of the impurity atom, but it can be mobilized by applying a small voltage across the crystal.

Silicon can also be doped with boron, an element whose atoms have three valence electrons. Each boron atom inserted in the silicon lattice creates a deficiency of one electron, a state that is called a hole. A hole also remains associated with an impurity atom under ordinary circumstances but can become mobile in response to an applied voltage. The hole is not a real particle, of course, but

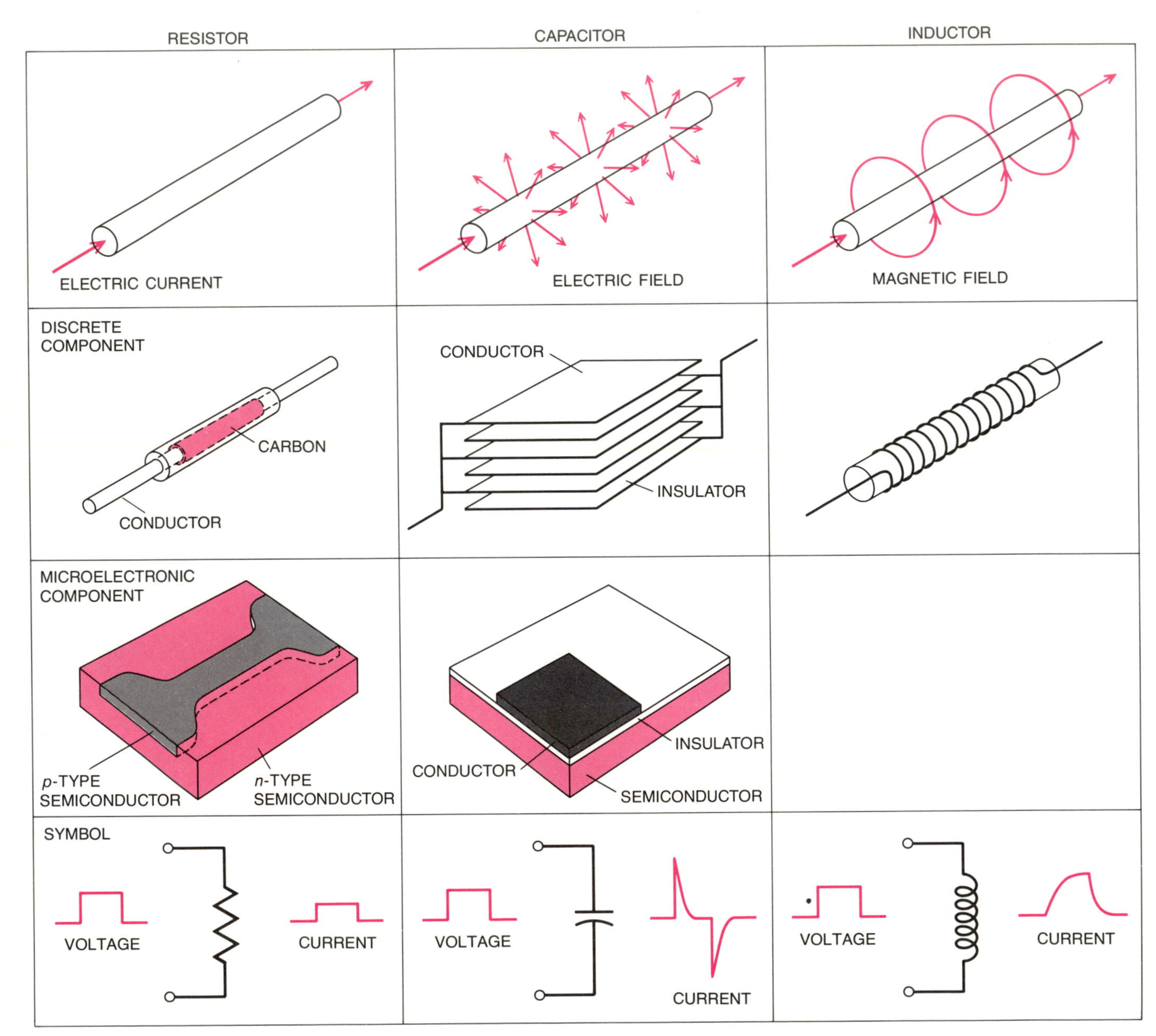

PASSIVE CIRCUIT ELEMENTS employed in electronic devices include resistors, capacitors and inductors. Resistance represents the energy dissipated by electrons as they move through the atomic structure of a conductor. Capacitance measures the energy stored in the electric field that surrounds a charged conductor. Inductance measures the energy stored in the magnetic field set up by an electric current. Resistors constructed as discrete components consist of carbon or some other substance that conducts electricity poorly, such as the alloy nichrome; a resistor that forms part of a microelectronic assembly consists of a thin ribbon of one type of semiconductor surrounded by semiconductor of the opposite type. A discrete capacitor is made of many interleaved conducting and insulating layers; a microelectronic capacitor is made by forming on the surface of a semiconductor crystal a thin layer of insulator followed by a layer of metal. Only small values of capacitance can be obtained in this way. A discrete inductor consists of a coiled wire, often wound on a core of ferromagnetic material; no satisfactory method for building microelectronic inductors has been devised. The symbols for each of the passive elements are shown at the bottom along with the signal currents that flow through them in response to an applied signal voltage.

merely the absence of an electron at a position where one would be found in a pure lattice of silicon atoms. Nevertheless, the hole has a positive electric charge and can carry an electric current. The hole moves through the lattice in much the same way that a bubble moves through a liquid medium. An adjacent atom transfers an electron to the impurity atom, "filling" the hole there but creating a new one in its own cloud of electrons; the process is then repeated, so that the hole is passed along from atom to atom.

Silicon doped with phosphorus or another pentavalent element is called an *n*-type semiconductor, the *n* representing the negative electric charge of the conduction electrons. Doping with boron or another trivalent element gives rise to a *p*-type semiconductor, the designation referring to the positive electric charge of the holes.

The simplest semiconductor device is a diode made up of adjoining *n*-type and *p*-type regions in a single crystal of silicon. When a positive voltage is applied to the *p*-type region and a negative voltage to the *n*-type region, countercurrents of electrons and holes are established. The holes in the *p*-type region are repelled by the positive charge applied to the *p* terminal and are attracted to the negative terminal, so that they flow across the junction. Electrons in the *n*-type region are propelled in the opposite direction. The large current that results is called the forward diode current.

If the connections to the diode are reversed, holes are pulled back toward the terminal of the *p*-type region, which now has a negative charge, and electrons are drawn back into the *n*-type region toward the positive terminal. No current flows across the junction. Actually a very small "reverse" current is always observed, carried by a few electrons found in the *p*-type silicon and by holes in the *n*-type. Such "minority" carriers are always present, but their concentration is low.

A diode is not capable of gain, and so it cannot serve as an active circuit element; on the other hand, it has a property that distinguishes it from other passive devices. Resistors, capacitors and inductors are all symmetrical devices: their effects on a signal are the same no matter what the polarity of the signal and no matter which way they are connected in a circuit. The most conspicuous property of the diode is its asymmetry: it presents a low resistance to a signal of one polarity and a high resistance to a signal of the opposite polarity.

A transistor can be made by adding a third doped region to a diode so that, for example, a *p*-type region is sandwiched between two *n*-type regions. One of the *n*-doped areas is called the emitter and the other the collector; the *p* region between them is the base. In structure,

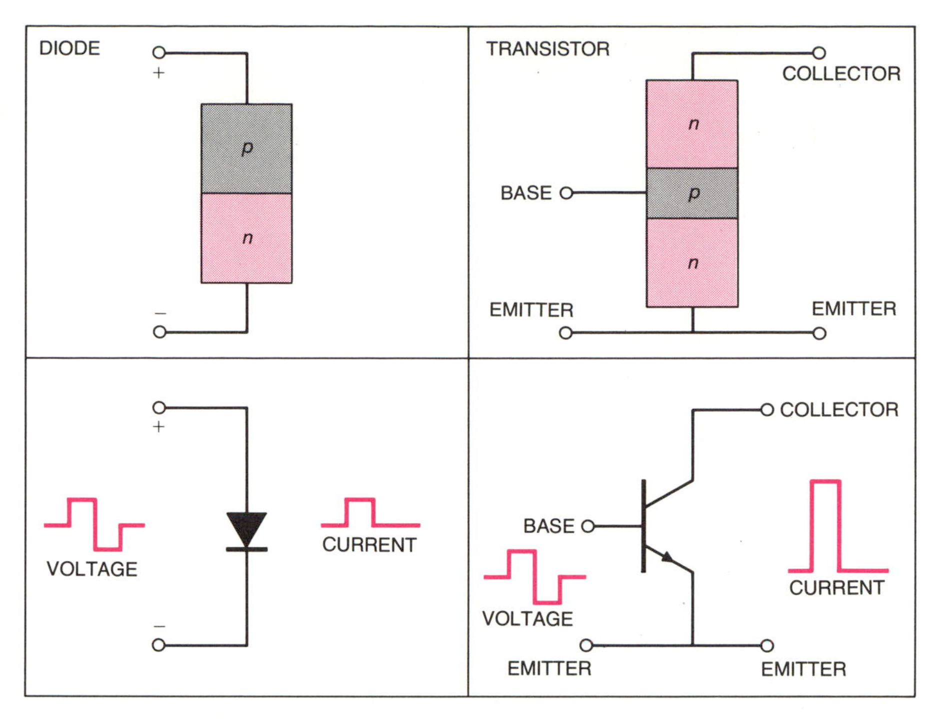

SEMICONDUCTOR DEVICES from which microelectronic circuits are constructed include the diode (*left*) and the transistor. Both devices are made of apposed regions of silicon doped with impurity elements; the two kinds of doped silicon are called *n* type and *p* type. The essential property of the diode is its asymmetry: connected as shown here, it transmits a signal of positive polarity but blocks a negative one. In this respect the transistor is also asymmetrical, but it has an additional property of greater importance. A transistor is capable of amplification; by drawing on an external power supply it can convert a low-power signal into a high-power one.

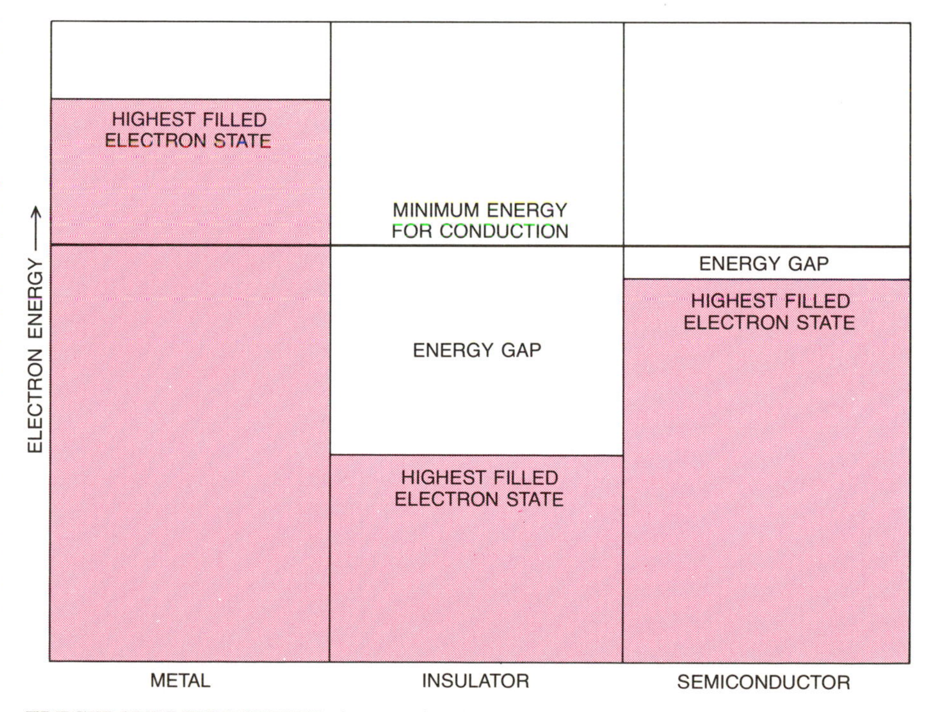

ELECTRONIC STRUCTURE of metals, insulators and semiconductors determines their electrical properties. Each of the electrons in a substance occupies a discrete, unique state, and the available states are filled in order, beginning with the one of lowest energy. Conduction requires electrons of relatively high energy. (For simplicity the minimum energy for conduction is shown here as if it were the same in all materials, although in fact there are substantial differences.) In a metal the highest filled electron state is at an energy exceeding the minimum for conduction, so that conduction electrons are always present; they wander freely through the atomic lattice of the metal. In an insulator there is a large gap between the highest electron energy and the energy needed for conduction; this gap can be interpreted as the energy needed to pull an electron free from the atom to which it is bound. There is also a gap in the electron structure of a semiconductor, but it is a small one. When they are undisturbed, electrons in a semiconductor remain in the vicinity of particular atoms, but they can be freed by a small expenditure of energy, such as that represented by the voltage applied to a diode or a transistor.

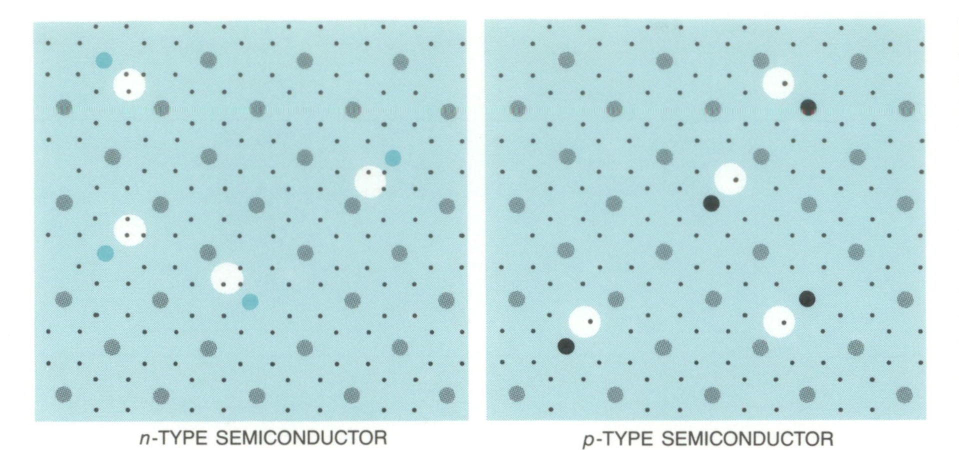

DOPING OF SILICON with impurity atoms alters the electronic structure of the substance in such a way that carriers of electric charge are easily freed from the atomic lattice. A silicon atom (*gray*) has four electrons in its valence, or outermost, shell, and in a pure crystal these electrons form pairs that are shared by adjacent atoms. As a result each atom is surrounded by eight electrons, an inherently stable configuration. An *n*-type semiconductor can be made by replacing a few atoms of silicon with those of an element, such as phosphorus (*color*), that has five electrons in its valence shell. The extra electron has no part to play in the bonds between the atoms of the crystal and so it readily becomes a mobile charge carrier. In a *p*-type semiconductor the impurity introduced is an element, such as boron (*black*), with three electrons in its valence shell. Each impurity atom gives rise to a deficiency of one electron, called a hole. A hole has a net positive charge, and under an applied voltage it can move from atom to atom through the crystal structure. The designations *n* and *p* refer respectively to the negative and positive polarities of the charge carriers. Here the concentration of impurity atoms is greatly exaggerated.

then, the transistor can be regarded as two diodes joined back to back on a single crystal of silicon. As might be expected from this analysis, the operation of the transistor depends on the relative voltages applied to the three regions.

As a point of reference the *n*-type emitter can be assigned a potential of zero volts; we shall then assume that the *p*-type base has a small positive voltage and the *n*-type collector has a larger positive voltage. With this arrangement of voltages the emitter and the base represent a diode operating in the forward direction: holes are drawn into the emitter and electrons are injected into the base. The voltages on the base and the collector, in contrast, are of the wrong polarities for conduction, and so only a negligible reverse current flows through that junction. The collector is not idle, however: electrons injected into the base by the emitter are transported to the collector by a diffusion process similar to the mixing of two gases. In a well-designed transistor almost all the injected electrons diffuse all the way through the base, and a large current flows from the emitter to the collector.

If the positive voltage on the base is reduced to zero or if the base is given a negative potential, the forward diode current from the emitter to the base is halted. Because electrons are no longer injected into the base, the emitter-to-collector current also stops. Hence the collector current is controlled by the base current. By changing the polarity of the base voltage the collector current can be switched on or off; between these two states the collector current is roughly proportional to the base current. If the doped regions of the transistor have the proper dimensions and arrangement, the collector current can be much larger than the base current, so that the transistor exhibits gain. The ratio of collector current to base current is commonly as high as 100, and current gains greater than 1,000 are attainable.

The most important dimensional constraint on the design of such a transistor is that the base must be narrow so that virtually all the electrons injected by the emitter are transported across the base to the collector. A narrow base also reduces the transit time for an injected electron and therefore increases the speed at which the transistor can change from one state to the other.

The transistor I have described is called an *npn* transistor, the term denoting the sequence of doped regions in the silicon. A complementary device, the *pnp* transistor, is made by inserting an *n*-type base between *p*-type regions that define the emitter and the collector. The principles of operation are the same for a *pnp* transistor, but all polarities are reversed, so that for normal operation the base and the collector must be negative with respect to the emitter. The polarities of the charge carriers within the device are also reversed: the current from emitter to collector consists of injected holes.

The *npn* and *pnp* transistors make up the class of devices called junction transistors. They are also known as bipolar transistors because charge carriers of both polarities are involved in their operation. It is the bipolar transistor that was invented in 1948 by John Bardeen, Walter H. Brattain and William Shockley of the Bell Telephone Laboratories.

A second basic kind of transistor was actually conceived almost 25 years before the bipolar devices, but its fabrication in quantity did not become practical until the early 1960's. This is the field-effect transistor. Of the several types that have been devised, the one that is common in microelectronics is the metal-oxide-semiconductor field-effect transistor. The term refers to the three materials employed in its construction and is abbreviated MOSFET.

In a typical MOSFET two islands of *n*-type silicon are created in a substrate of *p*-type material. Connections are made directly to the islands, one of which is called the source and the other the drain. On the surface of the silicon over the channel between the source and the drain a thin layer of silicon dioxide (SiO_2) is formed, and on top of the oxide a layer of metal is deposited, forming a third electrode called the gate. Silicon dioxide is an excellent insulator, and so the gate has no direct electrical connection with the semiconductor substrate. The gate is coupled to the silicon, however, by capacitance, that is, the electric field generated by any charge placed on the gate electrode can influence the motion of charge carriers in the semiconductor channel.

In a common mode of operation the source and the substrate are connected by an external conductor and both are held at a potential of zero volts. The drain is then given a positive voltage. No current flows between source and substrate because they are both grounded; between the drain and the substrate there is only the negligible reverse current of a diode. In the quiescent condition, with no voltage applied to the gate, the *p*-type channel under the gate contains a majority of holes and few electrons can be attracted to the positive potential of the drain. When a positive potential is applied to the gate, the electric field attracts a majority population of electrons to a thin layer at the surface of the crystal immediately under the gate. Because of the presence of numerous electrons in a region that is normally *p* type, the surface is said to be inverted. The inversion creates a continuous *n*-type channel from source to drain and large currents can flow. Like the bipolar transistor, the MOSFET is capable of amplification, although the gain is usually measured in terms of a voltage ratio instead of a current ratio.

It should be noted that whereas both electrons and holes participate in the base current of a bipolar transistor, essentially only one kind of charge carrier is present in the inverted channel of a MOSFET. In the device I have described these carriers are electrons, and the transistor is called an *n*-channel MOSFET or simply an *n*-MOS transistor. The complementary device can also be built; it consists of two islands of *p*-type material in an *n*-type substrate, with the same capacitance-coupled gate over the channel. All polarities are reversed in this device, and the charge carriers are holes rather than electrons. It is called a *p*-MOS transistor.

Still another kind of MOSFET is made by connecting two islands of *n*-type material with a thin but continuous *n*-type channel under the gate capacitor. Since a conduction channel is naturally present, current flows from the source to the drain in this device when the gate is not energized. A negative potential applied to the gate drives electrons from the channel and halts the current, turning off the transistor. Such a device is called a depletion-mode MOSFET, and of course the corresponding transistor with a *p*-type channel can also be constructed. Transistors whose conduction is turned on by a gate voltage are called enhancement-mode devices. All together, then, there are four types of MOS transistor: *n*-channel and *p*-channel types of both enhancement-mode and depletion-mode devices.

Dimensions are of importance in the fabrication of MOS transistors just as they are in bipolar technology. For a MOSFET the critical dimensions are the thickness of the oxide layer under the gate electrode and the distance separating the source and the drain regions; the sensitivity of the transistor's response to a gate voltage varies inversely as the thickness of the oxide layer.

The primary materials of a microelectronic device are areas of a silicon chip doped with various concentrations of *n*-type or *p*-type elements. In addition there are conductors, which can be made either of a metal, such as alumi-

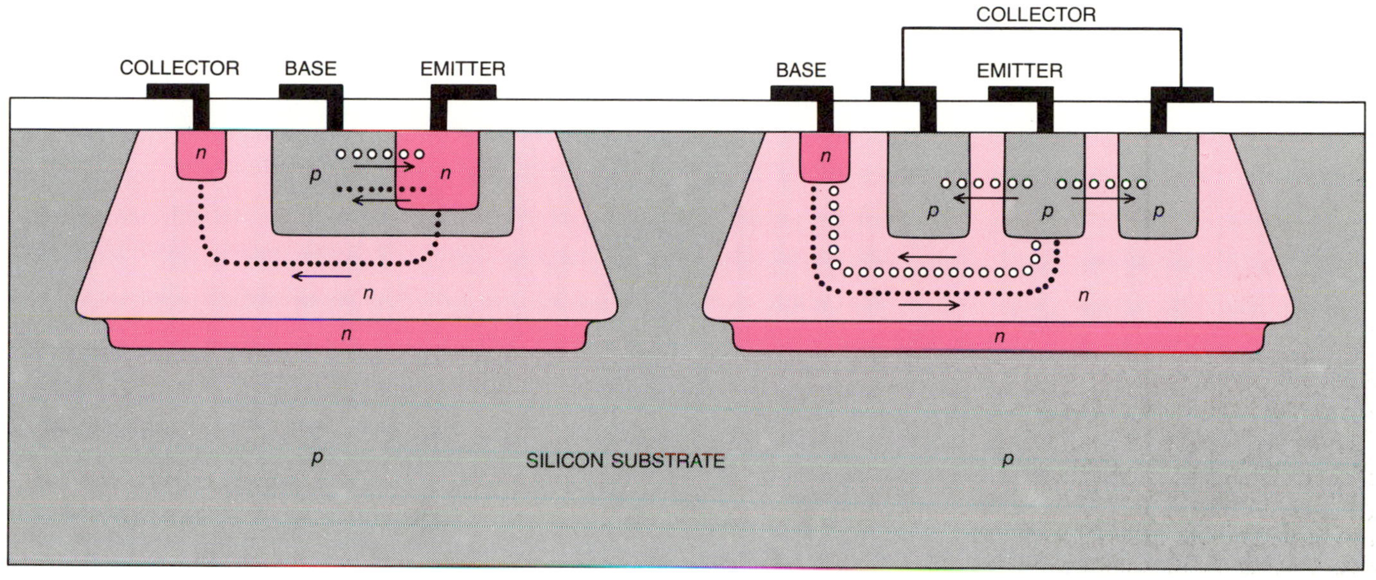

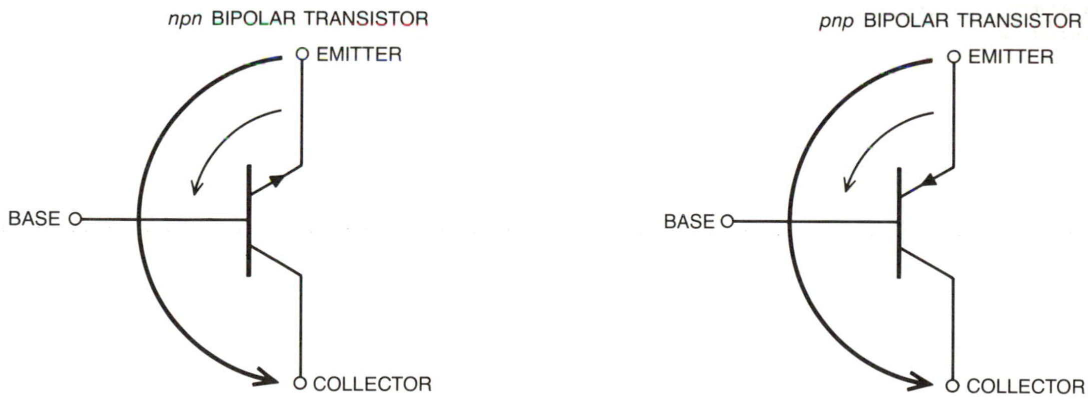

SILICON DIOXIDE (SiO_2)

p-TYPE SILICON

n-TYPE SILICON

HEAVILY DOPED *n*-TYPE SILICON

ALUMINUM CONDUCTOR

••••••• ELECTRONS

oooo HOLES

MICROELECTRONIC TRANSISTORS are fabricated in a single crystal of silicon through a series of operations that require access to only one surface of the silicon chip. In the example shown here the entire chip is doped with a *p*-type impurity, then islands of *n*-type silicon are formed. Smaller *p*-type and *n*-type areas are next created within these islands in order to define the three fundamental elements of the transistor: the emitter, the base and the collector. In an *npn* transistor (*left*) a positive voltage is applied to the base and the collector, and as a result holes flow from the base to the emitter and electrons are injected by the emitter into the base. Many of the injected electrons, however, migrate all the way through the base to reach the collector, and this emitter-to-collector current can be much larger than the emitter-to-base current. The device exhibits gain because a small signal applied to the base can control a large one at the collector. For simplicity in manufacturing, a *pnp* transistor (*right*) is generally constructed according to a different plan, in which the emitter-to-collector current is lateral rather than vertical. The principles of operation are the same except that all polarities are reversed. Connections to the circuit elements are made through aluminum conductors deposited over an insulating layer of silicon dioxide; some areas of *n*-type silicon are heavily doped to improve their conductivity. The large *n*-type islands are required to isolate the transistors. Because charge carriers of both polarities participate in the operation of these devices they are called bipolar transistors.

num, or of heavily doped polycrystalline silicon, which has a fairly high conductivity. Finally, silicon dioxide serves as a good insulator.

The fabrication of the circuit must be accomplished entirely from the surface. Areas to be doped, for example, are defined photolithographically, then the appropriate element is allowed to diffuse into the silicon structure. The process must be repeated several times in order to create all the needed doped regions, which are often nested one inside another. Areas of oxide are laid out in a similar way, and are formed from the substance of the chip itself, by heating it in the presence of oxygen. Conductors are deposited over a thick mat of oxide that covers the entire chip, breaking through the insulating barrier only where it is necessary to make electrical contact with the silicon.

A limited selection of passive circuit elements can be fabricated by these techniques. A resistor is formed by defining a thin ribbon of *p*-type silicon within an island of *n*-type material. Current flowing through the device is confined to the ribbon by maintaining it at a negative potential with respect to the *n*-type island so that only reverse current flows across the *pn* junction.

Doped silicon would be an unlikely choice of material for a resistor to be manufactured as a discrete component. It simply does not have enough of the desired property: electrical resistance per unit length. Practical values of resistance can be obtained in microelectronic circuits only because the precision of the fabrication technology makes it possible to form a ribbon of material with a very small cross-sectional area. Even so, resistors of large value are cumbersome.

Another disadvantage of the silicon resistor is that its ultimate value cannot be predicted with great precision. Circuits incorporating the resistors must therefore be designed to tolerate considerable variation. The manufacture of an entire integrated circuit in a single sequence of processes can compensate in part for such wide tolerances. Since all the resistors on a chip are formed at the same time, they all tend to depart from their specified values by roughly the same amount; moreover, the resistance values vary in concert with some of the properties of the transistors on the same chip. By arranging for the deviation of one component from its nominal value to be balanced by that of another, excellent performance of the circuit as a whole can be achieved.

The structure of a microelectronic capacitor is much like that of the gate electrode in a metal-oxide-semiconductor transistor. When the device is to function only as a capacitor and not as part of an active circuit element, the area of silicon under the electrode is doped very heavily in order to increase its conductivity. A thin layer of oxide is then formed, followed by a layer of aluminum. As in the case of resistors, this method of construction would not be employed in making discrete components. The silicon dioxide layer cannot be made as thin as some other insulators without the risk of a short circuit. Moreover, the total area of the conductors is very small; discrete capacitors are made up of many interleaved conductors, whereas the microelectronic equivalent is limited to a single pair of plates. For these reasons the values of capacitance available in microelectronic circuits are

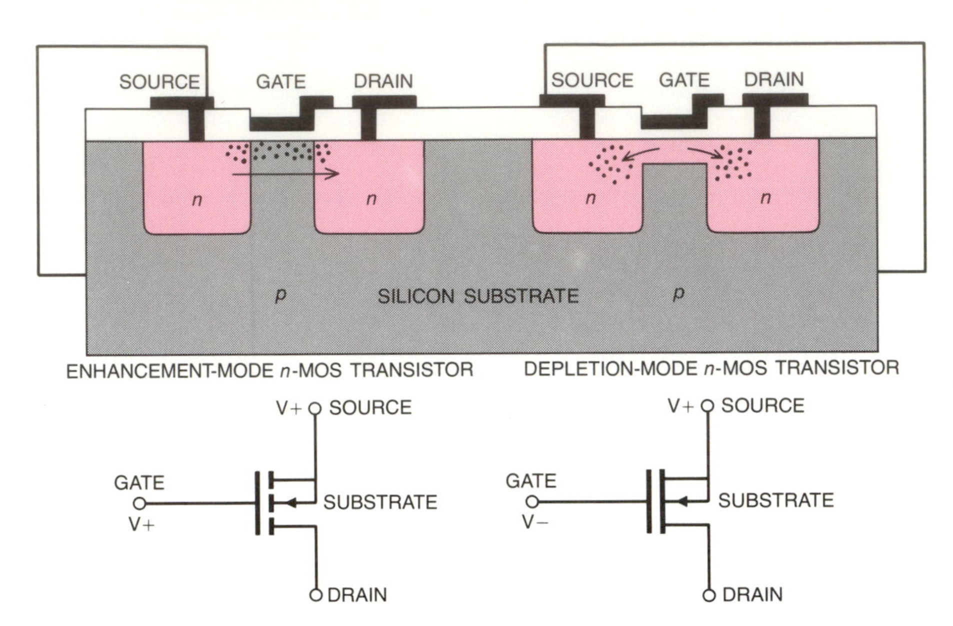

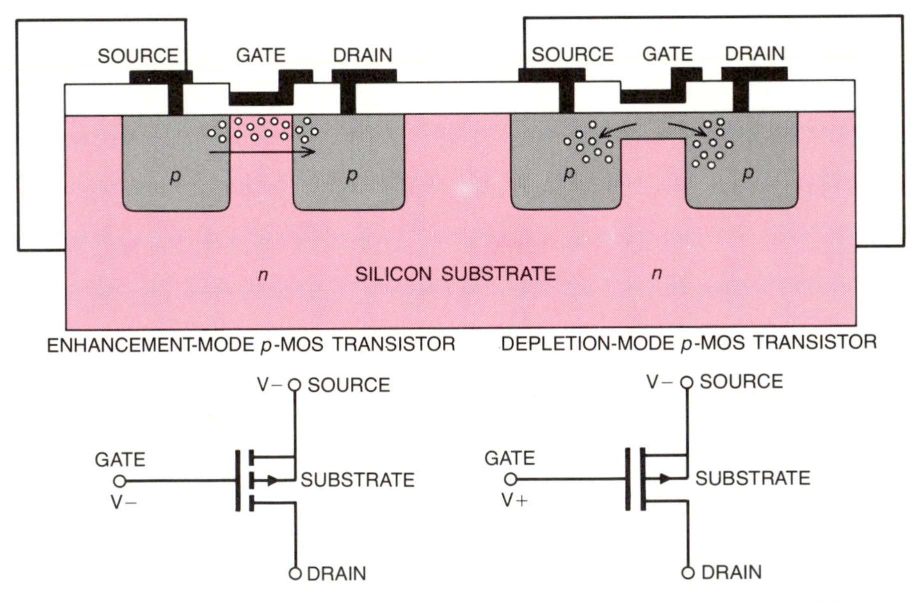

FIELD-EFFECT TRANSISTORS, made by the metal-oxide-semiconductor (MOS) technique, differ from bipolar ones in that only one kind of charge carrier is active in a single device. Those that employ electrons are called *n*-MOS transistors; those that employ holes are *p*-MOS transistors. In an *n*-MOS device (*top*) two islands of *n*-type silicon, called the source and the drain, are formed in a *p*-type substrate. Over the channel between the source and the drain is a metal electrode, the gate, that is prevented from making contact with the semiconductor by a thin layer of silicon dioxide. In an enhancement-mode transistor (*top left*) a positive potential at the drain exerts an attractive force on electrons available from the source, but the electrons cannot pass through the *p*-type channel, which is rich in holes. When a positive charge is applied to the gate, however, the resulting electric field attracts electrons to a thin layer at the surface of the channel and current flows from the source to the drain. In a depletion-mode *n*-MOS transistor (*top right*) there is a continuous channel of *n*-type silicon between the source and the drain, so that the transistor normally conducts; only when a negative voltage is applied to the gate is the current halted, since electrons are then expelled from the channel. By forming islands of *p*-type material in an *n*-type substrate the corresponding *p*-MOS devices (*bottom*) can be constructed; here the charge carriers are holes. Because MOS transistors require no isolation islands they can be packed more densely on a chip of silicon than bipolar transistors.

small, seldom more than a ten-thousandth the value of a typical capacitor in a discrete-component circuit.

If capacitors are difficult to shrink to the scale of microelectronic devices, inductors have so far proved impossible. The helical coils and ferromagnetic core materials of the discrete inductor are simply incompatible with silicon technology, and no substitute has been found. If an inductor is required, it must be supplied as a discrete component.

Improvements in the technology of fabrication will no doubt increase the selection of passive elements available to the designer of a microelectronic circuit. The potential for improvement is even greater, however, through the simple elimination of many passive components in favor of active ones.

The case of resistors is particularly dramatic: they can often be eliminated by direct substitution. A transistor can be regarded as a current-controlled or voltage-controlled resistor. It follows that if a suitable current or voltage is available to set the resistance, a transistor can be inserted into a circuit at almost every point where a resistor might be employed. This practice would never be adopted in a circuit made up of discrete components because discrete transistors are more expensive than discrete resistors. In a microelectronic circuit, on the other hand, the main cost of a component is the area of silicon it occupies, and that is at least as great for the resistor as it is for the transistor.

The replacement of other passive circuit elements with active ones involves more elaborate stratagems. In a sense the most elaborate of all is the sudden prominence in recent years of digital electronics. Many of the large capacitances and inductances of traditional electronic circuitry are employed to ensure the faithful reproduction of the continuously varying voltages and frequencies of analogue signals. Even in those circuits the large passive elements can often be replaced by a network of active ones, but the passive components are much more readily dispensed with in the processing of digital signals, which have only discrete levels. In most cases there are only two recognizable levels: high voltage and low (or zero) voltage. In such circuits transistors are operated as switches, which also have only two states—on or off—and in the switching mode they require fewer auxiliary devices.

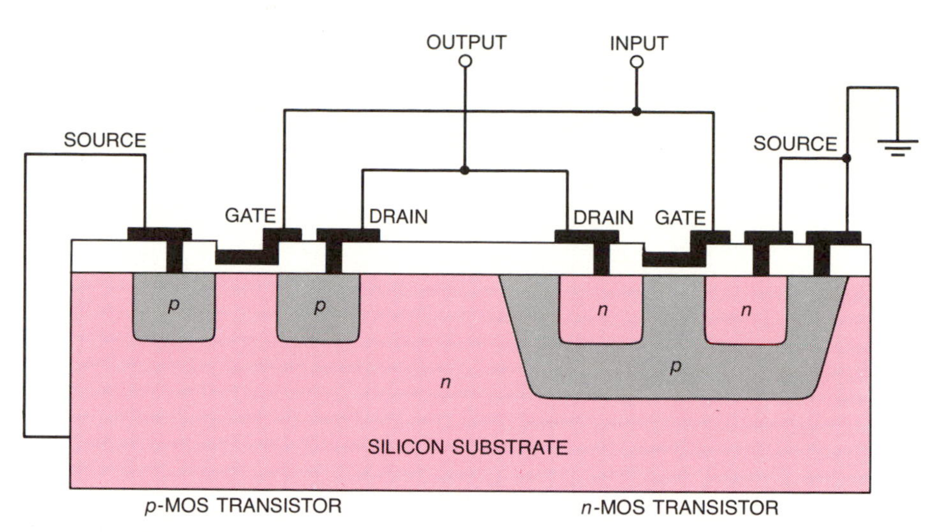

COMPLEMENTARY MOS DEVICE includes both *n*-MOS and *p*-MOS transistors on a single chip of silicon. If the circuit is fabricated in an *n*-type substrate, as it is here, then a *p*-channel transistor can be made in the normal way, but an *n*-channel device requires an island of *p*-type material. The need for such islands adds another step to the manufacturing process and also reduces the packing density, and thus the total number of transistors that can be fitted onto a single chip. Complementary MOS devices can be arranged to achieve low power consumption. In the circuit shown here the gates of both transistors are connected to a single input; since the two transistors require signals of opposite polarity for conduction they are never turned on at the same time and little current flows from the power supply ($V+$) to ground.

The two basic types of transistor, bipolar and MOSFET, divide microelectronic circuits into two large families. The bipolar devices were the first to be developed, and a great variety of bipolar technologies have evolved. Today the greatest density of circuit elements per chip can be achieved with the newer MOSFET technology, but that is not to say that bipolar circuits are likely to become obsolete.

In the construction of a typical bipolar integrated circuit the entire chip is first given an overall doping with a *p*-type contaminant. In this ocean of *p*-type silicon separate *n*-type islands are then created for each transistor (and for other circuit elements such as resistors as well). These islands serve the function of isolation; the circuit is designed in such a way that all of them are maintained at a positive voltage with respect to the substrate, so that no current except the small reverse diode current flows between them. In order to make an *npn* transistor a lake of *p*-type material is introduced into the *n*-type island, and finally a smaller island of *n*-type semiconductor is built within this lake. The small, innermost island is the emitter; the *p*-type lake is the base and the main, isolation island serves as the collector. The electrons that make up the collector current flow down from the emitter, across the base and into the comparatively large volume of the collector.

A *pnp* transistor of a similar design could be constructed in the same chip, but it would require adding one more doping operation to a schedule that already includes at least four. Once again an island of *n*-type material would be created in the *p*-doped sea. Then would be added in succession a lake of *p*-type semiconductor, a smaller island of the *n*-type material and finally within this small island a puddle of *p*-type silicon. Only the last three regions would participate in the operation of the transistor; the main island would serve for isolation only.

In order to avoid this complexity the *pnp* transistor is constructed according to a different plan, in which the current flow is lateral rather than vertical. In this design the *n*-type isolation island becomes the base and two separate *p*-doped regions are introduced. They are often formed in an annular arrangement, with the collector forming a ring around the emitter. The characteristics of the lateral *pnp* transistor are inferior to those of the vertical *npn* one because the base region separating the emitter from the collector is inevitably wider. As a result the current gain and the switching speed of the lateral transistor are reduced.

A metal-oxide-semiconductor transistor can also be fabricated at the surface of an ocean of *p*-type silicon, but in MOS circuits no isolation diodes are required. In order to build an enhancement-mode *n*-channel MOSFET it is sufficient to create two small islands of *n*-type material to serve as the source and the drain; then, of course, the oxide layer and the gate electrode must be formed on the surface of the chip over the channel. Isolation islands are not needed because the voltages applied to the source and the drain ensure that no current can flow between them and the substrate. The transistor isolates itself from all other elements on the chip.

An *n*-channel microelectronic device operating in the depletion mode is built in much the same way, except that a permanent channel of *n*-doped silicon is laid down between the source and the drain. MOS devices employing the *p* channel are most conveniently formed in a substrate that has been treated with

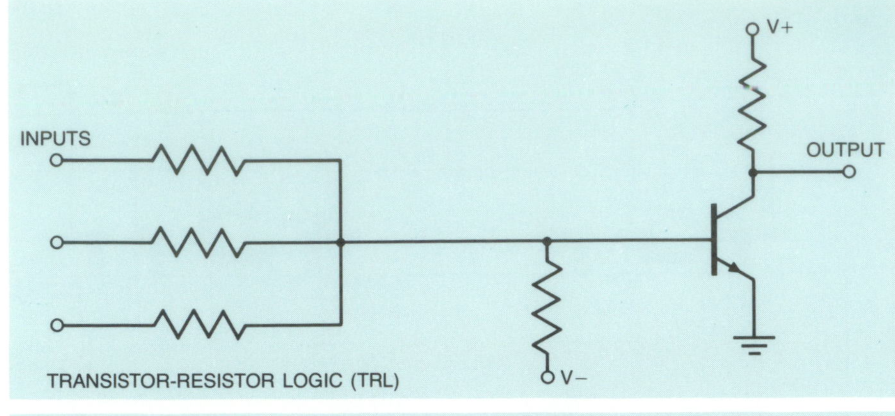

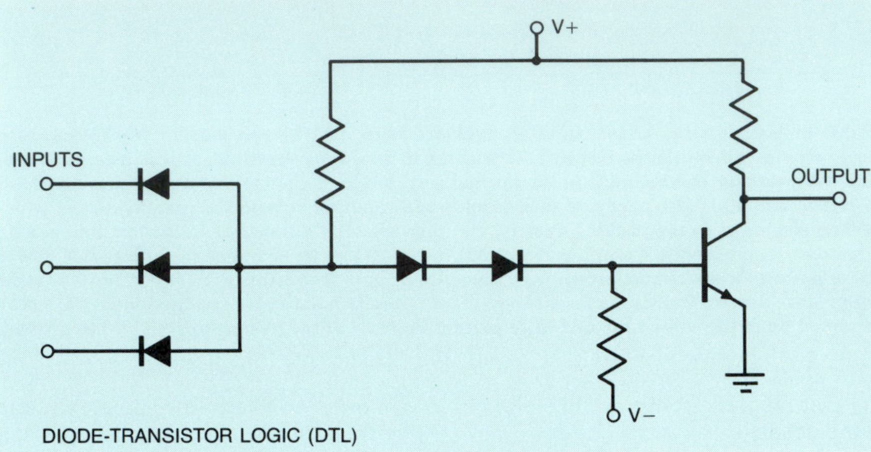

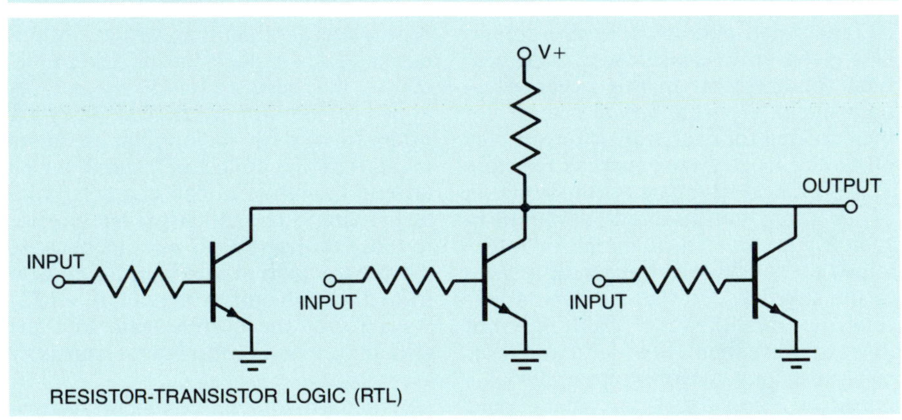

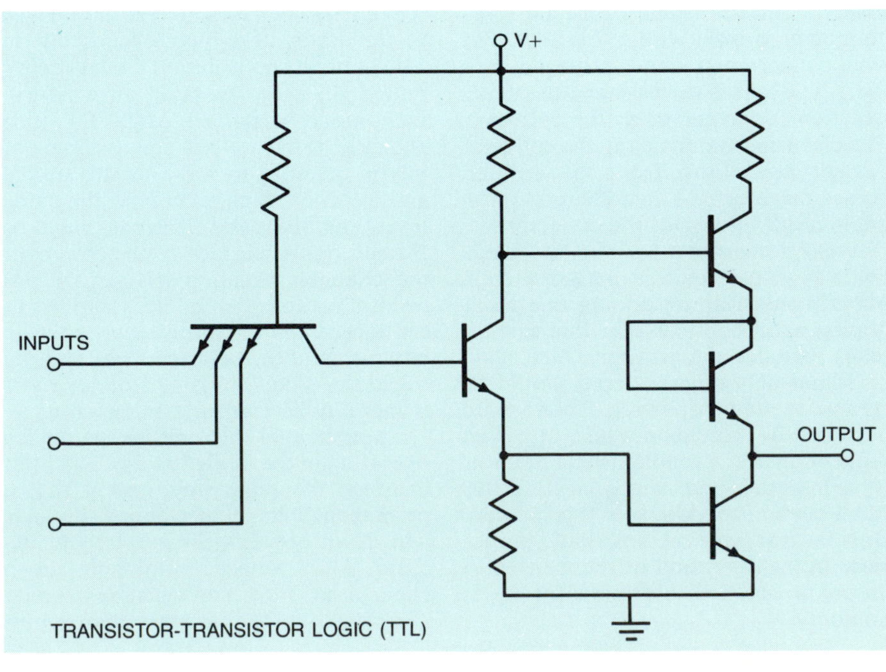

an *n*-type dopant. The polarities of all the semiconductor regions and the polarities of all the voltages applied to them must of course be reversed.

In microelectronic circuits employing metal-oxide-semiconductor technology the decision to employ *n*-channel or *p*-channel devices is usually a fundamental one because all the transistors on a single chip employ the same type of channel material. It is possible, however, to manufacture a chip containing both kinds of MOSFET, but only at the cost of increased complexity and lower density. Complementary MOS circuits are made, for example, by isolating all the *p*-channel devices within islands of *n*-type silicon. The complementary integrated circuits are more difficult to manufacture, but in some applications their convenience is worth the added cost. The particular advantage of the complementary MOS (CMOS) devices is the potential for low power consumption.

It is mainly because isolation islands are not required in conventional *n*-MOS and *p*-MOS devices that these transistors can be packed more densely than bipolar ones. The difference is a substantial one: roughly four times as many MOS transistors will fit on a given area of silicon as bipolar ones. For this reason MOS technology has come to dominate the manufacture of large-scale integrated circuits, in which tens of thousands of active elements are incorporated into a single device. A majority of one-chip microprocessors are built with metal-oxide-semiconductor technology, as are almost all semiconductor memory chips.

The most important virtue of bipolar circuits is higher speed of operation. The speed with which a bipolar transistor can change its state is determined mainly by the width of the base; the delay being measured depends on the time

DIGITAL LOGIC CIRCUITS employing bipolar semiconductor devices have evolved toward a state in which transistors are adopted for almost all functions. Digital logic operates with signals that have only two recognizable levels (such as high voltage and low voltage); logic circuits accept such signals as inputs and transform them according to fixed rules to generate an output. The first families of bipolar logic circuits were constructed from discrete components. In transistor-resistor logic (TRL) the number of resistors was maximized, since they were the cheapest devices. In diode-transistor logic (DTL) performance was improved by substituting semiconductor diodes for many of the resistors. Resistor-transistor logic (RTL) was the first microelectronic technology; a transistor was supplied for each input and only a few resistors with small values were required. Transistor-transistor logic (TTL) is today the commonest form of bipolar microelectronic technology. Transistors are abundant and are coupled directly together. The circuit illustrated here includes a device, the multiple-emitter transistor, that has no equivalent among discrete components.

required for electrons to diffuse from the emitter to the collector. In the vertical design generally employed for *npn* transistors the base can be quite narrow, substantially more so than the narrowest line that can be resolved photolithographically on the surface of the chip. In a MOS device the width of the gate plays a similar role, but because MOS transistors are lateral devices that width cannot be reduced below the limit of resolution. The mobility of the charge carriers also has an effect on the speed of a MOS transistor: electrons move faster than holes, and so *n*-channel devices tend to be faster than *p*-channel ones. Finally, the speed of a MOS device is limited by the capacitor made up of the gate, the oxide layer and the substrate, which requires a finite time for charging and discharging. Reducing the capacitance speeds operation, but it also reduces gain.

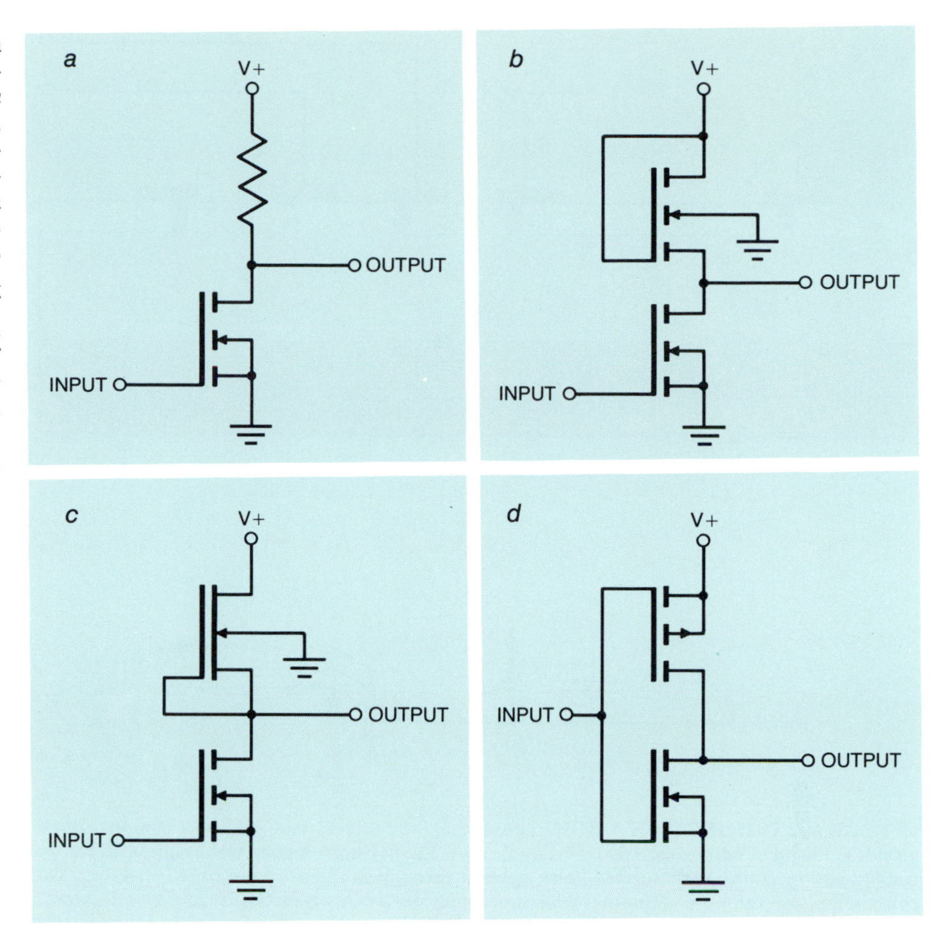

MOS LOGIC CIRCUITS were developed later than bipolar ones and for the most part have been built only in microelectronic form. Each of the circuits here performs the function of inverting a signal, so that if the input is high, the output is low, and vice versa. In each case the input is applied to the gate of an enhancement-mode *n*-MOS transistor; the circuits are distinguished by the choice of a load element needed to limit the current through this transistor. The simplest choice is a resistor (*a*), but it also gives the poorest performance. A second enhancement-mode *n*-MOS transistor (*b*) is the easiest to fabricate; a depletion-mode *n*-MOS transistor (*c*) gives the highest packing density. Finally, by adding an enhancement-mode *p*-channel device (*d*) the low power consumption of a complementary MOS circuit can be achieved.

Over the past two decades the development of microelectronic circuits has made high-quality transistors progressively cheaper and more abundant while providing only a meager selection of resistors and capacitors and no inductors at all. As a result circuit configurations have evolved that exploit the strengths of the technology and adeptly skirt its limitations. The trend can be traced in the evolution of digital logic circuits.

The fundamental units of electronic logic are circuits called gates (not to be confused with the gate electrode in an MOS transistor). Each kind of gate accepts a signal at its input terminals and transforms it according to a rule embodied in its internal wiring, the transformed signal appearing at the output terminals. Many gates accept more than one input and determine the output according to the combination of input signals. At the heart of every gate is at least one active circuit element, which acts as a switch; the switch controls the output of the circuit and is ultimately controlled by the input.

Among the families of logic circuits developed in the mid-1950's was one called transistor-resistor logic (TRL). The circuits were assembled entirely from discrete components and the number of resistors was maximized because resistors were the cheapest and most reliable devices. A typical gate with three inputs consisted of one transistor and five resistors.

In the late 1950's and early 1960's semiconductor diodes, also packaged as discrete components, became cheap enough to compete with resistors; the result was diode-transistor logic (DTL). The diodes were employed to isolate the input signals from one another and to shift voltage levels. A typical gate still relied on a single switching transistor, along with three resistors and five diodes. It offered both lower power consumption and higher speed than the equivalent transistor-resistor network.

The next generation of logic designs, prevalent in the early and mid-1960's, introduced the first microelectronic logic circuits. The family of devices was given the name resistor-transistor logic (RTL), but the resistors had values much smaller than those of earlier circuits and transistors were employed much more generously. In a typical gate each of three inputs was allotted its own switching transistor. Once again both power consumption and speed were improved. Integrated RTL circuits were manufactured with one gate or a few gates on a chip.

The fourth major logic family to be introduced remains today the commonest form of bipolar digital technology. It is called transistor-transistor logic (TTL), and as the term implies it delegates most functions to active circuit elements. TTL gates were the first commercially important ones to incorporate a microelectronic circuit element that could not have been assembled from discrete components. That element is the multiple-emitter transistor, in which two or three emitters share a common base and collector. In a TTL gate each of the emitters accepts an input signal. The multiple-emitter transistor controls a single switching transistor, and that in turn drives a network of three output transistors. Compared with the previous family of RTL gates, TTL circuits provide greater output power (so that more gates in the next stage of an array can be driven), less stringent tolerances in manufacturing and greater immunity to spurious voltages, or electrical noise. Integrated circuits containing several hundred TTL gates are now common.

Perhaps because metal-oxide-semiconductor technology developed later, its evolution has been less complicated. As a rule newly introduced logic families have not completely displaced older ones; rather, each of the families has found applications for which it is best suited.

The first MOSFET microelectronic circuits employed *p*-channel devices exclusively because they are the easiest to

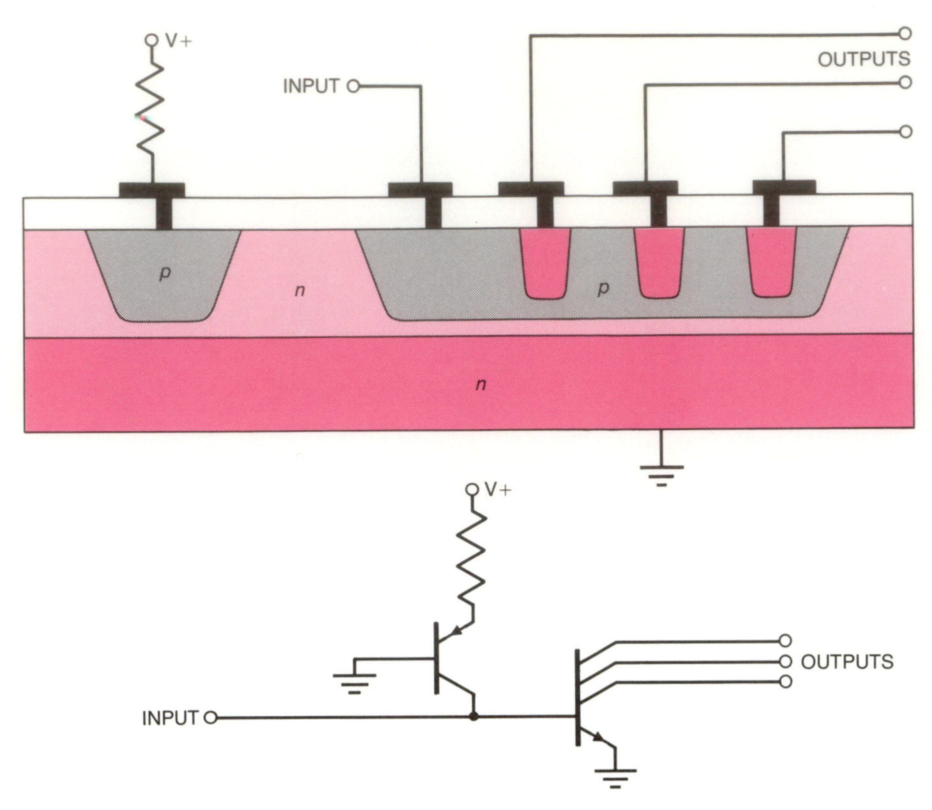

INTEGRATED-INJECTION LOGIC, often abbreviated I²L, compresses a complete logic circuit made up of two transistors into a single unit. The substrate serves as the emitter of an *npn* transistor (*right*) in which current flows upward through the base to multiple collectors. The substrate is also the base of another transistor, a *pnp* device in which the current flow is lateral. By this arrangement of elements isolation islands that are normally required in bipolar technology are eliminated and packing densities similar to those of MOS circuits can be attained.

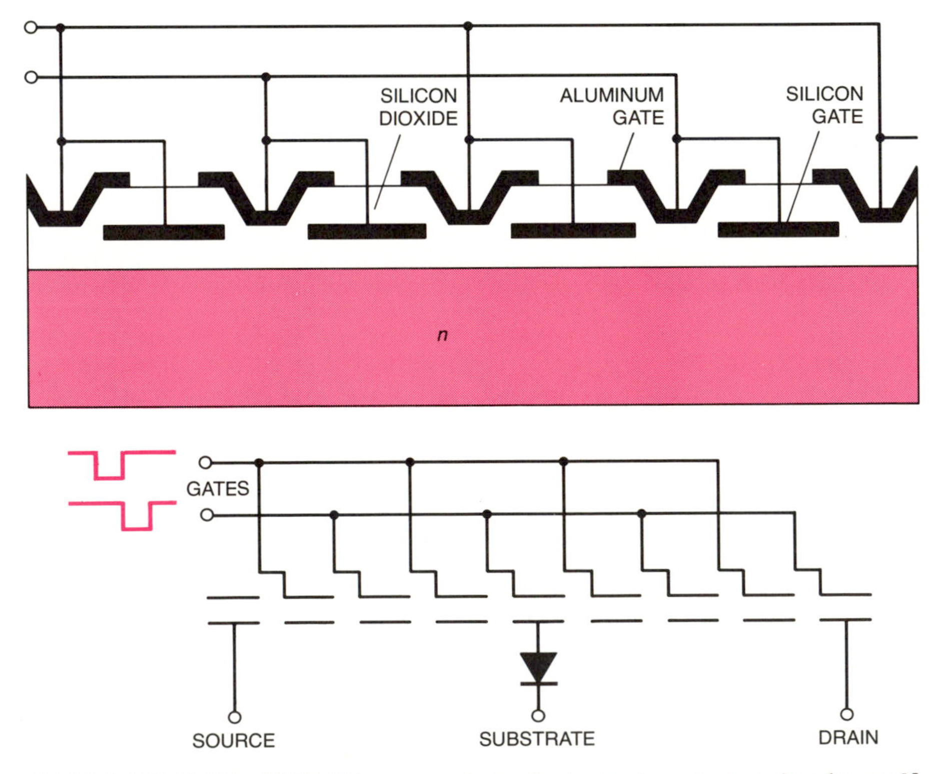

CHARGE-COUPLED DEVICE is a microelectronic circuit element whose function could not be duplicated by a practical assembly of discrete components. The device can be regarded as a stretched MOS transistor with a long string of gates (perhaps as many as 1,000) between the source and the drain. In the *p*-channel device shown here a charge packet, consisting of a concentration of holes, can be held in place for a short time by applying a steady negative voltage to one of the gates. If that voltage is then dropped and simultaneously the next gate in line is energized, the charge packet moves to a new position under the second gate. By applying pulses to alternate gates a sequence of charge packets can be transferred from source to drain.

make. They are also the slowest, because of the lower mobility of holes compared with electrons. Other technologies have largely replaced the *p*-MOS devices where high performance is needed, but the latter are still in widespread use in inexpensive electronic calculators, where speed of operation is not important and low cost is the foremost consideration.

In microprocessors and in semiconductor memories it is largely *n*-channel MOS circuits that have replaced the *p*-MOS devices. The varieties of *n*-MOS technology can be surveyed by considering a simple gate circuit with one input to an enhancement-mode *n*-channel field-effect transistor. The possible circuit configurations are distinguished by the element chosen as a load for this transistor.

The most straightforward choice and conceptually the simplest one is a resistor, but it also gives the poorest performance. Another possibility for the current-limiting load element is a similar enhancement-mode *n*-channel transistor, connected in series with the first one. This is the combination that is most easily fabricated and that is often chosen for low-cost applications. The highest packing density can be achieved with an *n*-channel depletion-mode device as the load element, and that is therefore the choice for demanding, large-scale integrated circuits, such as microprocessors. Finally, a *p*-channel enhancement-mode transistor can be employed as the load element. The gate is then no longer an *n*-MOS device but a complementary MOS (CMOS) circuit. Because only one of the transistors is conducting at any given time, except during the switching operation itself, there is little steady-state current through the gate and power consumption is low. An application in which CMOS technology excels is the electronic wrist watch, which must operate for long periods on the energy stored in a small battery.

Primarily through advances in fabrication methods and materials science the packing density of bipolar circuits has been considerably improved, and so has the speed of MOS devices. Nevertheless, a substantial gap remains between them. One candidate to fill this gap is a new bipolar technology called integrated-injection logic, usually abbreviated I²L.

The basic functional element in integrated-injection logic is not a single transistor but a pair of transistors formed as a unit in the silicon substrate. One of them has a vertical arrangement, but the current flows up to the collector rather than down as in the usual design. The substrate itself serves as the emitter of this transistor, and at the same time it forms the base of another, lateral transistor. Because of the voltages applied to the various doped regions no isolation islands are needed; the pair of "superintegrated" transistors forms a self-contained logic gate.

In speed I²L gates will probably not catch up with the fastest TTL circuits, although I²L is already faster than all the MOSFET technologies. On the other hand, the compact architecture of integrated-injection logic makes it a natural candidate for large-scale integration. Packing densities equal to those of *n*-MOS chips are a possibility.

The devices we have considered are in the mainstream of microelectronics development. The techniques learned in making gates for digital logic, however, have also been applied to the development of other kinds of microelectronic device. Circuits that process analogue signals are probably indispensable in devices such as the telephone and radios, and some of them can be made in microelectronic form. A typical low-power amplifier designed for assembly from discrete components has only one transistor but requires four resistors, two of them with large values, and three capacitors with large or very large values. An equivalent microelectronic device employs seven transistors but only one current-controlling resistor. In some cases a more productive way to handle analogue signals may be to convert them into digital form for processing, then convert them back again when necessary. The apparatus for the conversion can itself be fabricated in a microelectronic package.

Semiconductor technology has also provided a selection of transducers through which electronic devices can communicate with their environment. A novel example is a pressure gauge whose working element is a thin diaphragm of semiconducting silicon. The resistance of silicon changes as a response to mechanical strain, and so a change in the pressure across the diaphragm can be detected as a change in the resistance of the material.

A somewhat more familiar transducer is the photodiode and the fundamentally similar phototransistor. When a photon, or quantum of electromagnetic energy, is absorbed by the *p*-type region of a photodiode, the energy given up by the photon creates an electron and a hole. The electron migrates to the junction and crosses it, and the total current of such electrons is proportional to the flux of incident photons. The obverse device is the light-emitting diode, which is constructed so as to exploit collisions between electrons and holes. In each collision an electron fills a hole and both particles are effectively annihilated as charge carriers. Their energy appears in the form of a photon. In silicon, annihilations lead to the emission of infrared radiation but in the semiconducting compound gallium arsenide the light radiated is in the red portion of the visible spectrum. Arrays of gallium arsenide light-emitting diodes make up the numerical displays of many electronic calculators.

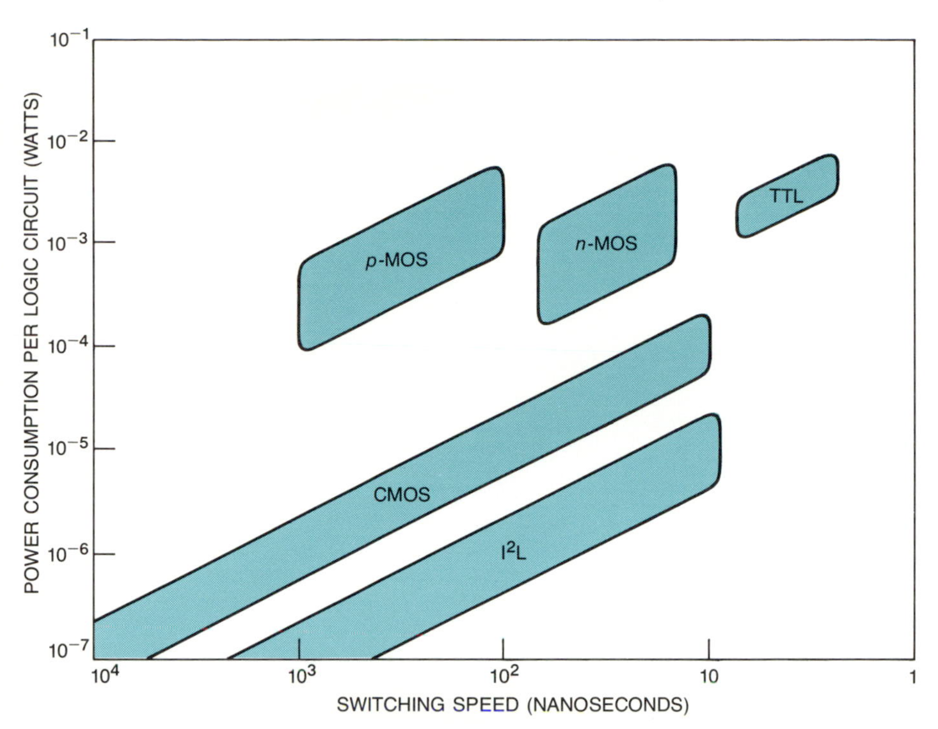

HIGH SPEED AND LOW POWER CONSUMPTION in a microelectronic device have a reciprocal relation: the circuit can be made to operate faster simply by applying more power. The product of the switching delay time and power consumption is an important figure of merit for a digital circuit: it represents the energy required for a single switching operation. The fastest logic circuits available today employ bipolar technology, as in the TTL devices, but these circuits also consume large amounts of power. In the MOS technologies *p*-MOS devices are being replaced by *n*-MOS ones because the latter operate faster without consuming more power. Complementary MOS circuits offer even lower power consumption without sacrifice of speed. I²L technology is attractive because even at its highest speed power consumption is quite low.

One recently invented microelectronic functional element has the distinction that no practical assembly of discrete components could emulate its operation; it is quintessentially microelectronic. Called a charge-coupled device, it can be regarded as an ingenious extension of the enhancement-mode MOS transistor. The charge-coupled device consists of a source and a drain separated by a long row of gates. A packet of charge injected by the source can be trapped under the first gate by a voltage applied to that electrode. If the voltage on the first gate is dropped and a voltage is simultaneously applied to the second one, the charge packet is attracted to and trapped under the second gate. Repeating the process hands the packet of charge down the line of gates one at a time until it reaches the drain.

A memory device in which digits are stored as long strings of bits can be fabricated by building a charge-coupled device with perhaps 1,000 gates. Since the packing density can be very high, many such devices can be fitted onto one chip. Charge packets can also be generated photoelectrically rather than being injected by the source, so that an array of charge-coupled devices can function as the heart of a television camera with digital output.

The time is long past when the designer of a microelectronic device could study the behavior of his proposed circuit by building a "breadboard" prototype from discrete components. Instead the circuit is represented by a mathematical model whose characteristics are investigated with the aid of a computer. The model in turn must be based on an accurate description of the electrical properties of the individual circuit elements. As the elements recede ever further from human scale, verifying those descriptions becomes increasingly difficult. As the dimensions of the elements shrink below one micrometer, physical mechanisms that are inconsequential in larger devices may become important and in some instances even dominant. Thus one requirement for continued progress that must not be neglected is the need for improved models of the circuit elements.

An equally systematic approach will be required in the fabrication of the next generation of integrated circuits. The models must reveal not only how the element works but also how it is to be made. Compared with models of electrical behavior, such process models are in a primitive state; fabrication remains an empirical art. Continued progress in microelectronics may depend to a significant extent on our ability to predict the properties of a transistor from a knowledge of the steps taken in its fabrication.

3

THE LARGE-SCALE INTEGRATION OF MICROELECTRONIC CIRCUITS

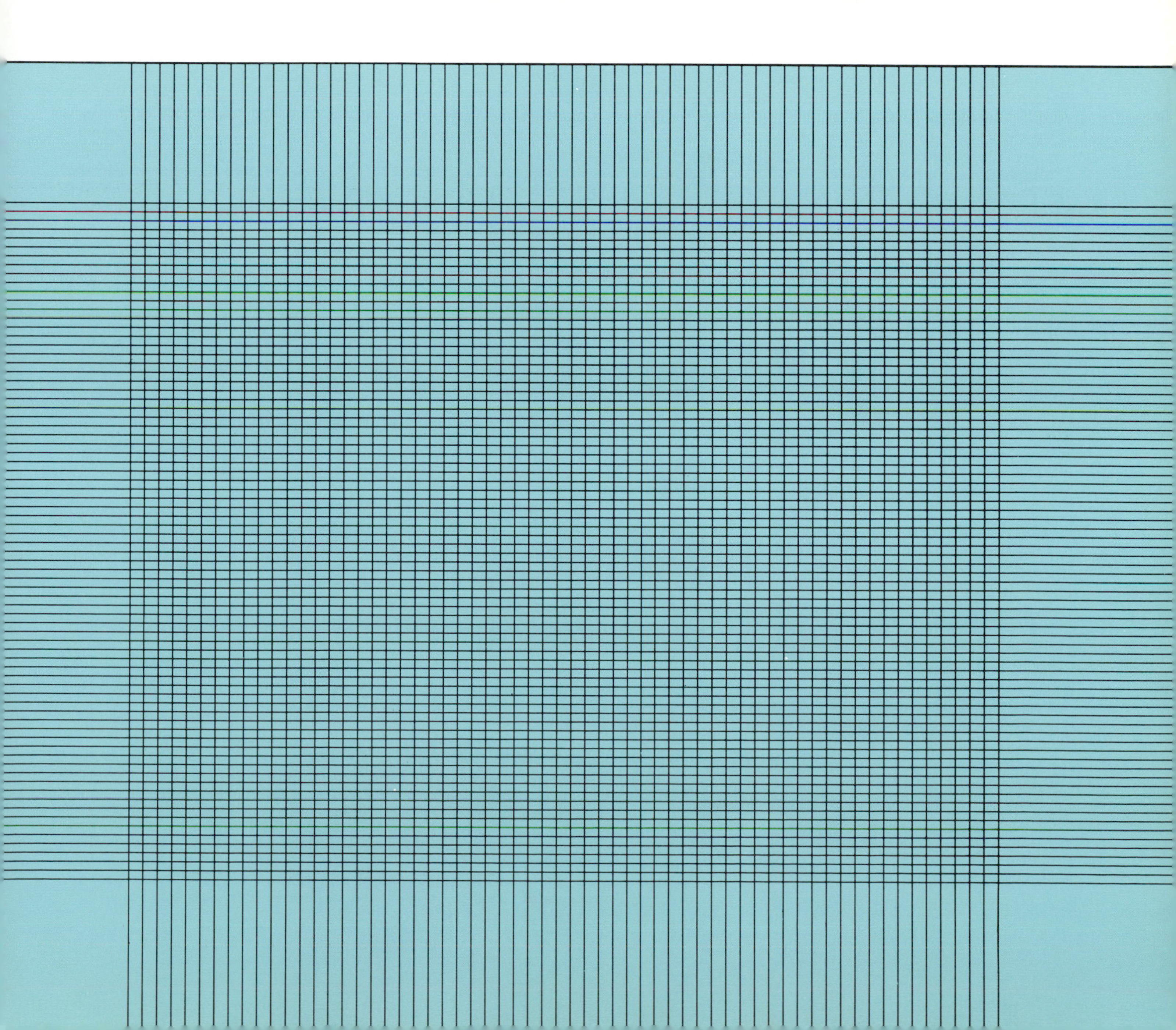

The Large-Scale Integration of Microelectronic Circuits

by WILLIAM C. HOLTON

When thousands of circuit elements are integrated on one chip, the integration is said to be large-scale. Many of these circuits not only are electrical but also follow the rules of Boolean logic

In the past 20 years the speed at which calculations can be performed has increased by six orders of magnitude, or roughly a million times. What is more, the cost of computation has fallen dramatically. Minicomputers available today for about $1,000 can nearly equal the capabilities of very large machines that cost as much as $20 million 15 years ago. By 1985 a medium-scale minicomputer may cost less than $100. The resulting increase in our ability to process information amounts to an intellectual revolution.

The need to improve the speed and computing capacity of machines was recognized in the 1940's, but performance was limited by the available electronic technology, which relied on vacuum tubes. Heat generated by the vacuum tubes resulted in their having a relatively short operating life, which placed an upper bound on the size of computers. The first completely electronic computer, ENIAC, was developed at the University of Pennsylvania in 1945; it had 18,000 vacuum tubes. Larger machines were clearly impractical; it would have taken almost 24 hours per day to find and replace the defective tubes.

The computer was saved from this premature end by the invention of the transistor in 1947. The reductions in size and cost that have been achieved in the past 10 years derive from the development of the integrated circuit in the late 1950's. Today the designing of computers and of other electronic devices is being transformed again by large-scale integration, the process whereby tens of thousands of transistors and their interconnections are manufactured simultaneously. Through this technology virtually all the logic elements of a digital computer can be fitted onto a chip of silicon no more than a quarter of an inch on a side.

The transistors of a digital circuit are operated as switches that generally have only two states: on or off, conducting or nonconducting. Similarly, the signals employed within the circuit have only two recognizable voltage levels, called simply high and low. It is because all the states of a digital circuit are confined to these two values that the logic and arithmetic functions of a computer are expressed in terms of the binary numbers, or numbers written to the base 2.

When a decimal number is read from right to left, each digit is understood to be multiplied by a progressively higher power of 10; indeed, the digits are often referred to as the ones, the tens, the hundreds and so forth. Binary numbers employ a similar positional notation except that the multipliers are powers of 2, so that the digits represent ones, twos, fours, eights and so on. Through this counting system any quantity can be represented as a string of 0's and 1's.

Arithmetic in the binary system is in many respects simpler than the corresponding operations with decimal numbers. Addition involves little more than counting, with provisions for carrying a binary digit, or bit, forward to the next higher power of 2. Multiplication is also a straightforward procedure. Since each digit of the multiplier must be either a 0 or a 1, each partial product formed must be equal either to zero or to the multiplicand. Thus the rule for multiplication is simply to write down the multiplicand, shifted one place to the left for each bit in the multiplier that is a 1, and sum those numbers.

Subtraction can be performed in the usual way, including the borrowing of bits from the next higher column, but it can also be accomplished in another way that is often more conveniently mechanized. By "inverting" all the bits of a binary number, that is, by changing all the 0's to 1's and all the 1's to 0's, a number called the complement is formed, which has some of the characteristics of a negative number. Numbers can therefore be subtracted by converting the subtrahend into its complement form and performing a specified scheme of addition. Finally, division can be done by counting how many times one number can be subtracted from another, with the count being the quotient. An important conclusion that can be drawn from these procedures is that all arithmetic operations on binary numbers can be reduced to addition. It follows that a computer can get along without special circuits for subtraction, multiplication and division provided it has the ability to add.

Perhaps even more important than the convenience of binary arithmetic is the facility with which binary numbers can express propositions in symbolic logic. In 1938 Claude E. Shannon of the Massachusetts Institute of Technology pointed out that switching circuits of a kind that were then built with electromechanical relays could be employed to evaluate logic statements. In other words, the dualities of on or off, high voltage or low, 1 or 0 could be made to stand for the duality of true or false.

The logic system employed by digital

ELECTRONIC ADDER in the photomicrograph on the opposite page is a small portion of a microprocessor formed entirely in the surface of a single chip of silicon. The microprocessor, an example of large-scale integration, is the TMS 9900, made by Texas Instruments Incorporated. The organization of the adder can be perceived in the repetitive pattern that divides the area shown into four vertical columns. Each column is a circuit, made up of 31 transistors, that operates on one digit, or bit, of the binary numbers to be added. All together there are 16 such columns, so that the microprocessor can add numbers 16 bits wide. The adder is employed in all arithmetic calculations. Green areas on the chip are mainly the source and drain regions of metal-oxide-semiconductor transistors; the gates of the transistors are pink. Mauve color appears in certain areas coated with silicon dioxide, and the silver lines are aluminum conductors.

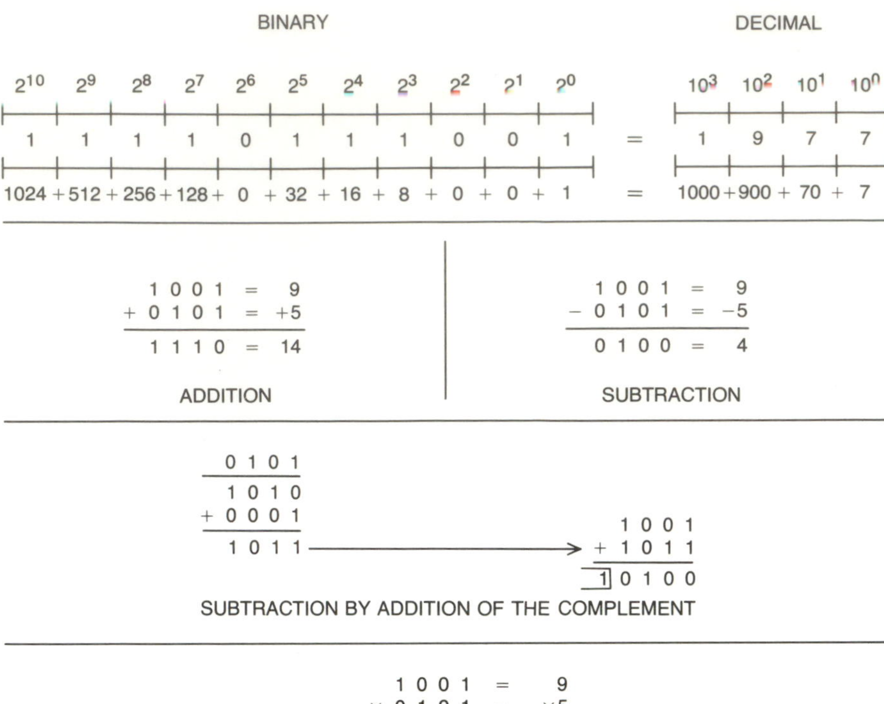

BINARY NUMBERS can represent any quantity as a string of 0's and 1's. Read from right to left, the bits of a binary number are understood to be multiplied by progressively higher powers of 2, just as the digits of a decimal number are multiplied by powers of 10. Binary addition involves little more than counting, with provisions for carrying a bit to the next higher power of 2. Subtraction can be done in the usual way, with borrowing from the next higher power of 2, or it can be done by means of a number called the complement, formed by changing all 0's to 1's and 1's to 0's, then adding 1 to the result. The difference between two numbers is found by adding the complement of the subtrahend to the minuend. Multiplication is much simplified because the digits of the multiplier are always 0 or 1; hence each partial product is equal either to zero or to the multiplicand. The partial products are written down, shifted an appropriate number of places to the left and added. Division, which is not illustrated here, can be done by repetitive subtraction. All these operations consist of procedures that can readily be performed by machine. Moreover, all the arithmetic operations can be reduced to addition.

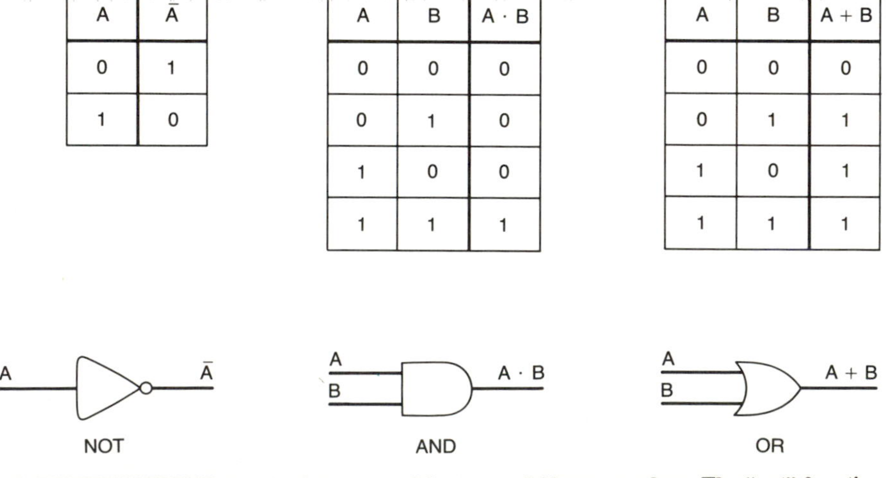

A	$\bar{A}$
0	1
1	0

A	B	A · B
0	0	0
0	1	0
1	0	0
1	1	1

A	B	A + B
0	0	0
0	1	1
1	0	1
1	1	1

LOGIC FUNCTIONS can also be expressed in terms of binary numbers. The "not" function "inverts" a binary digit, changing 0 to 1 and 1 to 0. The "and" and "or" functions accept two bits as inputs (here designated *A* and *B*) and generate an output bit determined by the values of the inputs. The function *A* "and" *B* (written in symbols $A \cdot B$) is a 1 only if both *A* and *B* are 1's. The "or" function (written $A + B$) generates a 1 output if either *A* or *B* is a 1 or if both inputs are 1's. At the bottom are symbolic representations of "gates" that perform these logic operations.

computers was devised by the British mathematician George Boole almost a century before the first electronic computer was built. The Boolean algebra provides rigorous procedures for deciding whether a statement is true or false provided only that the statement can be expressed in terms of variables that have just two possible values. Here we shall arbitrarily assume that "true" in Boolean logic is represented by a binary 1 or by a high-voltage state.

Through Boolean algebra logical analysis can be performed with just three functions, called "not," "and" and "or." The "not" function changes a binary bit to its opposite value: it converts a 0 into a 1 or a 1 into a 0. The "and" and "or" functions determine a single output bit from the values of two or more input bits. The "and" function is true only if all inputs are true. The "or" function, on the other hand, is true if one or more of its inputs are true.

The electronic representations of these functions are called logic gates. The "not" gate accepts a single bit as input and changes its value, or inverts it. If the input is a high voltage, the output is low; when the input changes, so does the output. The "and" gate accepts multiple input signals and produces a high output voltage only when a high voltage is simultaneously applied to all the inputs. The output of an "or" gate is high as long as a high voltage appears at one or more of its inputs.

Logic gates are basic functional units for both arithmetic and logic operations. An exercise that effectively demonstrates how the gates can be combined is the construction of a circuit that adds binary numbers. The circuit must accept as inputs the two bits to be added along with a carry bit, which could be either a 0 or a 1, from the next lower power of 2. It must produce as outputs a sum bit and a carry bit for the next higher power of 2.

The first step in the design of the circuit is the construction of a truth table that gives the desired output for every possible combination of inputs. There are eight combinations of the three binary inputs. The procedure for generating the sum bit consists in finding all the combinations that require a 1 as an output, then arranging gates to yield that output. If all three of the input bits are 0's, then of course the sum bit is also a 0. If exactly one of the input bits is a 1 while the other two are 0's, then a 1 output is required. There are three such combinations—the addend, the augend or the carry bit could be a 1—and circuitry must be provided for each of them. If two of the inputs are 1's and the third is a 0, then the sum bit is a 0, since in binary arithmetic 1 plus 1 equals 10. (The carry bit required will be generated in a separate group of gates.) Finally, all

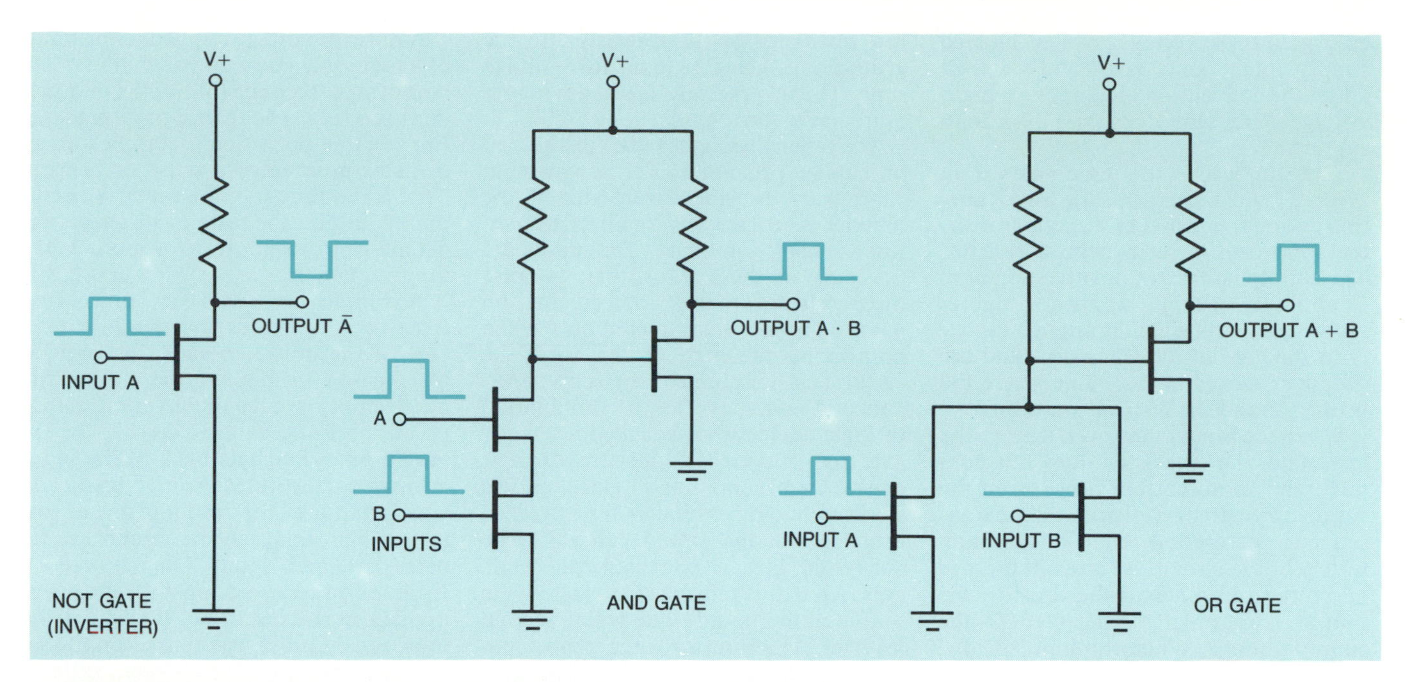

ELECTRONIC LOGIC GATES evaluate arithmetic and logic expressions in which binary values are represented by voltages. By convention a binary 1 is represented by a high voltage and a 0 by a low voltage. The gates shown here are constructed from metal-oxide-semiconductor transistors. The simplest is the "not" gate, or inverter. When the input to this gate is in the low state, the transistor does not conduct and only a negligible current flows from the supply voltage (*V*+) through the resistor and transistor to ground. As a result there is little voltage drop across the resistor and the output is in effect connected to the supply voltage. When a high signal is applied to the input, the transistor conducts and the comparatively large current flowing through the circuit produces a considerable voltage drop across the resistor. The output voltage is now near ground and is therefore in the low state. The "and" gate has two input transistors connected in series; current flows through them only when both receive a high signal simultaneously. In order to restore the proper polarity of the signal the output of the two series transistors is followed by an inverter. In an "or" gate the input transistors are arranged in parallel, so that a high signal applied to either one of them results in conduction. Again, an inverter is required to change the polarity of the output.

three of the inputs can be 1's, in which case the output is again a 1 (and a carry bit must again be generated). All together, then, four of the eight combinations of inputs require a 1 output: three combinations where exactly one input is in the 1 state and one combination in which all three inputs are 1's.

A circuit for the calculation of the sum bit can be built in segments, each segment representing one of the four "true" combinations of inputs. We can begin with the combination in which both the addend and the augend are 0's but the carry bit from the previous column is a 1. This condition must give rise to a 1 output, and so the addend and the augend are first passed through "not" gates, converting them into 1's, and are then applied to the inputs of an "and" gate. The output of this gate and the carry bit (in noninverted form) serve as inputs to another "and" gate, which gives the proper result.

Two more identical arrangements of gates are required. In one the addend and the carry bit are inverted by "not" gates and then combined with the augend; in the other the augend and the carry bit are inverted and combined with the addend. Finally, a fourth array of gates is needed to produce a true output when all three inputs are true. In this case "not" gates are unnecessary; the addend and the augend are applied to an "and" gate, whose output is combined with the carry bit in a second "and" gate. These four systems of gates will give the correct sum bit for all combinations of input bits. It remains only to tie the outputs of the four groups together with "or" gates so that the entire system of gates has a single output line that always gives the correct sum bit.

The generation of a carry bit for the next higher column is somewhat simpler than the calculation of the sum bit. The truth table for the carry bit indicates the bit should be a 1 whenever at least two of the inputs are 1's. It therefore suffices to combine the inputs in pairs. An "and" gate is provided for the addend and the augend, for the addend and the carry bit (from the next lower order) and for the augend and the carry bit. The outputs of these "and" gates are then combined by an "or" gate. If at least two of the inputs are true, then at least one of the "and" gates will have a true output and a carry bit of 1 will be generated.

When the adder is constructed, it has three input terminals (for the addend, the augend and the input-carry bit), which are applied to all the appropriate gates simultaneously. The two outputs are the sum bit and the carry bit for the next higher order. In this design the complete adder has three "not" gates, 11 "and" gates and five "or" gates. It is a simple matter to cascade such circuits to add binary numbers of more than one bit. The carry-output line of one stage is simply connected to the carry-input terminal of the next stage.

In a computer the symbols of Boolean algebra are reduced to the hardware of transistors and other electronic devices. In a microcomputer this transformation is particularly elegant because the symbolic logic is mapped almost directly onto the surface of a silicon chip.

We shall consider electronic realizations of the logic gates in only one semiconductor technology, that of enhancement mode *n*-channel metal-oxide-semiconductor (*n*-MOS) transistors [see "Microelectronic Circuit Elements," by James D. Meindl, page 12]. These transistors employ a positive power-supply voltage, and they conduct when a positive signal is applied to the gate. Only the simplest circuit configuration for each gate will be discussed here. These are not necessarily the best configurations, but the refinements commonly added to real computer circuits do not change any of the basic operating principles.

The three elements of an *n*-MOS transistor are the source, the gate and the drain. The gate (which should not be confused with a logic gate) controls the current flowing between the source and the drain. When a positive potential of about two volts is applied to the gate, the transistor conducts; when the voltage is reduced to zero, conduction ceases. Again these voltages can be labeled simply "high" and "low," and we shall adopt the convention of equating a high-voltage state with a binary 1 or a logical "true."

The simplest of the logic gates is an inverter, which performs the logic function "not." It consists of a single transistor connected in series with a resistor. The supply voltage (a positive potential of about five volts) is applied to the resistor. The input signal is applied directly to the gate of the transistor, and the output is sensed at the point where the transistor and the resistor are joined.

When the input to the inverter is in the low state, the transistor does not conduct, and so no current flows from the supply through the resistor and the transistor to the ground point, or common return of the circuit. As a result there is no voltage drop across the resistor: the output is essentially connected to the supply voltage. When the input is low, the output is high.

If a high signal is now applied to the gate, the transistor conducts. With a large current flowing through the circuit there is a substantial voltage drop across the resistor. The output of the gate is now effectively connected, through the low resistance of the transistor, to the ground; it is therefore in the low-voltage state. This is precisely the behavior required of a "not" gate.

A	B	C	SUM BIT	CARRY BIT
0	0	0	0	0
0	0	1	1	0
0	1	0	1	0
0	1	1	0	1
1	0	0	1	0
1	0	1	0	1
1	1	0	0	1
1	1	1	1	1

TRUTH TABLES for binary addition give the calculated values of the sum bit and the carry bit for all possible values of the input. The inputs are the two bits to be added (*A* and *B*) and the carry bit (*C*) from the next lower power of 2. The rules for addition specify that the sum bit is a 1 if exactly one of the inputs is a 1 or if all three are 1's; otherwise the sum bit is a 0. The carry bit (for the next higher power of 2) is a 1 if at least two of the inputs are 1's.

The other logic gates are constructed on similar principles. For a two-input "and" gate the single transistor of the inverter is replaced by a pair of transistors connected in series. The inputs are the gates of these transistors; because the transistors are in series a conduction path to ground is completed only when both inputs are high. In this form the circuit acts simultaneously as an "and" gate and as an inverter, a combination of logic devices that is called a "nand" gate (for "not-and"). Logic arrays can be constructed from "nand" gates, and in some technologies that is the standard practice, but the circuit can easily be converted into a conventional "and" gate. All that is required is to follow the output of the "nand" gate with a second inverter, which restores the proper polarity to the signal. When both inputs to the gate are high, the output of the "nand" portion of the circuit is low; that forces the output of the inverter—and of the circuit as a whole—into the high-voltage state.

An "or" gate is built in the same way except that the two input transistors are arranged in parallel rather than in series. As a result of this change a conduction path to ground exists if either of the inputs is high. As in the "and" gate, an inverter must be appended to the circuit if a high output is to follow from a high input. Without the inverter the circuit is a "nor" gate (for "not-or").

By simply ganging together logic gates as they have been described, a one-bit binary adder could be built from about 50 transistors and almost as many resistors. Actually an adder need not be that complex. Refinements to the design that allow one transistor to serve more than one function reduce the number of elements to 17 transistors and four resistors, or merely 21 transistors.

Even in its most simplified form the adder circuit is a logic array of considerable complexity. Moreover, it can only add numbers that are one bit wide. Microprocessors commonly work with binary numbers from eight to 16 bits wide; for a complete adder the circuit would have to be replicated a corresponding number of times.

The logical and arithmetic manipulation of binary numbers is not the only function of the basic gate circuits. Another essential function in computers is the storage of information. This can be accomplished by combining gates into circuits that are variously called flip-flops, latches, registers and memory cells, the name applied in a particular instance being determined in large part by the purpose the circuit is being made to serve.

A pulse of voltage applied to the input of a logic gate passes through it in a few tens of nanoseconds (billionths of a second) and leaves no record of its passage. For reliable operation a complex array of gates must have a sense of history. That is the function of a latch: a circuit that maintains the state of its output indefinitely in response to a momentary input signal.

A simple latch can be constructed from two "or" gates and two inverters. One of the inputs to each "or" gate is reserved as an input to the circuit; the other inputs receive the inverted output of the opposite gate. Referring the inverted output signals back to the input terminals constitutes positive feedback, which stabilizes the circuit in one of two states. The inputs to the circuit are labeled "set" and "reset." The output is a 0 if the pulse received most recently was applied to the reset line; it is a 1 if the most recent pulse was sensed on the set line. This circuit, which is often called a flip-flop because of the manner in which it changes state, can be fabricated with as few as two transistors.

A somewhat more versatile device called a register latch can be made by preceding the inputs to a flip-flop with a simple circuit consisting of two "and" gates and an inverter. Of the two inputs to this circuit one is reserved for a control signal and the other receives data. As long as the control signal is low, data pulses are blocked and have no effect on the state of the output. When the control line is high, the output changes to reflect the momentary status of the data and retains that state even after the control pulse is removed. A single latch can store one bit of information, and so a complete register requires one latch for each bit that is to be stored.

In microprocessors registers are employed for the temporary storage of data, of partial results, of instructions and of the addresses where other data or instructions are to be found. A special register designed to count continuously from zero to its maximum capacity is universally employed to step through a sequence of instructions.

A number of other logic devices are required for a functioning computer. For example, a comparator determines whether two binary numbers are equal or unequal. The individual bits of the numbers to be compared are applied to the inputs of "and" gates, and the outputs of those gates are combined in such a way that a single high bit is produced only if all the bits are identical. A decoder accepts a binary number as an input and activates one output or in some cases more than one output depending on the value of the number. A decoder that accepts a three-bit number could select one of eight lines, since there are eight binary numbers of three bits each (ranging from 000 to 111).

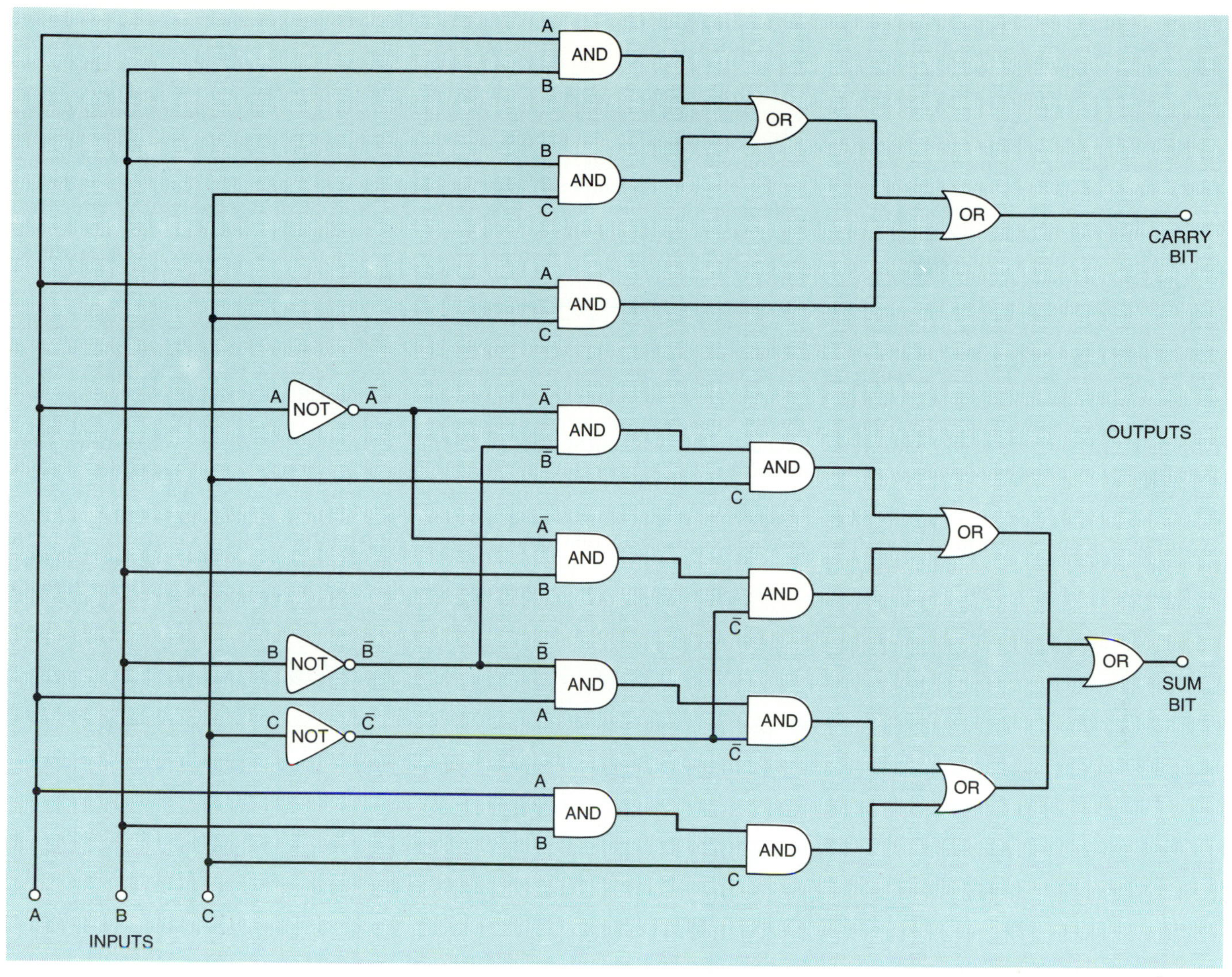

BINARY ADDER provides an array of logic gates for each combination of input bits (*A*, *B* and *C*) that requires an output. The four rows of gates at the bottom calculate the sum bit; the three rows at the top calculate the carry bit for the next higher power of 2. As an example, suppose both of the bits to be added (*A* and *B*) are 0's but the carry bit (*C*) from the preceding column is a 1; the truth table on the opposite page indicates that this combination of inputs must yield a sum bit of 1 and a carry bit of 0. The sum bit is generated in the fourth row from the bottom, where *A* and *B* are applied to an "and" gate in inverted form, so that both appear as binary 1's. The output of this gate, which is a 1, is combined with the carry bit in a second "and" gate, and the 1 output is passed through a series of "or" gates to the output of the adder. None of the gates for the calculation of the carry bit responds to this combination of inputs, and so the carry bit is a 0; if two or more of the inputs were 1's, then a carry bit of 1 would be issued. A binary adder requires a logic array like this one for each bit of the numbers to be added, with the carry-output line of each stage connected to the carry-input line of the next stage.

Any function that can be stated in terms of binary variables can be represented by an array of logic gates. We have seen how a circuit for addition can be constructed using about a dozen gates for each bit of the numbers to be added. By applying the same principles much more elaborate operations could be performed. For example, we might design a logic array that would square a number, then multiply the square by a coefficient, then add to that result the product of the original number and another coefficient and finally add a constant to the grand total. Such a circuit would evaluate quadratic expressions. It would be a complicated and expensive device, however, incorporating at least three multipliers (each in itself a complex circuit) along with adders, registers and other components. What is more, it could handle only quadratic expressions; to deal with cubic equations, another even more cumbersome circuit would be needed. If a computer were required to have a separate array of logic gates for every function it could perform, a machine of any versatility would be prohibitively large and costly.

Real computers make do with relatively few logic elements by effectively changing the interconnections between the devices and thereby creating at will any needed logic array. A calculation may call for several additions, for example, but the computer requires only one adder circuit, which is supplied in turn with each new set of operands. In order to operate a logic array in this flexible manner a more elaborate system of timing and control circuits is needed. Something else is required as well, something that is not present at all in the logic circuits we have considered so far: a program, or set of commands for changing the configuration of the logic array.

The concept of a stored program was introduced in 1833 by Charles Babbage as a feature of a "calculating engine" he proposed to build. His machine, which would have employed mechanical counting gears to do decimal arithmetic, would have incorporated a number of principles that are basic to the design of electronic computers. Although the idea was sound, the calculating engine was never built. The modern form of the stored program was introduced by John von Neumann, who added a significant refinement. He specified that both the data to be processed and the instructions for processing them be written in the same notation. By this contrivance instructions can be manipulated by the machine as if they were data. A program can therefore alter another program or even itself.

It is worthwhile considering in some detail how the logic elements of a computer are controlled by the program. For this purpose we shall invent a hypothetical microprocessor and a program for adding two binary numbers.

An organizational principle of almost all microprocessors is that the various parts of the machine communicate with one another through a system of conductors called a bus. The bus is simply a set of parallel conducting paths connecting various areas of the microprocessor chip and extending through connecting pins to a set of parallel conductors outside the chip.

All the logic devices that make up the computer are connected to the bus, but they do not all operate simultaneously. The memory, where both the program and the data are stored, communicates this information to the machine through the bus. The arithmetic and logic unit (ALU), as its name implies, contains devices such as adders for arithmetic and logic operations on data received over the bus. Several registers are connected to the bus for the temporary storage of operands and results; one of these registers, called the accumulator, is closely associated with the ALU. Another register is commonly dedicated to storing the memory addresses where needed data or instructions are to be found. Still another register, the program counter, always contains the address of the next instruction to be executed. Finally, decoding and controlling circuitry is attached to the bus, but it can also communicate with elements of the machine through separate control lines. Clock signals are provided to all parts of the machine over conduction paths independent of the bus.

The bus simplifies communication within the computer. Instead of installing a private line from each part of the machine to every other part, all the devices share one or more common buses. The cost of this simplification is that only one device can use any given bus at a time. Once again, more elaborate timing and control circuitry is required. The control is exercised by effectively disconnecting from the bus all devices except those that are transmitting or receiving data at a given time.

When the microprocessor is under the control of a program, it oscillates between two cycles: the instruction-fetch cycle and the instruction-execution cycle. Although the timing of events within these cycles varies from one microprocessor to another, we shall assume that in our machine the cycles are further divided so that the machine has a total of four states. In the first state an address is sent to memory and in the second an instruction is returned. In the

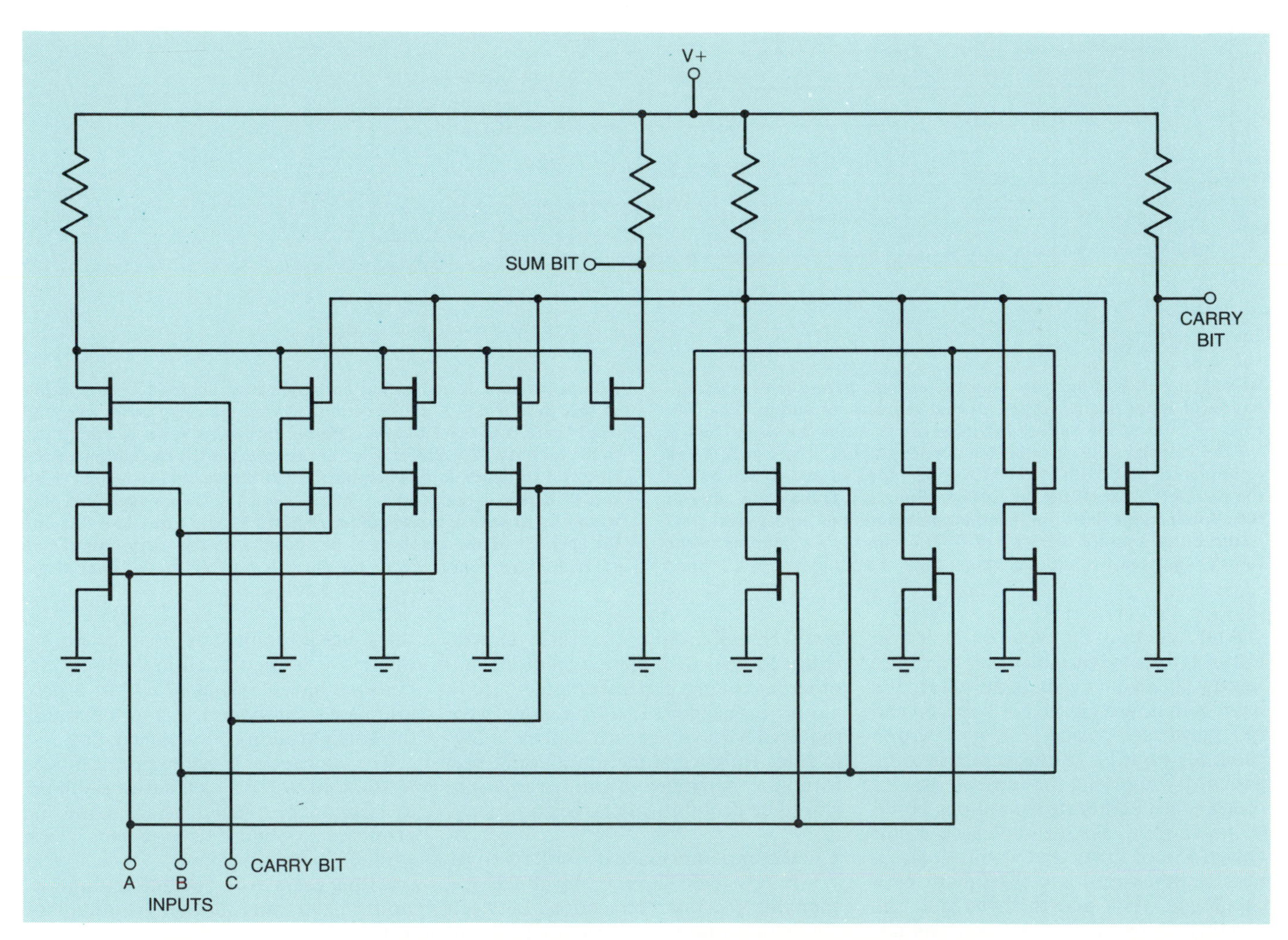

ADDER CIRCUIT is the functional equivalent of the logic array on the preceding page, but its structure is simplified by allowing some circuit elements to serve more than one function. The carry bit is generated in a straightforward manner by the seven transistors at the right. If any two of the inputs are in the high-voltage state, then one of the three pairs of transistors wired in series will conduct; if at least one of those pairs is conducting, then the common output of the three pairs is forced to the low-voltage state. This low signal is inverted by the transistor at the extreme right to produce a high, or binary 1, carry bit. The intermediate result of the carry-bit calculation is also employed in calculating the sum bit. In the array of seven transistors just to the left of center each of the inputs is combined with the inverted form of the carry bit, and the common output of these gates is also followed by an inverter. The result of this operation is that the sum bit is a 1 if one of the input bits is a 1 and the carry bit is a 0; examination of the truth tables for addition shows that this procedure always gives the correct sum. Finally, the three transistors wired in series at the extreme left generate a 1 output (through an inverter) if all three inputs are 1's. The circuit employs 17 transistors and four resistors, but in an integrated circuit the resistors could be replaced by transistors.

third state another address is sent to memory; in the fourth state an operand is returned from memory and processed in accordance with the instruction that has been received.

We shall assume that the program is properly stored in memory beginning at location 0 and that the program counter has been reset to 0. When the machine is started, it immediately enters the first state: the contents of the program counter are put on the bus and are interpreted as an address in memory. This start-up procedure is possible because the clock, which defines the four states, controls a system of gates that allows only the program counter to transmit signals to the bus during the first state. Only the memory system is able to receive signals from the bus during this period.

The address issued by the program counter is 0 since the counter was assumed to have been reset to that value. This binary number must be decoded by gates that form part of the memory system, so that a signal is applied to the memory cells with the address 0. The number of bits in the address determines the maximum number of memory locations. An eight-bit address can select only from among 256 locations, since there are only 256 binary numbers of eight bits each. A 16-bit address gives access to 65,536 locations. (Additional memory can be addressed if extra control signals are provided.)

The signal provided by the address-decoding circuitry to location 0 is a "read" signal; as a response to it the contents of that memory location are latched onto the memory output lines. When the machine leaves the first state and enters the second, the program counter and the address decoder are disconnected from the bus. At the same time the output of the memory is connected to it, again through gates controlled by the system clock. The contents of memory location 0 therefore appear on the bus during the second clock period. It will be remembered that the signals on the bus during this period are to be interpreted as an instruction, and so the clock must also activate the instruction-decoding and control circuitry. In the control unit the instruction is temporarily stored in a latch.

This entire sequence of events has served merely to fetch one instruction from memory and load it into the instruction register. What happens in the remainder of the machine cycle is now determined by the contents of that register. The decoding of the instruction is a crucial step in the operation of the machine. As in the decoding of memory addresses, the number of bits determines the maximum number of instructions, but that is rarely a serious limitation: few microprocessors have a repertory of more than 100 instructions.

The decoding of the instruction can be done in several ways. One method employs a "hard-wired" decoder, in which an array of gates selects a unique combination of active output lines for each possible combination of bits in the operation code. These outputs are the control lines that connect and disconnect various devices from the bus and that perform other functions as well, such as incrementing counters and initiating computations in the ALU. Instructions can also be decoded by applying them to a read-only memory, where a fixed pattern of bits is permanently stored when the processor is manufactured. Each of the operation codes is then interpreted as an address in the read-only memory; the output of the memory is latched and applied to the control lines.

We shall assume that the instruction fetched from memory location 0 is a binary number that has been assigned the meaning "Fetch the next word in memory and load it into the accumulator." This instruction might be executed in the following way. The instruction decoder (whatever its mechanism) could be constructed in such a way that the pattern of bits representing this instruction would activate five control lines. One of these control lines increments the program counter, so that it contains the address of the next location: 1. The other four enable the program counter, the accumulator, the address decoder and the memory output lines to be connected to the bus. Not all those devices, however, are connected at the same time, even though they all receive control signals simultaneously. When the clock enters its third period, marking the beginning of the instruction-execution cycle, the machine enters a state that is reserved for sending an address to memory. Hence only the source of the address (in this case the program counter) and the destination (the address decoder) are activated during this period by both a clock signal and a control signal.

In the fourth clock state the program counter and the address decoder are isolated from the bus, and the memory outputs and the accumulator are connected to it. As a result the number stored at

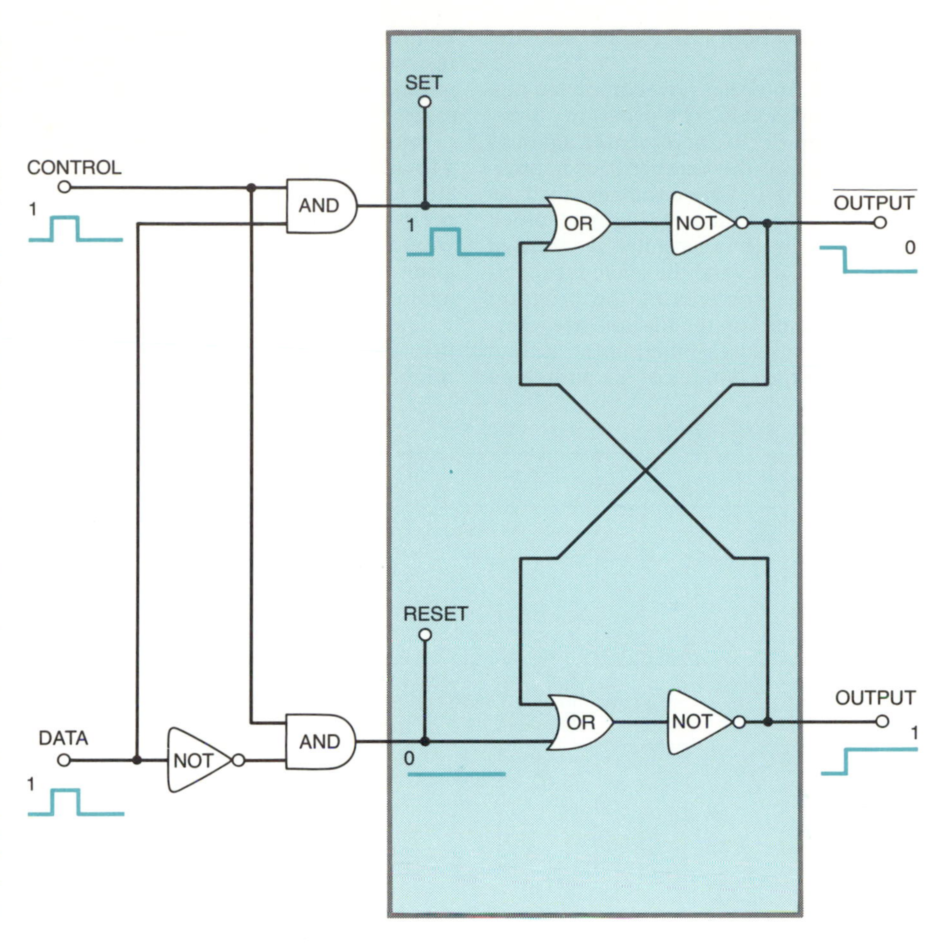

LATCH, OR "FLIP-FLOP," CIRCUIT maintains a record of momentary signals. In its simplest form (*colored box*) the circuit has inputs designated "set" and "reset." With the reset input low a brief high pulse applied to the set input forces the output to the high-voltage state, which is then maintained indefinitely by feedback from the output. With the set line low a high pulse at the reset terminal resets the output to zero. A second output terminal provides an inverted signal. This circuit can be preceded by a system of gates to make a controlled register latch. One of the inputs is now designated "control," and no signals can reach the flip-flop unless the control signal is high. In the presence of a high control signal, data can be applied to the remaining input. All high data pulses are steered to the set input and all low data pulses go to the reset input, so that the output always reflects the state of the most recent data signal.

memory location 1 is loaded into the accumulator.

The computer has now completed one full machine cycle. The program counter is automatically incremented again so that it contains the number 2, or in binary notation 010, where the next instruction is to be found. When the clock "ticks" again, the computer is returned to its initial state and the cycle is repeated. First the contents of the program counter are put on the bus and are interpreted as an address, then in the second clock state the contents of the addressed memory location are interpreted as an instruction. That second instruction might call for the contents of one of the registers to be issued as an address in memory during the third clock state, with the data returned being loaded into still another register during the fourth period. The second machine cycle would then be complete and the program counter would be incremented again (to 3, or binary 011).

The instruction fetched during the first part of the next machine cycle might call for adding the contents of a register to the contents of the accumulator. In this case there is no need to fetch an operand from memory during the third clock period and so that state could be ignored or perhaps deleted from the schedule. In order to execute the instruction the control circuitry must activate the adder in the ALU and arrange for the operands to be supplied by the correct registers. Results of operations within the ALU are ordinarily retained in the accumulator.

Subsequent instructions might call for comparing the sum formed by this oper-

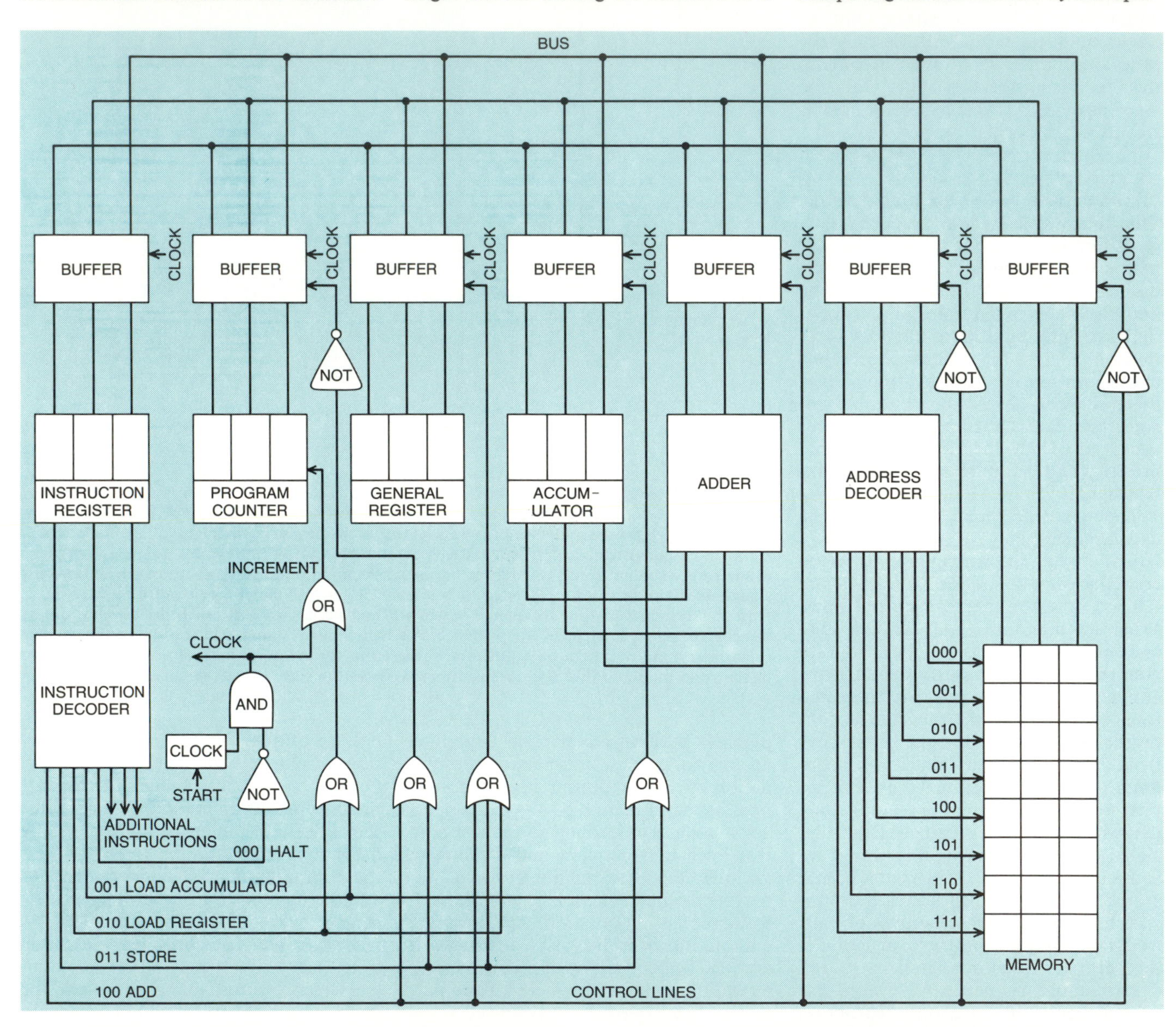

HYPOTHETICAL COMPUTER is an array of electronic logic elements whose interconnections can be altered by instructions stored in the computer itself. Instructions, operands and the "addresses" that designate locations in the computer memory are all expressed as binary numbers. In this machine the binary numbers are just three bits wide; since there are only eight possible combinations of three bits, operands can have only eight values, the machine can have only eight memory locations and it is limited to a repertory of eight instructions. The major pathway for communication in the machine is the "bus," a set of parallel conductors with connections to almost all the devices. These devices include four registers: the program counter, which ordinarily contains the address of the next instruction; the instruction register, where the binary code for each operation is retained; a general register for the temporary storage of operands or addresses, and the accumulator, a register with connections both to the bus and to the arithmetic and logic unit (ALU) of the computer. In this case the ALU is merely an adder, but in real machines it is capable of several operations. Data are written into the memory or read from it at a location specified by the address decoder. The decoder is an array of logic gates that activates one of eight address lines in response to a three-bit binary number. Instructions are decoded in the same way. Each of these devices is isolated from the bus by a "buffer," which connects the device to the bus only on receipt of a control signal from the instruction decoder along with a simultaneous signal from the system clock, which issues a continuous train of pulses to all parts of the machine. Information appearing on the bus could represent an operand, an instruction or an address; how it is interpreted depends on the state of the timing and control signals. The capabilities of a computer this small are trivial, but it reflects principles of organization common in real machines, including microprocessors.

ation with a value stored in memory. If the two numbers were different, no action would be taken, and the machine would proceed to the next instruction in the sequence. If the numbers were identical, however, a signal would be issued resetting the program counter to some preselected value. Execution of the program would then resume at this new memory location. Operations of this kind are called conditional branch instructions, since they enable the machine to choose one of two paths through a program depending on the results of a calculation. Other operation codes could instruct the machine to read data from "peripheral,"or external, devices, or to provide data as output to those devices. Ultimately the machine must come to a "halt" instruction, which stops all operations.

The machine described above is not equivalent to any particular microcomputer, but it incorporates features common to many of them (and to many large computers as well). The alternate fetching and execution of instructions, for example, is an almost universal mode of operation. A more fundamental feature is the use of the same language—the binary numbers—for all signals in the system. A pattern of bits appearing on the bus of our microprocessor can represent an address in memory, an instruction or an operand. How the pattern is interpreted depends on the state of the machine and thus depends critically on the timing and control signals, which are in turn determined by the instruction codes.

MACHINE CYCLE	CLOCK PERIOD	PROGRAM COUNTER	INFORMATION ON THE BUS		MEANING OF INSTRUCTIONS
1 FETCH	1	000	CONTENTS OF PROGRAM COUNTER	000	LOAD CONTENTS OF NEXT MEMORY LOCATION INTO GENERAL REGISTER
	2	000	INSTRUCTION "LOAD REGISTER"	010	
EXECUTE	3	001	CONTENTS OF PROGRAM COUNTER	001	
	4	001	DATA	111	
2 FETCH	1	010	CONTENTS OF PROGRAM COUNTER	010	LOAD CONTENTS OF NEXT MEMORY LOCATION INTO ACCUMULATOR
	2	010	INSTRUCTION "LOAD ACCUMULATOR"	001	
EXECUTE	3	011	CONTENTS OF PROGRAM COUNTER	011	
	4	011	DATA	001	
3 FETCH	1	100	CONTENTS OF PROGRAM COUNTER	100	ADD CONTENTS OF GENERAL REGISTER TO CONTENTS OF ACCUMULATOR AND RETAIN THE RESULT IN THE ACCUMULATOR
	2	100	INSTRUCTION "ADD"	100	
EXECUTE	3	100	(IDLE)	—	
	4	100	CONTENTS OF GENERAL REGISTER	111	
4 FETCH	1	101	CONTENTS OF PROGRAM COUNTER	101	STORE THE CONTENTS OF THE ACCUMULATOR IN MEMORY AT AN ADDRESS SPECIFIED BY THE CONTENTS OF THE GENERAL REGISTER
	2	101	INSTRUCTION "STORE"	011	
EXECUTE	3	101	CONTENTS OF GENERAL REGISTER	111	
	4	101	CONTENTS OF ACCUMULATOR	000	
5 FETCH	1	110	CONTENTS OF PROGRAM COUNTER	110	
	2	110	INSTRUCTION "HALT"	000	
EXECUTE	3	110	—	—	
	4	110	—	—	STOP ALL OPERATIONS

PROGRAM for the three-bit computer calls for the addition of two numbers. The system clock divides each machine cycle into four periods; the first two periods are devoted to fetching an instruction, the last two to executing it. In the first clock period the contents of the program counter (initially 000) are put on the bus and interpreted as an address in memory; in the second period the contents of the designated memory location are latched into the instruction register. Events in the third and fourth periods depend on the operation specified by the instruction. In the first machine cycle the operation code is 010, which has been assigned the meaning "Fetch the contents of the next location in memory and load this number into the general register." In response to this pattern the instruction decoder activates a control line that increments the program counter, so that the counter points to the next item in memory, and enables the general register to receive the stored value. In the second machine cycle another operand is loaded into the accumulator, then in the third cycle the number in the register is added to the one in the accumulator. The numbers added are 111 and 001, and since the machine has no provisions for overflow the result is 000. In the fourth machine cycle the value in the general register, which was initially an operand, is treated as an address, where the result of the addition is stored.

Large-scale integration has transformed the computer from a machine that must be built from many components into a component that can be incorporated into larger systems. A number of microprocessors now on the market include on a single chip all the circuitry we have discussed with the exception of random-access memory and in most cases the clock [see "Microprocessors," by Hoo-Min D. Toong, page 66]. A few single-chip devices even include a limited amount of memory.

The first generation of commercially produced microelectronic devices are now referred to as small-scale integrated circuits. They included only a few gates. Even where multiple gates were included in one package they were not always connected internally; the circuitry defining a logic array had to be provided by external conductors.

Devices with more than about 10 gates on a chip but fewer than about 200 are medium-scale integrated circuits. Many of these chips contain complete functional blocks of a computer. For example, chips are available that contain complete registers, decoders, comparators or counters. Perhaps the upper boundary of medium-scale integrated-circuit technology is marked by chips

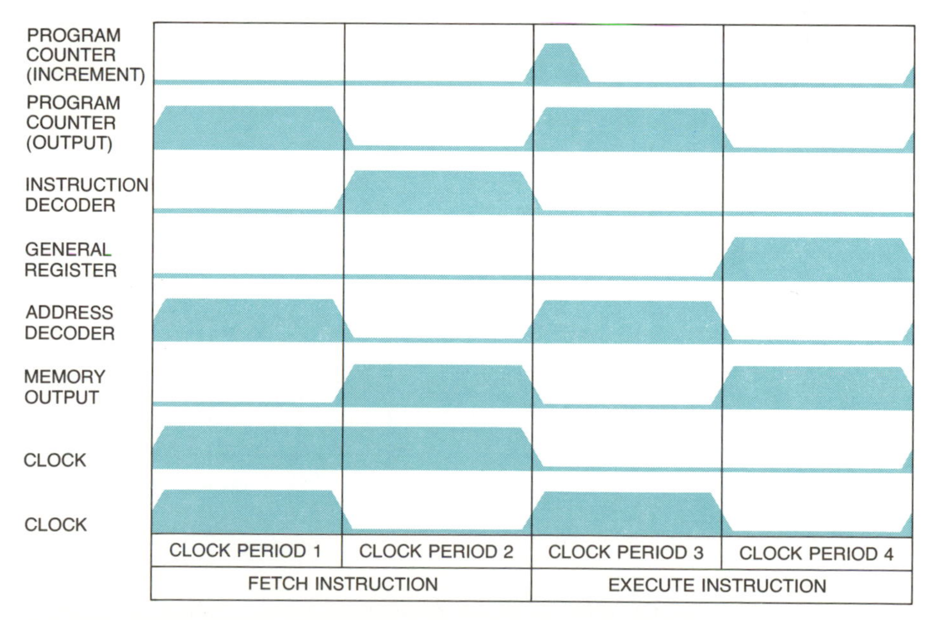

TIMING DIAGRAM gives the sequence of events during the first machine cycle in the execution of the program at the top of the page. Two clock signals divide the machine cycle into four clock periods. In the first period only the program counter and the address decoder are connected to the bus, and in the second period only the memory output and the instruction decoder are connected. In the third and fourth periods the machine is under the control of both the clock and the instruction decoder. Together they increment the program counter, then connect both it and address decoder to bus. In fourth period the memory output and the general register are active. Finally, as the clock returns to the first state, the program counter is incremented again.

that contain a complete arithmetic and logic unit. This unit accepts as inputs two operands and can perform any one of a dozen or so operations on them depending on the state of several control lines. The operations include addition, subtraction, comparison, logical "and" and "or" and shifting one bit to the left or right.

The greatest variety of medium-scale integrated circuits is available in the bipolar semiconductor technology called TTL, for transistor-transistor logic. Even with the development of large-scale integration, TTL devices have retained their utility. Many microprocessors are designed so that their signal voltages and power-supply requirements are compatible with those of TTL circuits. Hence TTL devices can be employed as auxiliary components in a mi-

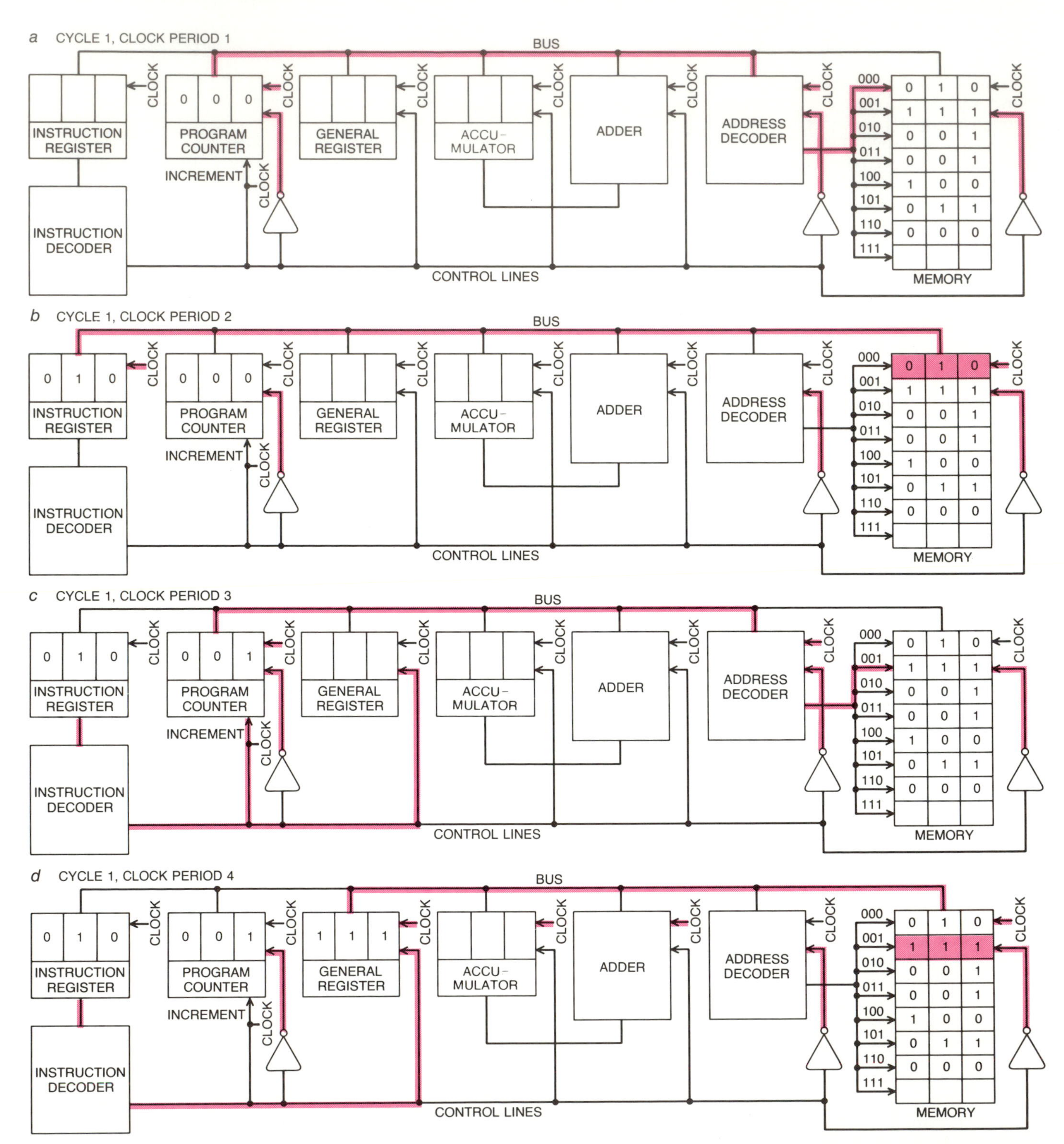

EXECUTION OF A PROGRAM by the three-bit computer is traced in a sequence of simplified diagrams. Initially the program is stored in the first seven memory locations and the program counter is set to 000. During the first two clock periods no instruction has yet been decoded and so the machine must operate entirely under the control of the system clock. This mode of operation is made possible by inverters in the control lines to the program counter, the address decoder and the memory output, which ensure that these devices can be activated by the clock in the absence of a control signal. In the first clock period (*a*) the program counter and the address decoder are connected to the bus, with the ultimate result that location 000 in memory is designated for reading. The contents of this location are latched into the instruction register during the second period (*b*). It is not until the third clock period that control signals from the instruction decoder become available. The control line that is energized immediately increments the program counter and provides a signal enabling the general register to be connected to the bus; in addition this control line fails to disconnect the program counter, the address de-

crocomputer system for the control of memory or of input and output terminals and for other functions such as generating timing signals.

For the most part large-scale integrated circuits have been fabricated not with the bipolar transistors of TTL but with metal-oxide-semiconductor transistors, which can be made smaller and packed closer together. The change in scale can be measured crudely by counting the number of transistors that can be fitted onto a chip. Small-scale integrated circuits have on the order of 10 transistors per chip; medium-scale circuits have on the order of 100. Recent microprocessors made through large-scale integration have between 10,000 and 20,000 transistors. In the orderly geometric array of elements in semiconductor memory chips the number of transis-

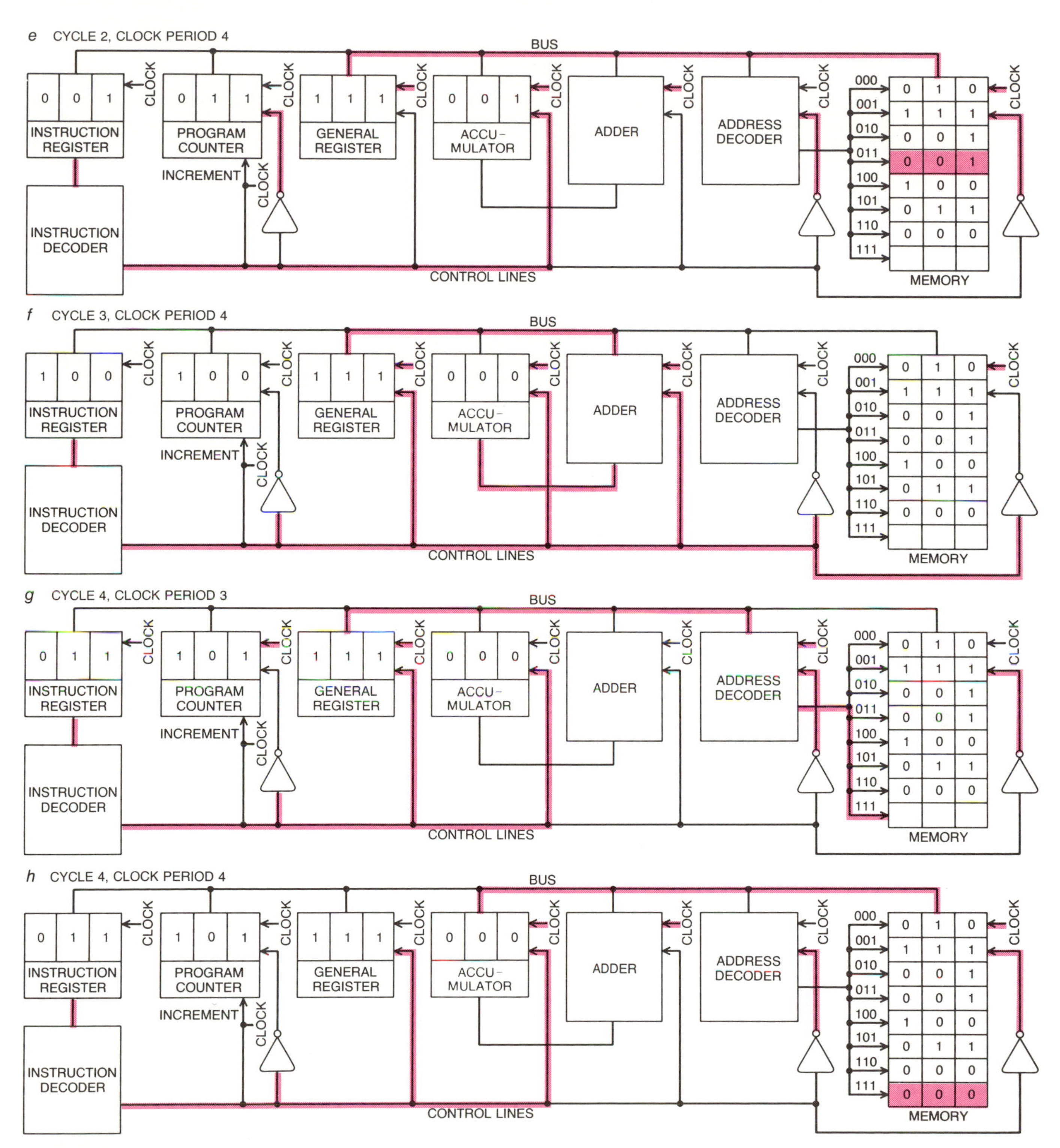

coder and the memory output. Not all these devices are activated simultaneously, however. Since the third clock period (*c*) is reserved for sending the address of an operand to memory, the clock activates only the program counter and the address decoder during this period. In the fourth period (*d*) the program counter and the address decoder are turned off by the clock, and the memory output and the general register are activated. The instruction decoded during the second machine cycle (*e*) calls for loading another value from memory into the accumulator. Addition is performed during the third cycle (*f*). Since this operation requires no memory access, signals are issued by the instruction decoder that disable the address decoder and the memory output; the program counter is also disabled. At the same time the general register, the accumulator and the adder are activated. In the third clock period of the fourth cycle (*g*) the contents of the general register are put on the bus, and since the only other active device is the address decoder this value is interpreted as an address in memory. The contents of the accumulator, where the result of the addition was retained, are stored at this location during the fourth period (*h*).

tors is now approaching 100,000 [see "Microelectronic Memories," by David A. Hodges, page 54].

Three factors have contributed to the rapid growth in the number of circuit elements per chip. One factor is improvement in techniques for growing large single crystals of pure silicon. By increasing the diameter of the wafers—the disks of silicon on which chips are manufactured—more chips can be made at one time, reducing the unit cost. Moreover, the quality of the material has also been improved, reducing the number of defects per wafer. This has the effect of increasing the maximum practical size of a chip because it reduces the probability that a defect will be found within a given area. The chip size for large-scale integrated circuits has grown from less than 10,000 square mils (thousandths of an inch) to 70,000.

A second factor is improvement in optical lithography, the process whereby all the patterns that make up a circuit are ultimately transferred to the surface of the silicon. By developing optical systems capable of resolving finer structures, the size of a typical transistor, as measured by the gate length, has been reduced from a few thousandths of an inch in 1965 to two ten-thousandths of an inch today. Finally, refinements in circuit structure that make more efficient use of silicon area have led to a hundredfold increase in the density of transistors on the chip.

Continued evolution of the microcomputer will demand further increases in packing density. As the limit of optical resolution is now being reached, new lithographic and fabrication techniques will be required. Circuit patterns will have to be formed with radiation having wavelengths shorter than those of light, and fabrication techniques capable of greater definition will be needed.

The present scale of microelectronic structures requires an edge acuity of 2,500 angstroms, which is roughly the wavelength of the ultraviolet radiation employed in making the circuits. To form sharp images of still smaller structures, patterning by means of a steered electron beam can be employed. The highest potential resolution is offered by X-ray lithographic systems, which have an effective wavelength of about 10 angstroms.

Electron-beam and X-ray patterning could conceivably reduce the linear dimensions of a transistor by another factor of 50. The area occupied by the transistor would therefore be reduced more than 1,000 times. If the capabilities of the present generation of microprocessors are remarkable, the prospects for this next generation of very-large-scale integrated circuits (VLSI) are extraordinary. A processor would take up a trivial portion of such a chip; millions of bits of memory could be included as well.

Although the immediate impediments to building more complex microprocessors lie in the fabrication process, the task of design is hardly trivial. The design process begins with the determination of the overall organization of the computer. The "width" of the machine, or in other words the number of bits in a standard word, is a fundamental consideration; so is the amount of memory to be addressed, which dictates the number of address bits that must be provided. Among a multitude of other considerations, attention must be given to the total number of connections to be made to the chip: each connection requires a metallic pad at the edge of the chip, and there is limited room for these pads.

Once the architecture, or organization, of the machine has been decided on, logic diagrams must be prepared for all the functions that are to be included. The logic diagrams must then be reduced to their electronic equivalents, and finally the resulting circuit diagrams must be expressed in terms of electronic devices and interconnection paths between them that can be manufactured at a reasonable cost within the allotted area of silicon [see "The Fabrication of Microelectronic Circuits," by William G. Oldham, page 40].

The properties of individual transistors and other components depend critically on their dimensions. If the circuit is to function reliably, these must remain within prescribed limits in spite of variations in the manufacturing process. Timing is equally critical: when a signal is issued by one part of the computer, another part must be ready to receive it. One obvious approach to these problems is to build a prototype of the processor out of existing components. In recent years it has become far commoner, however, to build a simulated prototype with the aid of a computer. Mathematical models of the individual circuit elements are constructed and combined to form models of entire systems. In essence a large computer is made to emulate the behavior of a smaller one.

When a design is complete, the computer itself can prepare the master drawings that will be employed to define the circuit pattern on the silicon. Other computers supervise the fabrication process, and still another one tests the completed chips, automatically marking the defective ones. There is no more telling demonstration of the need for evolutionary development in the creation of a complex technology. The integrated circuits being designed and manufactured today could not be made without the assistance of those made in past years.

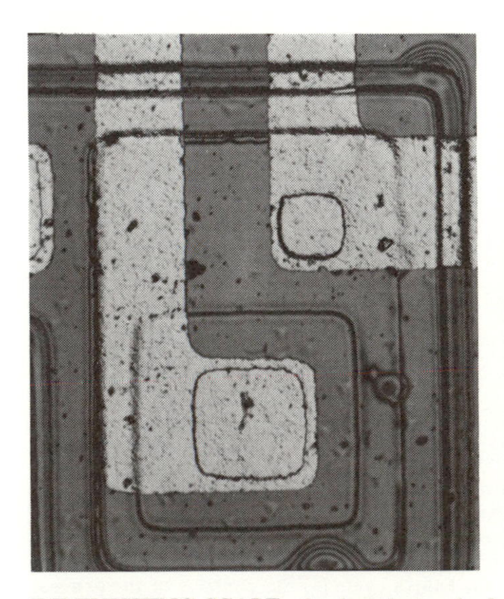
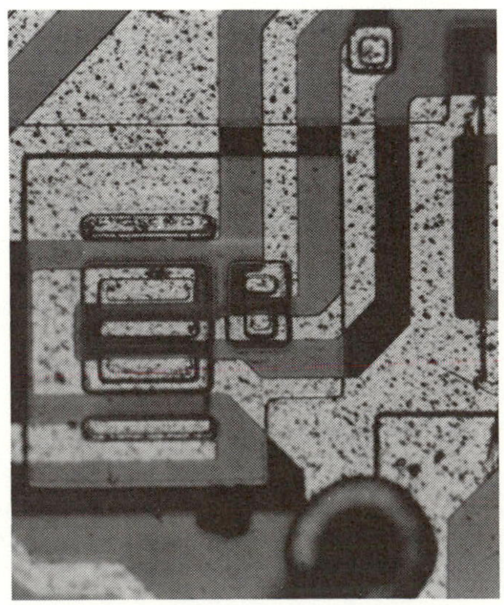
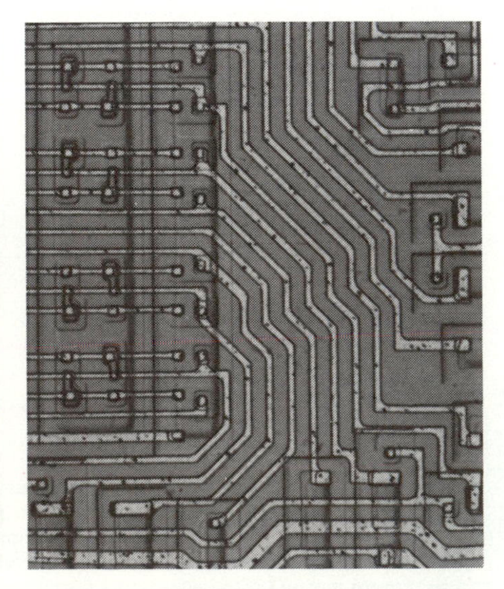

DIMINISHING SCALE of microelectronic devices is apparent in three chips seen at the same magnification: about 300 times. At the left is a chip manufactured in 1961; a single transistor fills the field of view. The chip in the middle was made in 1965; portions of several active elements are visible. The large-scale integrated circuit at right, manufactured in 1975, crowds more than a dozen transistors into the same area of silicon. All these devices employ bipolar semiconductor technology; higher density can be achieved with MOS transistors.

4

THE FABRICATION OF MICROELECTRONIC CIRCUITS

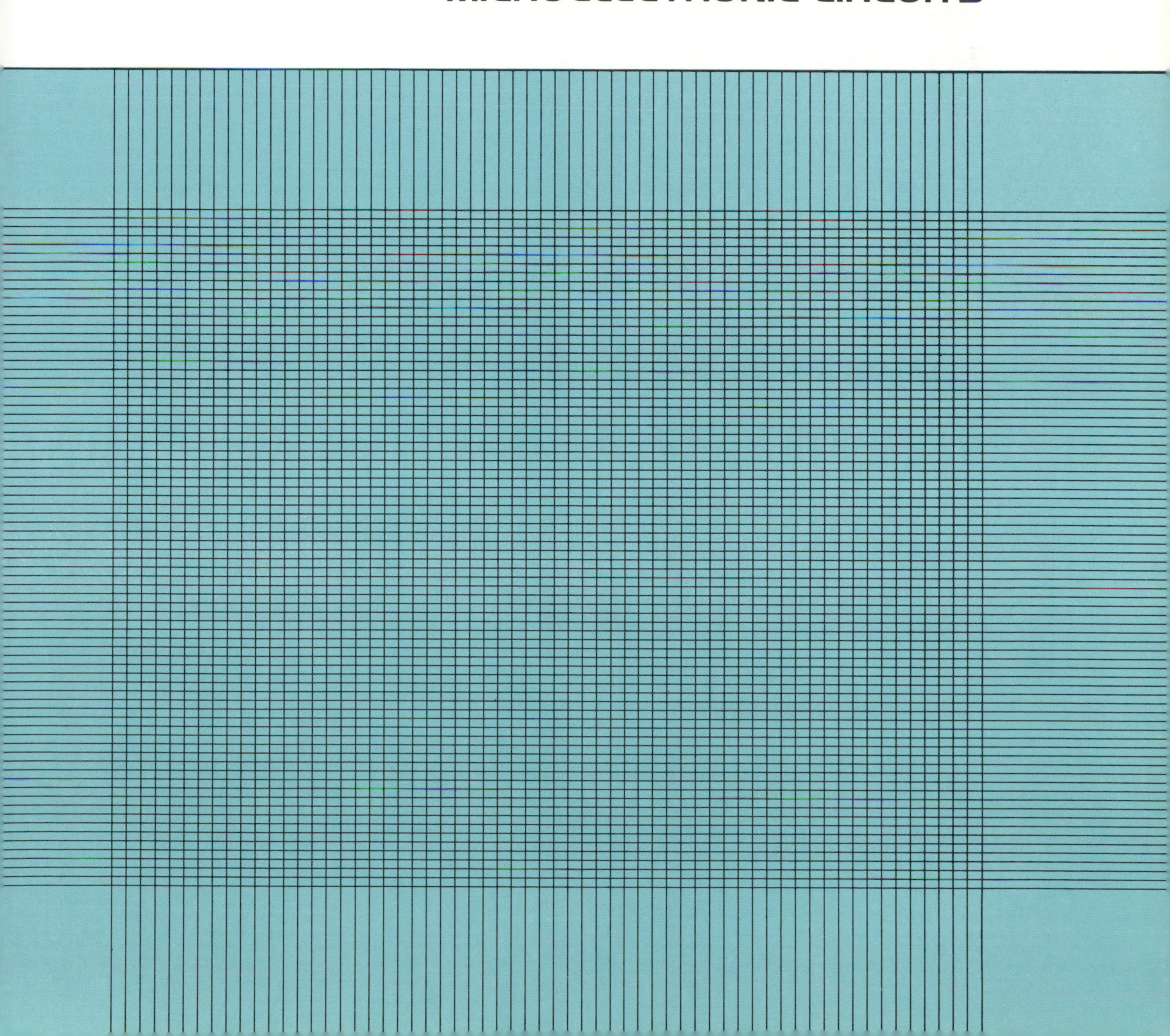

The Fabrication of Microelectronic Circuits

by WILLIAM G. OLDHAM

The patterns of integrated circuits are made in the large and then are put on the chip photographically. The objective of fabrication methods is the production of chips at the lowest cost per function

The manufacture of large-scale integrated circuits has as its primary goal the lowest possible cost per electronic function performed. The main features of the fabrication processes adopted by the microelectronics industry can be best understood in terms of this goal. These features include the fabrication of many circuits at a time (an extraordinary example of mass production), the reduction of the circuits to the smallest possible size and the maximum simplification of the processing technology.

The dramatic reduction in the cost of microelectronic circuits achieved in the past few years has not resulted from any major new breakthrough in fabrication technology. Indeed, most of the basic manufacturing processes involved have been widely adopted in the industry for five years or more. The recent sharp drop in fabrication cost has been achieved during a period of general economic inflation. The cost of processing a "wafer" of silicon, the substrate on which the microelectronic circuits are made, has risen moderately, but the area of the wafers has increased more rapidly, approximately doubling every four years. Thus the processing cost per unit area has actually decreased. Meanwhile the space required for a given electronic function has shrunk by a factor of two every 18 months or so. This reduction in size has come not only from great ingenuity in designing simple circuits and simple technological processes for making them but also from the continuing miniaturization of the circuit elements and their interconnections. Moreover, the gradual elimination of defects in various manufacturing steps has resulted in a significant decrease in the net cost of fabrication. With a lower frequency of defects the yield of good circuits on a given wafer increases.

The current pace of developments in the manufacture of microelectronic circuits suggests that the progress in reducing production costs will continue. Virtually every stage of fabrication—from photolithography to packaging—is either in the midst of a significant advance or on the verge of one.

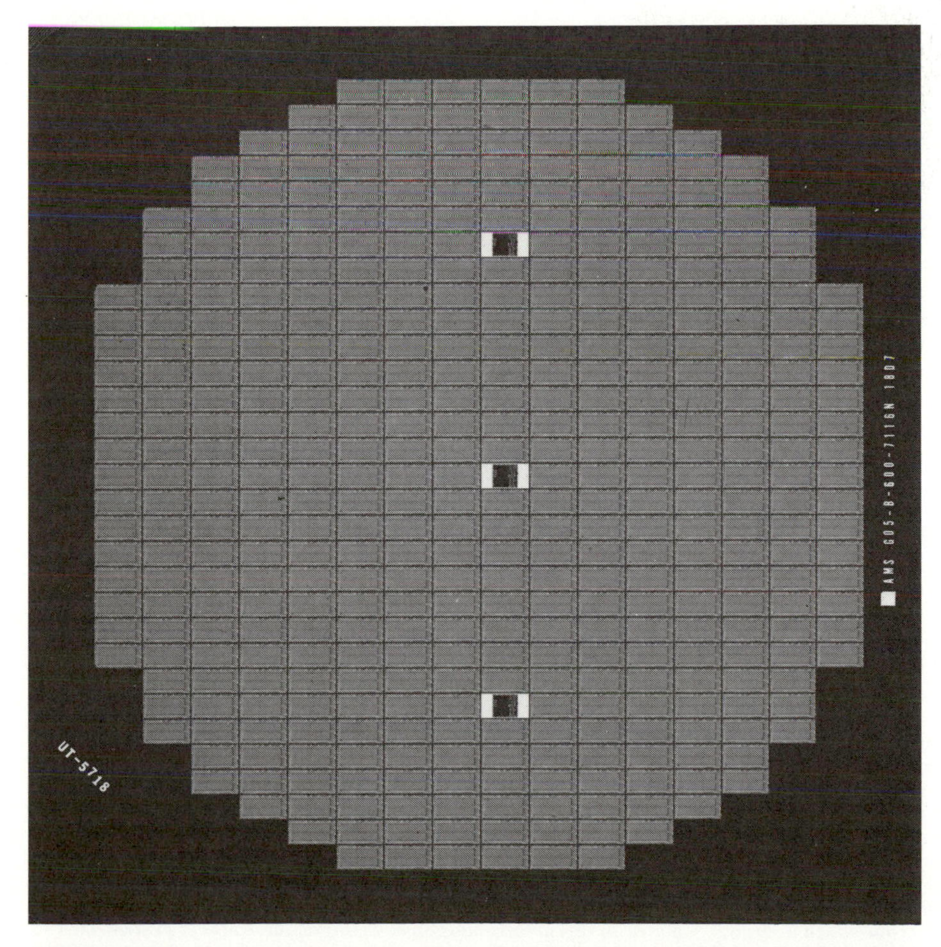

PHOTOMASK, a glass plate bearing the pattern for a single layer of integrated circuitry etched in a chromium film on its surface, is shown in silhouette in these two photographs, one reproduced actual size (*above*) and the other enlarged 10 diameters (*opposite page*). The photomask is used in the manufacture of a 16,384-bit random-access memory, a product of Intersil Inc.

A large-scale integrated circuit contains tens of thousands of elements, yet each element is so small that the complete circuit is typically less than a quarter of an inch on a side. The pure, single-crystal silicon wafers that bear the circuits are much larger: currently three or four inches in diameter. One of the key economies in the manufacture of microelectronic circuits is the simultaneous fabrication of hundreds of circuits side by side on a single wafer. An even greater scale of mass production is attained in several stages of manufacture by processing as many as 100 wafers together in a batch. Hence the cost of labor and equipment is shared by thousands of circuits, making possible the extremely low per-circuit cost that is characteristic of microelectronics.

After a wafer has passed through the fabrication stage it is sectioned into individual dice, or chips, each of which is a complete microelectronic circuit. Not all the circuits will work. Defects in a wafer cannot be avoided, and a single defect can ruin an entire circuit. For example, a scratch only a few micrometers long can break an electrical connection. It is impossible both physically and economically to repair the defective circuits; they are simply discarded.

As one might expect, the larger the die, the greater the chance for a defect to appear and render the circuit inoperative. The yield (the number of good circuits per wafer) decreases with the size of the dice, both because there are fewer places on a wafer for larger dice and because larger circuits are more likely to incorporate a defect.

At first it might appear most economical to build very simple, and therefore very small, circuits on the grounds that more of them would be likely to be good ones. It is true that small circuits are inexpensive; simple logic circuits are available for as little as 10 cents each. The costs of testing, packaging and assembling the completed circuits into an electronic system, however, must also be taken into account. Once the circuits are separated by breaking the wafer into dice, each die must be handled individually. From that point on the cost of any process such as packaging or testing is not shared by hundreds or thousands of circuits. In fact, for medium-scale integrated circuits packaging and testing costs often dominate the other production costs.

A typical microelectronic system con-

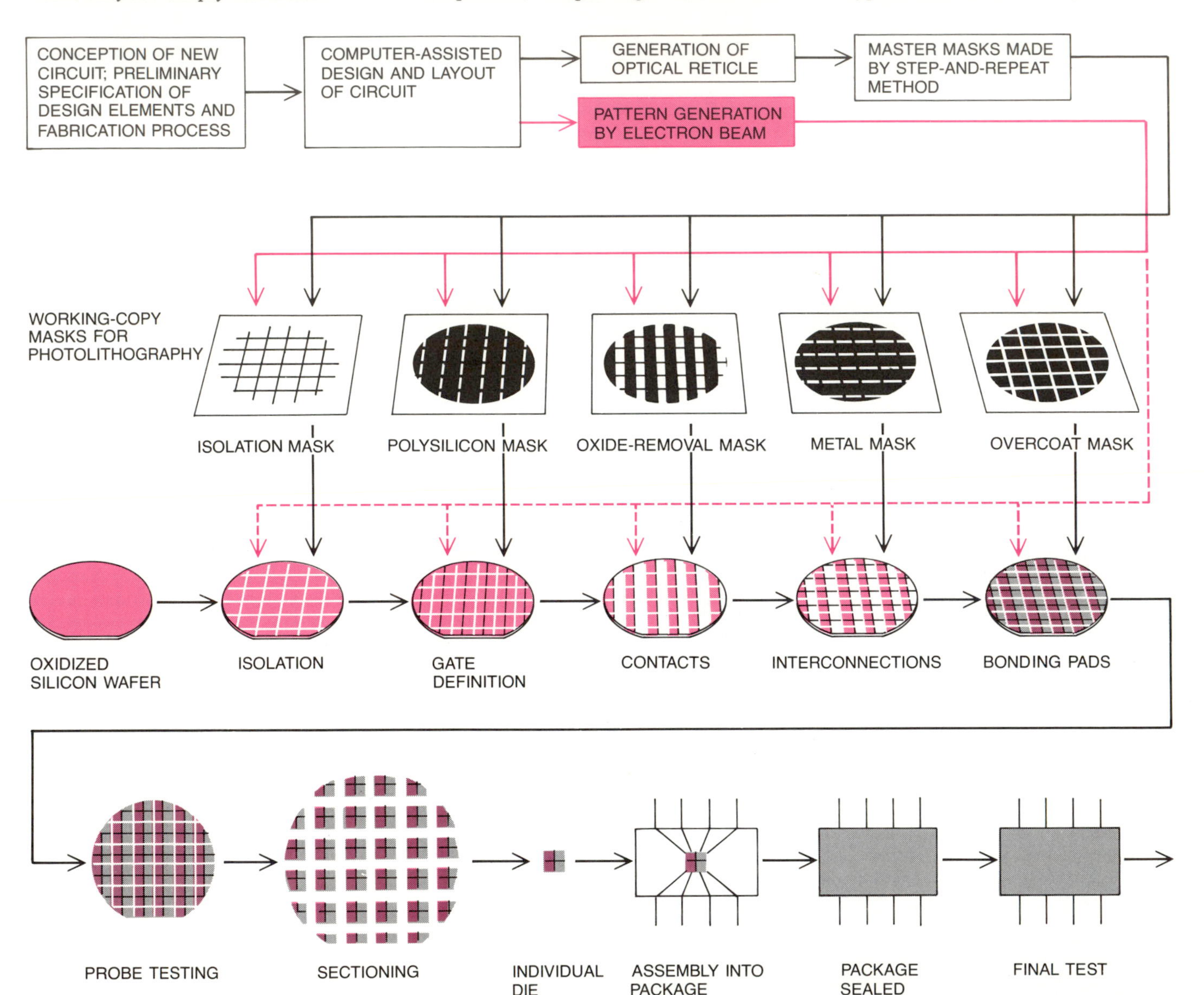

MANUFACTURE of a large-scale integrated circuit is outlined in this schematic diagram. The design of the circuit is often carried out with the aid of a computer, which helps to determine the most space-conserving layout of the circuit elements. The resulting layout is used to prepare a set of photomasks, each containing the pattern for a single layer. Usually this is done by first generating a tenfold enlargement of each layer, called a reticle, which is checked, corrected and regenerated until it is perfect. A photographically reduced image of the reticle is next reproduced hundreds of times in a "step and repeat" process to yield a set of final-size master masks, from which a large number of working plates are copied. (The electron-beam approach to pattern generation, represented here in color, is beginning to replace the optical method for producing the masks, thereby making it possible to eliminate two photographic steps and write the pattern directly on the working mask from the information stored in the computer memory; in time it is expected that electron-beam lithography will also be introduced directly into the fabrication of the circuits themselves.) The silicon "wafers" that serve as substrates for the circuits are obtained by sawing a long single crystal of silicon into thin slices, which are polished, cleaned and oxidized in preparation for the first patterning step. (A "flat" is ground along the length of the crystal so that each wafer will have a reference edge parallel to a natural crystal plane.) Following the five-mask fabrication process depicted in this greatly simplified example, the wafer is probe-tested to determine the good circuits, and the defective circuits are marked with an ink spot. Individual "dice" bearing good circuits are then selected from the sectioned wafer and assembled into packages. A final test is performed to ensure that the packaged circuit works properly.

tains both large-scale and medium-scale integrated circuits, and the cost of designing and constructing the system rises rapidly as the number of circuits increases. To minimize the total cost of the system one would like ideally to use either a small number of very powerful circuits (requiring that each circuit be large) or a large number of very cheap circuits (requiring that each circuit be small). In many complex systems the minimum total cost is achieved with the more expensive, larger circuits: ones costing closer to $10 each than to $1. If less powerful, cheaper circuits were to be used, many more would be needed for constructing the system, and testing and assembly costs would tend to build up. On the other hand, if too much electronic function is packed into a circuit, the large size of the die would result in such a low yield per wafer that the cost per circuit would become prohibitive.

Assuming a typical selling price of $10 for a large-scale integrated circuit, one can work backward and estimate optimum die sizes. (The selling price of the circuit must of course be high enough to recover not only the direct costs of manufacture but also the costs of research and development, marketing and general overhead; thus it is reasonable to assume that the direct manufacturing cost is a good deal less, say about $5.) It costs roughly $100 just to process a silicon wafer, regardless of the size of the dice; when testing and packaging are included, the total manufacturing cost is perhaps doubled. Hence if the manufacturing cost of the particular integrated circuit in question is assumed to be $5, the optimum die size is one that yields about 40 good circuits per wafer. At present it is possible to achieve such a yield for dice that measure approximately five millimeters on a side. It is interesting to note that in this example a rather low percentage of the circuits on the wafer are good. The yield is only 40 good circuits out of the 250 that can be fabricated on a single wafer 100 millimeters in diameter.

The structure of an integrated circuit is complex both in the topography of its surface and in its internal composition. Each element of such a device has an intricate three-dimensional architecture that must be reproduced exactly in every circuit. The structure is made up of many layers, each of which is a detailed pattern. Some of the layers lie within the silicon wafer and others are stacked on the top. The manufacturing process consists in forming this sequence of layers precisely in accordance with the plan of the circuit designer.

Before examining how these layers are formed it will be helpful to take an overall look at the procedure by which an integrated circuit is transformed from a conception of the circuit designer to a physical reality. In the first stage

MICROCIRCUIT IS LAID OUT with the aid of a computer. The portion of the large-scale integrated circuit being laid out in this demonstration is studied by the circuit designer on the screen of a cathode-ray tube. By typing in commands or redrawing with a "light pen," circuit elements can be added, subtracted or moved. The terminal seen here is made by Applicon Inc.

LAYOUT IS CHECKED by scrutinizing a photomask image that is 500 times larger than the actual circuit. Usually a reticle representing a tenfold enlargement of the pattern for a single layer of the device is generated for this purpose on a glass photographic plate five inches square. In this case an image of the reticle that has been magnified an additional 50 times is checked to verify that every circuit element is correctly placed and is the right size. The master masks are prepared by projecting a reduced image of the reticle onto another photographic plate by means of a lens system capable of extremely high resolution. The circuit seen on the light table here is the same as the one shown in the first two photographs in this article.

WAFER-FABRICATION FACILITY needs to be kept extremely clean in order to avoid contaminating the manufacturing environment with even the tiniest dust particles. Special clothing is worn by the workers, and the air is continuously filtered and recirculated to keep the dust level to a minimum. This photograph of a "clean room" was made at a plant operated by Zilog Inc. in Cupertino, Calif.

of the development of a new microelectronic circuit the designers who conceive of the new product work at specifying the functional characteristics of the device. They also select the processing steps that will be required to manufacture it. In the next stage the actual design of the device begins: the size and approximate location of every circuit element are estimated. Much of this preliminary design work is done with the aid of computers.

A computer can simulate the operation of the circuit in much the same way that electronic television games simulate the action of a table-tennis game or a space war. The circuit designer monitors the behavior of the circuit voltages and adjusts the circuit elements until the desired behavior is achieved. Computer simulation is less expensive than assembling and testing a "breadboard" circuit made up of discrete circuit elements; it is also more accurate. The main advantage of simulation, however, lies in the fact that the designer can change a circuit element merely by typing in a correction on a keyboard, and he can immediately observe the effect of the modification on the behavior of the circuit.

The final layout giving the precise positions of the various circuit elements is also made with the aid of a computer. The layout designer works at a computer terminal, placing and moving the circuit elements while observing the layout magnified several hundred times on a cathode-ray-tube display. The layout specifies the pattern of each layer of the integrated circuit. The goal of the layout is to achieve the desired function of each circuit in the smallest possible space. The older method of drawing circuit layouts by hand has not been entirely replaced by the computer. Many parts of a large-scale integrated circuit are still drawn by hand before being submitted to the computer.

At each stage of this process, including the final stage when the entire circuit is completed, the layout is checked by means of detailed computer-drawn plots. Since the individual circuit elements can be as small as a few micrometers across, the checking plots must be greatly magnified; usually the plots are 500 times larger than the final size of the circuit.

The time required to complete the task of circuit design and layout varies greatly with the nature of the circuit. The most difficult circuits to design are microprocessors, and here the design and layout can take several years. Other devices, such as static memories with a largely repetitive pattern, can be designed and laid out more quickly, in some cases in only a few months.

When the design and layout of a new circuit is complete, the computer memory contains a list of the exact position of every element in the circuit. From that description in the computer memory a set of plates, called photomasks, is prepared. Each mask holds the pattern for a single layer of the circuit. Since the circuits are so small, many can be fabricated side by side simultaneously on a single wafer of silicon. Thus each photomask, typically a glass plate about five inches on a side, has a single pattern repeated many times over its surface.

The manufacture of the photomasks is an interesting story in itself. Typically the process consists in first generating from the computer memory a complete pattern for each layer of the circuit. That is done by scanning a computer-controlled light spot across a photographic plate in the appropriate pattern.

This primary pattern, called the reticle, is checked for errors and corrected or regenerated until it is perfect. Typically the reticle is 10 times the final size of the circuit. An image of the reticle that is one-tenth its original size is then projected optically on the final mask. The image is reproduced side by side hundreds of times in a process called "step and repeat." The constraints on both the mechanical system and the photographic system are demanding; each element must be correct in size and position to within about one micrometer. The original plate created by the step-and-repeat camera is copied by direct contact printing to produce a series of submasters. Each submaster serves in turn to produce a large number of replicas, called working plates, that will serve for the actual fabrication process. The working

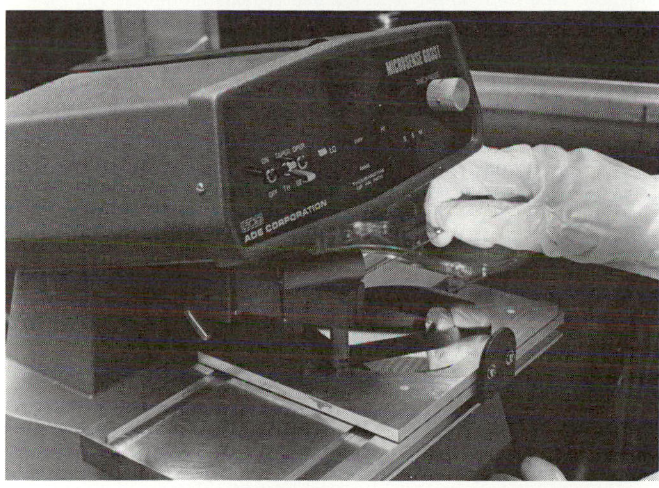

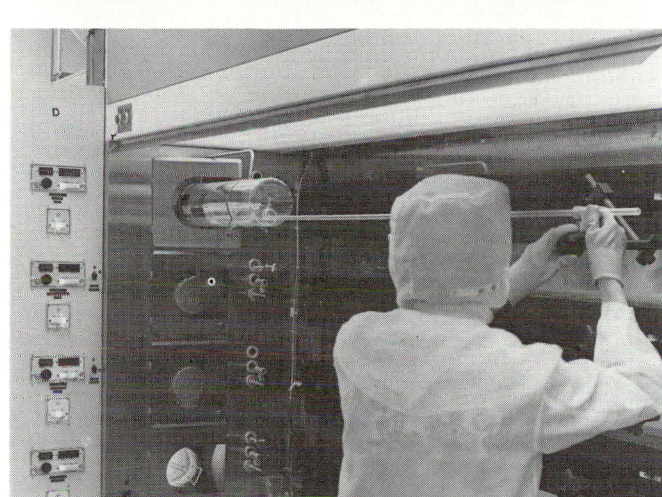

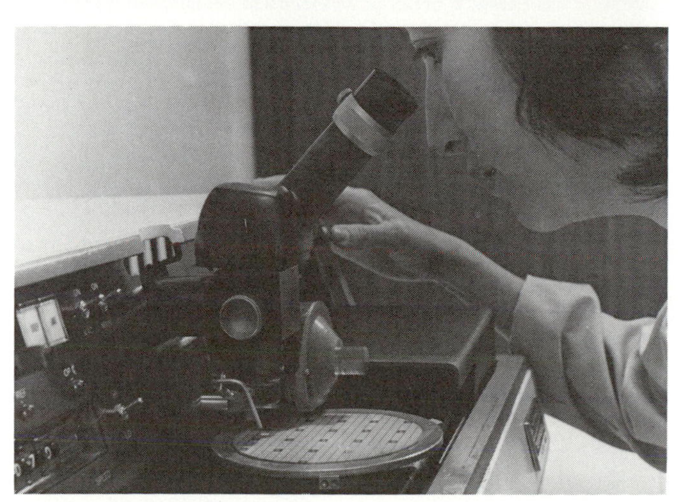

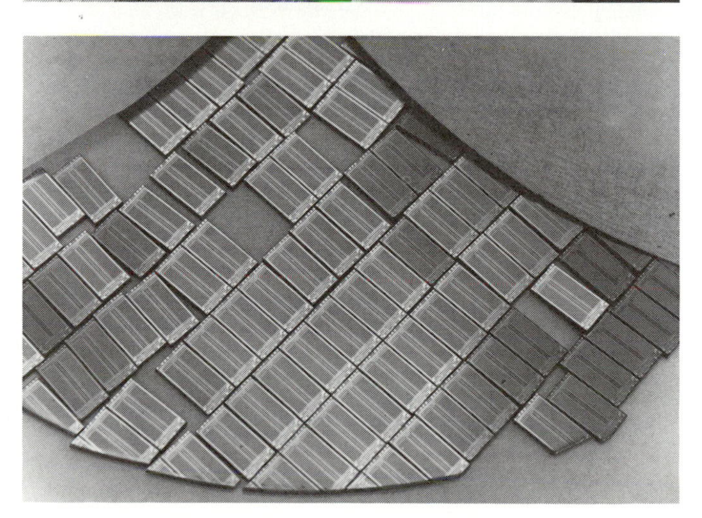

SIX STAGES in the fabrication of a microelectronic circuit are depicted in this sequence of photographs: growing a single-crystal ingot of "doped" silicon by withdrawing a small "seed" crystal of silicon from a quartz crucible containing pure molten silicon to which the desired dopant has been added (*top left*); slicing the precisely machined silicon crystal into thin wafers with a diamond-edged circular saw (*top right*); testing the smoothness of a polished wafer in a precision measuring device (*middle left*); forming a thin surface layer of silicon dioxide on a batch of wafers, shown being inserted into the oxidation furnace in a quartz "boat" (*middle right*); sawing grooves on the surface of the wafer prior to breaking the wafer into individual dice (*bottom left*); breaking the wafer into dice, each bearing a complete circuit (*bottom right*). The first three photographs were made in the Siltec Corporation's plant in Menlo Park, Calif. The last three were made at the Zilog facility. Numerous other processing steps, not shown here, are involved in building up the various circuit layers.

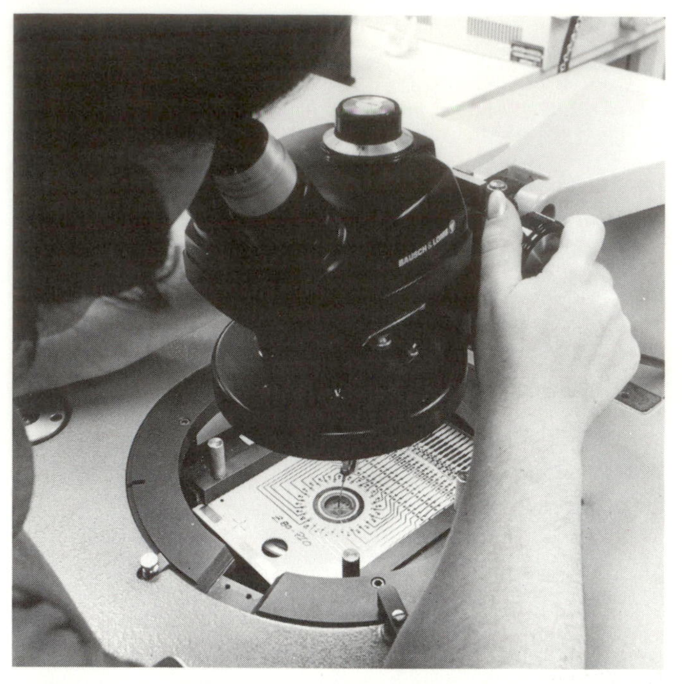

ELECTRICAL TESTING to distinguish good circuits from defective ones is carried out before the silicon wafer is broken into separate dice. An appropriate probe card carrying an array of extremely fine needlelike contacts is used for this purpose. The computer-controlled testing machine positions itself over each die, establishes contact with the circuit pads, tests the circuit and, if the circuit is defective, marks it with an ink spot to indicate that it should be discarded. The machine then steps to the next die and repeats the operation. Photograph at left shows an operator monitoring testing through a microscope; photograph at right shows closeup view of circuit being tested.

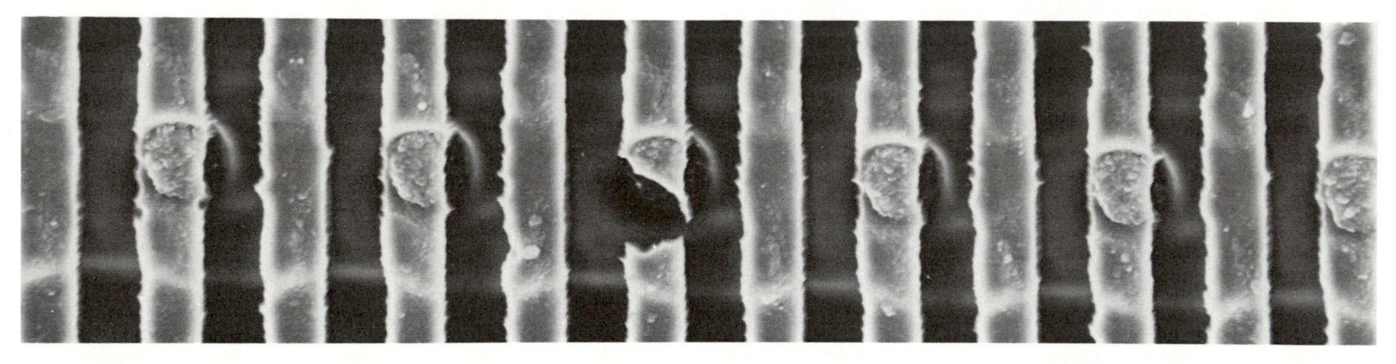

SINGLE DEFECT can ruin an entire microelectronic circuit. This scanning electron micrograph, made by William Roth of the Coates and Welter Instrument Co., shows a small portion of a 16,384-bit memory. A dust particle on the photomask used to define the aluminum layer has caused a break in an electrical connection. Aluminum strips are six micrometers wide. Magnification is 1,600 diameters.

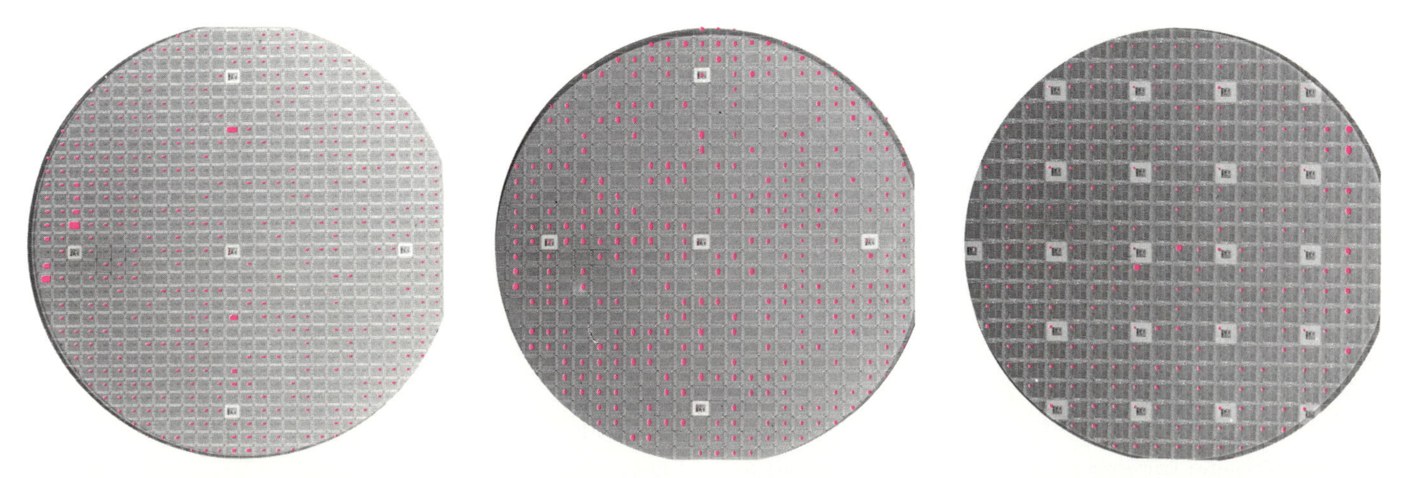

EFFECT OF CIRCUIT SIZE ON YIELD is demonstrated for microcircuits of three different sizes. The wafers bearing the circuits all measure 100 millimeters in diameter, and all were manufactured in the same Zilog fabrication facility. The circuits on the wafers have been tested and the defective dice marked with an ink spot. Wafer at left contains 340 good circuits, each 10.2 square millimeters in area. Wafer at center contains 240 good circuits, each 13.7 square millimeters in area. Wafer at right contains 80 good circuits, each 22.9 square millimeters in area. (In practice the yield varies greatly from wafer to wafer even for the same type of circuit; these three wafers were selected to be representative of the general trend.) The number of good circuits on a wafer decreases rapidly with increasing circuit size, both because there are fewer places on a wafer for larger dice and because larger circuits are more likely to incorporate a defect.

plate may be either a fixed image in an ordinary photographic emulsion or a much more durable pattern etched in a chromium film on a glass substrate.

A complete set of correct masks is the culmination of the design phase of the development of the microelectronic circuit. The plates are delivered to the wafer-fabrication facility, where they will be used to produce the desired sequence of patterns in a physical structure. This manufacturing facility receives silicon wafers, process chemicals and photomasks. A typical small facility employing 100 people can process several thousand wafers per week. Assuming that there are 50 working circuits per wafer, such a plant can produce five million circuits per year.

The inside of a wafer-fabrication facility must be extremely clean and orderly. Because of the smallness of the structures being manufactured even the tiniest dust particles cannot be tolerated. A single dust particle can cause a defect that will result in the malfunction of a circuit. Special clothing is worn to protect the manufacturing environment from dust carried by the human operators. The air is continuously filtered and recirculated to keep the dust level at a minimum. Counting all the dust particles that are a micrometer or more in diameter, a typical wafer-fabrication plant harbors fewer than 100 particles per cubic foot. For the purpose of comparison the dust level in a modern hospital is on the order of 10,000 particles per cubic foot.

The circuit manufacturer often buys prepared wafers of silicon ready for the first manufacturing step. The low price (less than $10) of a prepared wafer belies the difficulties encountered in its manufacture. Raw silicon is first reduced from its oxide, the main constituent of common sand. A series of chemical steps are taken to purify it until the purity level reaches 99.9999999 percent. A charge of purified silicon, say 10 kilograms, is placed in a crucible and brought up to the melting point of silicon: 1,420 degrees Celsius. It is necessary to maintain an atmosphere of purified inert gas over the silicon while it is melted, both to prevent oxidation and to keep out unwanted impurities. The desired impurities, known as dopants, are added to the silicon to produce a specific type of conductivity, characterized by either positive (*p* type) charge carriers or negative (*n* type) ones.

A large single crystal is grown from the melt by inserting a perfect single-crystal "seed" and slowly turning and withdrawing it. Single crystals three to four inches in diameter and several feet long can be pulled from the melt. The uneven surface of a crystal as it is grown is ground to produce a cylinder of standard diameter, typically either three inches or 100 millimeters (about four

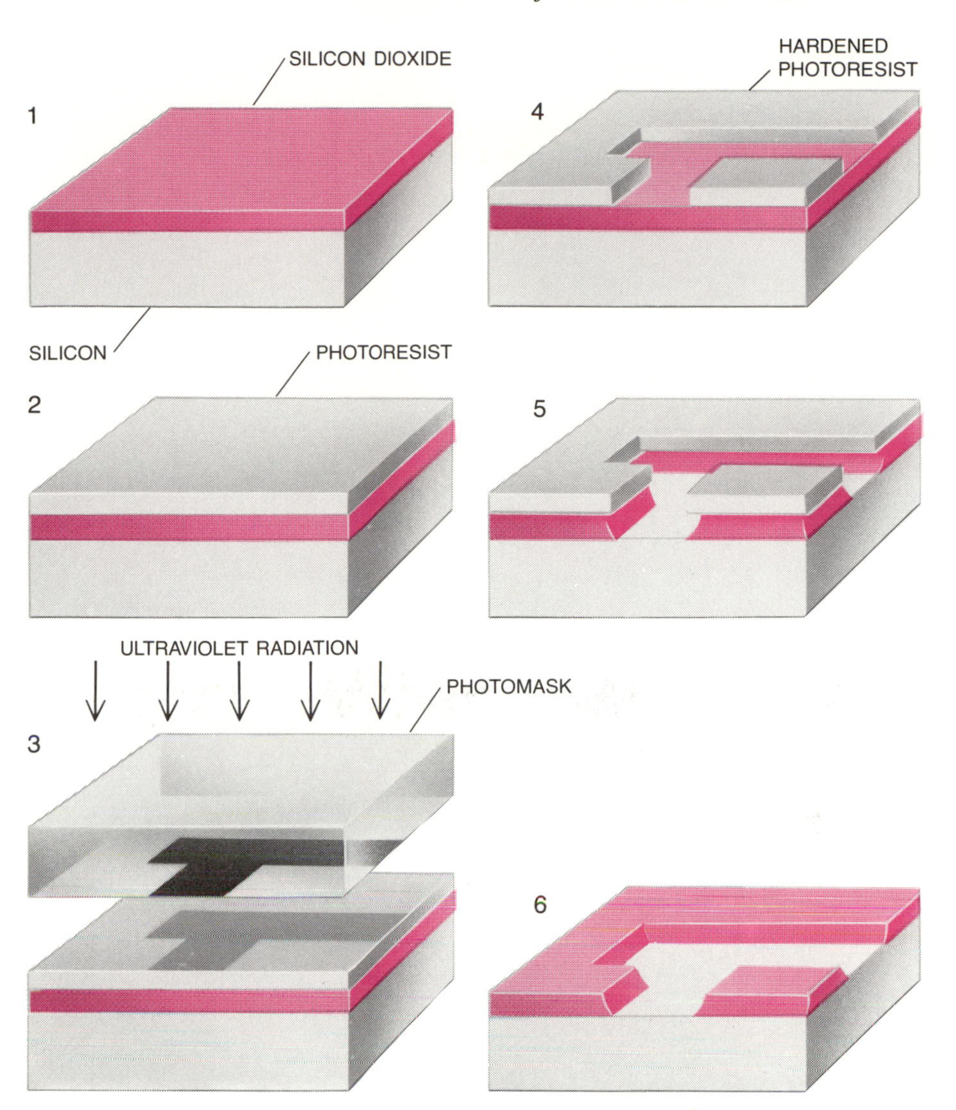

PHOTOLITHOGRAPHY is the process by which a microscopic pattern is transferred from a photomask to a material layer in an actual circuit. In this illustration a pattern is shown being etched into a silicon dioxide layer (*color*) on the surface of a silicon wafer. The oxidized wafer (*1*) is first coated with a layer of a light-sensitive material called photoresist (*2*) and then exposed to ultraviolet light through the photomask (*3*). The exposure renders the photoresist insoluble in a developer solution; hence a pattern of the photoresist is left wherever the mask is opaque (*4*). The wafer is next immersed in a solution of hydrofluoric acid, which selectively attacks the silicon dioxide, leaving the photoresist pattern and the silicon substrate unaffected (*5*). In the final step the photoresist pattern is removed by means of another chemical treatment (*6*).

TWO-LAYER STRUCTURE corresponding to step 5 in the top illustration on this page is seen in this scanning electron micrograph. The sample was cleaved to display this cross section just after the oxide layer was etched. Silicon appears black. Magnification is 10,000 diameters.

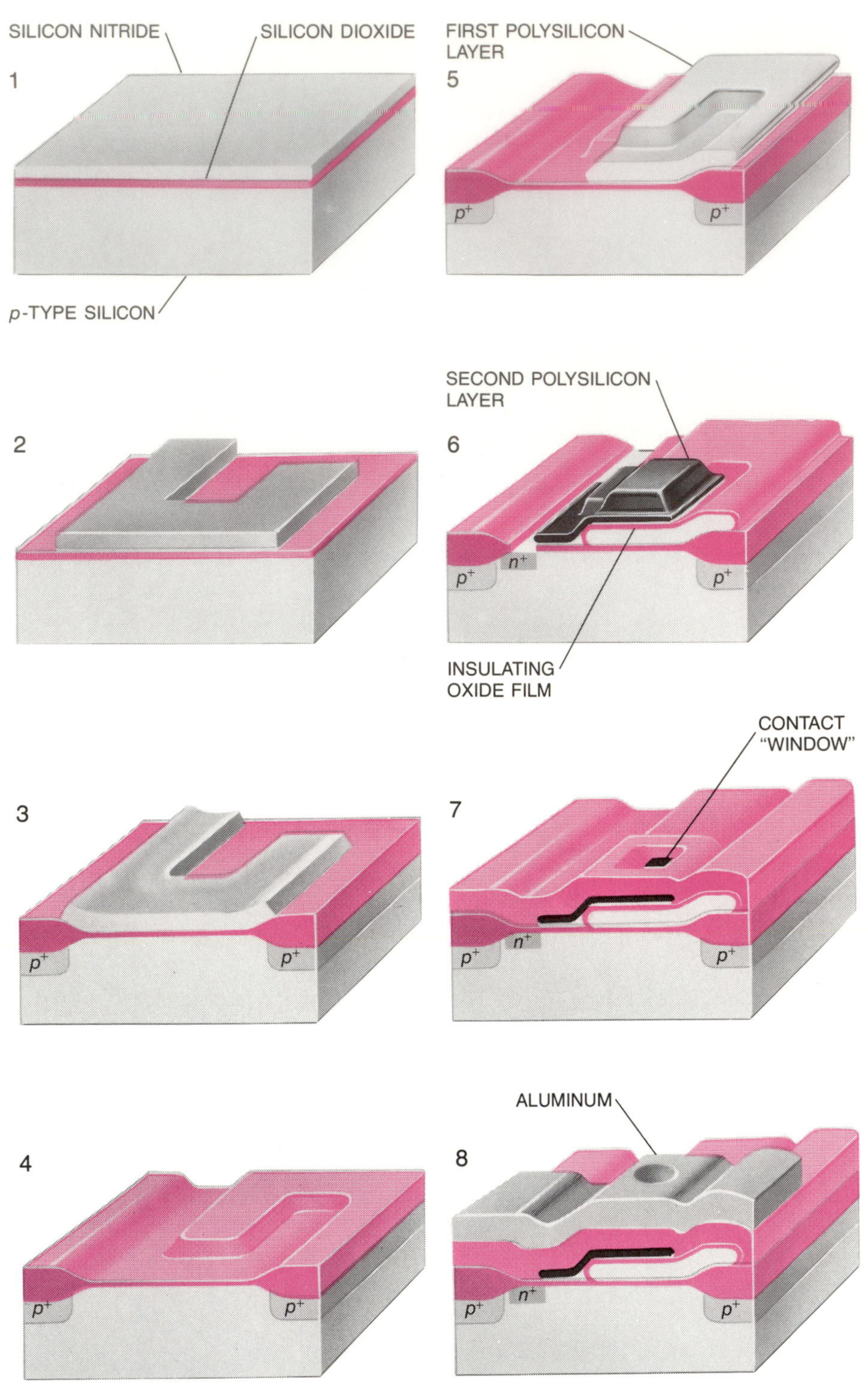

COMPLETE FABRICATION SEQUENCE for a two-level *n*-channel (denoting negative charge carriers) polysilicon-gate metal-oxide-semiconductor (MOS) circuit element requires six masking steps. The first few process steps involve the selective oxidation of silicon with the aid of a film of silicon nitride, which acts as the oxidation mask. A thin film of silicon dioxide is grown over the entire wafer, and a layer of silicon nitride is deposited from a chemical vapor (*1*). The layer is selectively removed in a conventional photolithographic step in accordance with the pattern on the first mask (*2*). A *p*-type dopant (for example boron) is implanted using the silicon nitride film as a mask, followed by an oxidation step, resulting in a thick layer of silicon dioxide in the unmasked areas (*3*). The silicon nitride is then removed in a selective etchant that does not attack either silicon or silicon dioxide (*4*). Since silicon is consumed in the oxidation process, the thick oxide layer is partly recessed into the silicon substrate. The first layer of polycrystalline silicon is then deposited and patterned in the second masking step (*5*). A second insulating film of oxide is grown or deposited, followed by the deposition of the second polysilicon layer, which is in turn patterned in the third masking step (*6*). A short etch in hydrofluoric acid at this stage exposes certain regions to an implantation or diffusion of *n*-type dopant. A thick layer of silicon dioxide is deposited next, and contact "windows" are opened with the fourth mask (*7*). Finally a layer of aluminum is deposited and patterned in the fifth masking operation (*8*). The wafer will also receive a protective overcoating of silicon dioxide or silicon nitride (not shown); the fact that openings must be provided in this overcoating at the bonding pads accounts for sixth masking step. Vertical dimensions are exaggerated for clarity.

inches). The crystal is mounted in a fixture and cut into wafers with a thin high-speed diamond saw. In the finishing step the wafers are first smoothed on both sides by grinding and then are highly polished on one side. The final wafer is typically about half a millimeter thick. The final steps must also be carried out in an absolutely clean environment. There can be no defects, polishing damage, scratches or even chemical impurities on the finished surface.

The dominant role of silicon as the material for microelectronic circuits is attributable in large part to the properties of its oxide. Silicon dioxide is a clear glass with a softening point higher than 1,400 degrees C. It plays a major role both in the fabrication of silicon devices and in their operation. If a wafer of silicon is heated in an atmosphere of oxygen or water vapor, a film of silicon dioxide forms on its surface. The film is hard and durable and adheres well. It makes an excellent insulator. The silicon dioxide is particularly important in the fabrication of integrated circuits because it can act as a mask for the selective introduction of dopants. Convenient thicknesses of silicon dioxide can be grown at temperatures in the range between 1,000 and 1,200 degrees C. The exact thickness can be accurately controlled by selecting the appropriate time and temperature of oxidation. For example, a layer of oxide a tenth of a micrometer thick will grow in one hour at a temperature of 1,050 degrees C. in an atmosphere of pure oxygen. A layer five times thicker will grow in the same time and at the same temperature in steam.

An important aspect of the oxidation process is its low cost. Several hundred wafers can be oxidized simultaneously in a single operation. The wafers are loaded into slots in a quartz "boat," separated by only a few millimeters. The high-temperature furnace has a cylindrical heating element surrounding a long quartz tube. A purified stream of an oxygen-containing gas passes through the tube. The boats of wafers are loaded into the open end and slowly pushed into the hottest part of the furnace. The temperature in the process zone is controlled to an accuracy of better than one degree C. Often the entire procedure is supervised by a computer. A small process-control computer monitors the temperature, directs the insertion and withdrawal of the wafers and controls the internal environment of the furnace.

The fabrication of integrated circuits requires a method for accurately forming patterns on the wafer. The microelectronic circuit is built up layer by layer, each layer receiving a pattern from a mask prescribed in the circuit design. The photoengraving process known as photolithography, or simply masking, is employed for the purpose.

The most basic masking step involves the etching of a pattern into an oxide. An oxidized wafer is first coated with photoresist, a light-sensitive polymeric material. The coating is laid down by placing a drop of the photoresist dissolved in a solvent on the wafer and then rapidly spinning the wafer. A thin liquid film spreads over the surface and the solvent evaporates, leaving the polymeric film. A mild heat treatment is given to dry out the film thoroughly and to enhance its adhesion to the silicon dioxide layer under it.

The most important property of the photoresist is that its solubility in certain solvents is greatly affected by exposure to ultraviolet radiation. For example, a negative photoresist cross-links and polymerizes wherever it is exposed. Thus exposure through a mask followed by development (washing in the selective solvent) results in the removal of the film wherever the mask was opaque. The photoresist pattern is further hardened after development by heating.

The wafer, with its photoresist pattern, is now placed in a solution of hydrofluoric acid. The acid dissolves the oxide layer wherever it is unprotected, but it does not attack either the photoresist or the silicon wafer itself. After the acid has removed all the silicon dioxide from the exposed areas the wafer is rinsed and dried, and the photoresist pattern is removed by another chemical treatment.

Other films are patterned in a similar way. For example, a warm solution of phosphoric acid selectively attacks aluminum and therefore can serve to pattern an aluminum film. Often an intermediate masking layer is needed when the photoresist cannot stand up to the attack of some particular etching solution. For example, polycrystalline silicon films are often etched in a particularly corrosive mixture containing nitric acid and hydrofluoric acid. In this case a film of silicon dioxide is first grown on the polycrystalline silicon. The silicon dioxide is patterned in the standard fashion and the photoresist is removed. The pattern in the silicon dioxide can now serve as a mask for etching the silicon film under it, because the oxide is attacked only very slowly by the acid mixture.

Photolithography is in many ways the key to microelectronic technology. It is involved repeatedly in the processing of any device, at least once for each layer in the finished structure. An important requirement of the lithographic process is that each pattern be positioned accurately with respect to the layers under them. One technique is to hold the mask just off the surface and to visually align the mask with the patterns in the wafer. The machine that holds the wafer and mask for this operation can be adjusted to an accuracy of one or two micrometers. After perfect alignment is achieved the mask is pressed into contact with the wafer. The mask is then flooded with ultraviolet radiation to expose the photoresist. The space between the wafer and the mask is often evacuated to achieve intimate contact; atmospheric pressure squeezes the wafer and the mask together. According to whether a high vacuum or a moderate one is used, the process is called "hard" or "soft" contact printing. In another variation, proximity printing, the mask is held slightly above the wafer during the exposure.

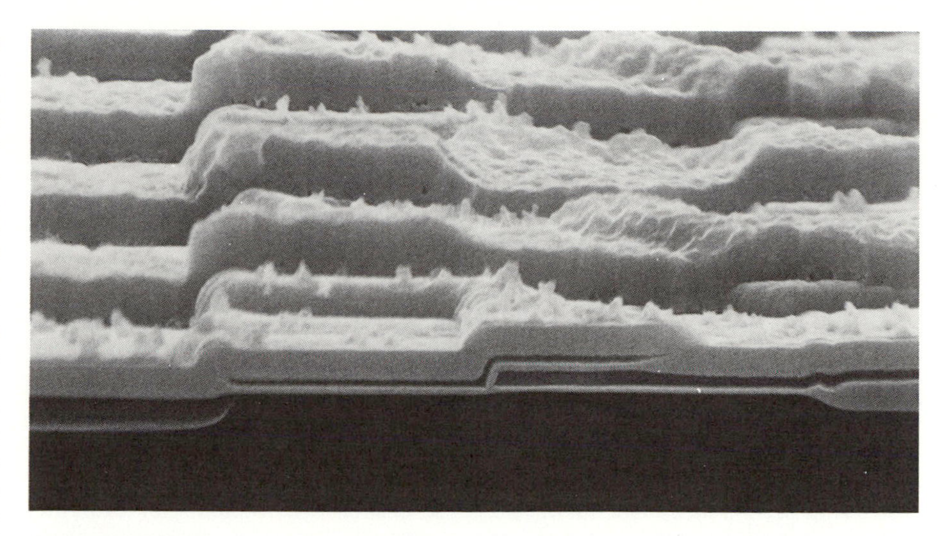

CROSS SECTION of a 16,384-bit microelectronic memory circuit manufactured by means of the two-level *n*-channel polysilicon-gate MOS method is seen in this scanning electron micrograph. The structure corresponds to the final step in the illustration on the opposite page.

The variations in the masking process arise from the need to print very small features with no defects in the pattern. If the mask were to be positioned very far from the surface, diffraction of the ultraviolet radiation passing through the mask would cause the smaller features to blur together. Thus hard contact would be preferred. On the other hand, small particles on the wafer or the mask are abraded into the mask when it is pressed against the wafer. Hence the masks can be used for only a few exposures before the defects accumulate to an intolerable level. A masking technique is chosen that is appropriate to the particular technology. Depending on the flatness of the wafer and the mask, and on the type of mask employed, a technique is chosen that gives reasonable mask life and sufficient resolution to print the smallest circuit elements in the device.

A recent trend has been toward the technique known as projection alignment, in which the image of the mask is projected onto the wafer through an optical system. In this case mask life is virtually unlimited. It is only in the past few years, however, that optics capable of meeting the photolithographic requirements for fabricating integrated circuits have become available. The fact that the wafers increase in size every few years is a continuing problem, and the task of designing optics capable of forming an accurate image over the larger area is becoming more difficult. Recent projection aligners, however, circumvent the extreme difficulty of constructing a lens capable of resolving micrometer-sized features over an area of many square inches. A much smaller area, on the order of one square centimeter, is exposed, and the exposure is repeated by either stepping or scanning the image over the wafer.

Active circuit elements such as metal-oxide-semiconductor (MOS) transistors and bipolar transistors are formed in part within the silicon substrate. To construct these elements it is necessary to selectively introduce impurities, that is, to create localized *n*-type and *p*-type regions by adding the appropriate dopant atoms. There are two techniques for selectively introducing dopants into the silicon crystal: diffusion and ion implantation.

If silicon is heated to a high temperature, say 1,000 degrees C., the impurity atoms begin to move slowly through the crystal. Certain key impurities (boron and phosphorus) move much more slowly through silicon dioxide than they do through silicon itself. This important fact enables one to employ thin oxide patterns as impurity masks. For example, a boat of wafers can be placed in a furnace at 1,000 degrees in an atmosphere containing phosphorus. The phosphorus enters the silicon wherever it is unprotected, diffusing slowly into the bulk of the wafer. After enough impurity atoms have accumulated the wafers are removed from the furnace, and solid-state diffusion effectively ceases. Of course, every time the wafer is reheated the impurities again begin to diffuse; hence all the planned heat treatments must be considered in designing a process to achieve a specific depth of diffusion. The important variables controlling the depth to which impurities diffuse are time and temperature. For example, a layer of phosphorus one micrometer deep can be diffused in about an hour at 1,100 degrees.

To achieve maximum control most diffusions are performed in two steps. The predeposit, or first, step takes place in a furnace whose temperature is selected to achieve the best control of the amount of impurity introduced. The temperature determines the solubility of the dopant in the silicon, just as the temperature of warm water determines the solubility of an impurity such as salt. After a comparatively short predeposit treatment the wafer is placed in a second furnace, usually at a higher temperature. This second heat treatment, the "diffusion drive-in" step, is selected to achieve the desired depth of diffusion.

In the formation of *pn* junctions by solid-state diffusion the impurities diffuse laterally under the oxide mask about the same distance as the depth of the junction. The edge of the *pn* junction is therefore protected by a layer of silicon dioxide. This is an important feature of the technique, because silicon dioxide is a nearly ideal insulator, and many of the electronic devices will not tolerate any leakage at the edge of the junction.

Another selective doping process, ion implantation, has been developed as a means of introducing impurities at room temperature. The dopant atoms are ionized (stripped of one or more of their electrons) and are accelerated to a high energy by passing them through a potential difference of tens of thousands of volts. At the end of their path they strike the silicon wafer and are embedded at various depths depending on their mass and their energy. The wafer can be selectively masked against the ions either by a patterned oxide layer, as in conventional diffusion, or by a photoresist pattern. For example, phosphorus ions accelerated through a potential of 100,000 volts will penetrate the photoresist to a depth of less than half a micrometer. Wherever they strike bare silicon they penetrate to an average depth of a tenth of a micrometer. Thus even a one-micrometer layer of photoresist can serve as a mask for the selective implantation of phosphorus.

As the accelerated ions plow their way into the silicon crystal they cause considerable damage to the crystal lattice. It is possible to heal most of the damage, however, by annealing the crystal at a moderate temperature. Little diffusion takes place at the annealing temperature, so that the ion-implantation conditions can be chosen to obtain the desired distribution. For example, a very shallow, high concentration of dopant can be conveniently achieved by ion implantation. A more significant feature of the technique is the possibility of accurately controlling the concentration of the dopant. The ions bombarding the crystal each carry a charge, and by measuring the total charge that accumulates the number of impurities can be precisely determined. Hence ion implantation is used whenever the doping level must be very accurately controlled. Often ion implantation simply replaces the predeposit step of a diffusion process. Ion implantation is also used to introduce impurities that are difficult to predeposit from a high-temperature vapor. For example, the current exploration of the use of arsenic as a shallow *n*-type dopant in MOS devices coincides with the availability of suitable ion-implantation equipment.

A unique feature of ion implantation is its ability to introduce impurities through a thin oxide. This technique is particularly advantageous in adjusting the threshold voltage of MOS transistors. Either *n*-type or *p*-type dopants can be implanted through the gate oxide, resulting in either a decrease or an increase of the threshold voltage of the device. Thus by means of the ion-implantation technique it is possible to fabricate several different types of MOS transistors on the same wafer.

The uppermost layers of integrated circuits are formed by depositing and patterning thin films. The two most important processes for the deposition of thin films are chemical-vapor deposition and evaporation. The polycrystalline silicon film in the important silicon-gate MOS technology is usually laid down by means of chemical-vapor deposition. Silane gas (SiH_4) decomposes when it is heated, releasing silicon and hydrogen. Accordingly when the wafers are heated in a dilute atmosphere of silane, a uniform film of polycrystalline silicon slowly forms on the surface. In subsequent steps the film is doped, oxidized and patterned.

It is also possible to deposit insulating films such as silicon dioxide or silicon nitride by means of chemical-vapor deposition. If a source of oxygen such as carbon dioxide is present during the decomposition of silane, silicon dioxide is formed. Similarly, silicon nitride is grown by decomposing silane in the presence of a nitrogen compound such as ammonia.

Evaporation is perhaps the simplest method of all for depositing a thin film, and it is commonly employed to lay down the metallic conducting layer in most integrated circuits. The metallic charge to be evaporated, usually aluminum, is placed in a crucible, and the wafers to be coated are placed above the crucible in a movable fixture called a planetary. During evaporation the wafers are rotated in order to ensure the maximum uniformity of the layer. The motion of the planetary also wobbles the wafers with respect to the source in order to obtain a continuous aluminum film over the steps and bumps on the surface created by the preceding photolithographic steps. After a glass bell jar is lowered over the planetary device and a high vacuum is established the aluminum charge is heated by direct bombardment with high-energy electrons. A pure aluminum film, typically about a micrometer thick, is deposited on the wafer.

In the fabrication of a typical large-scale integrated circuit there are more thin-film steps than diffusion steps. Therefore thin-film technology is probably more critical to the overall yield and performance of the circuits than the diffusion and oxidation steps are. In a recent development a thin film is even employed to select the areas on a wafer that are to be oxidized. The compound silicon nitride has the property that it oxidizes much more slowly than silicon. A layer of silicon nitride can be vapor-deposited, patterned and used as an oxidation mask. The surface that results is

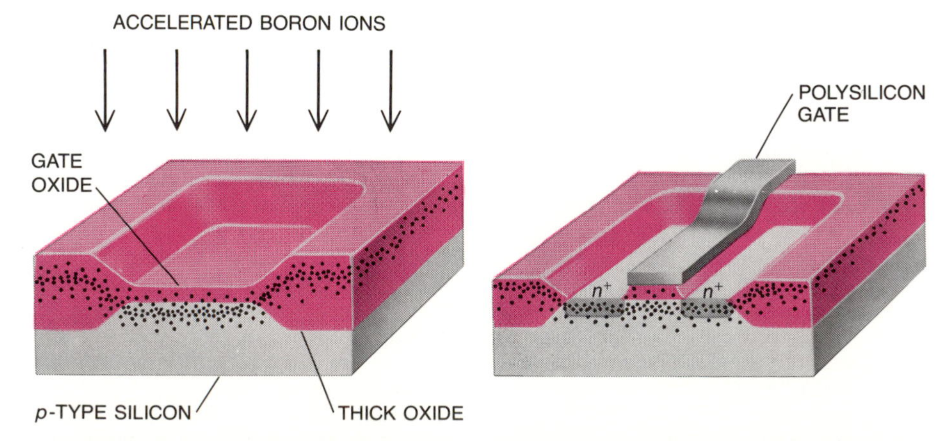

ION IMPLANTATION is employed to place a precisely controlled amount of dopant (in this case boron ions) below the gate oxide of an MOS transistor. By choosing a suitable acceleration voltage the ions can be made to just penetrate the gate oxide but not the thicker oxide (*left*). After the boron ions are implanted polycrystalline silicon is deposited and patterned to form the gate regions of the transistor. A thin layer of the oxide is then removed, and the source and drain regions of the transistor are formed by the diffusion of an *n*-type impurity (*right*).

much flatter than the surface if the thick oxide is grown everywhere and selectively removed. For n-channel MOS devices there is the additional advantage that an ion-implantation step involving boron can be added just before the oxidation step, relying on the nitride pattern as a mask. This procedure results in a heavily doped p-type region located precisely under the oxide, which acts as an obstacle to the formation of channels from adjacent elements in the device.

This "channel stopper" diffusion step is necessary in high-performance n-channel MOS technology. Without selective oxidation a special masking step would have to be added. The spacing between elements would then necessarily be larger; hence selective oxidation leads to greater circuit density. Bipolar integrated circuits also benefit greatly from the use of selective oxidation. By replacing the conventional diffused isolation with oxide isolation the space taken up by one bipolar transistor is reduced by more than a factor of four.

The wafer-fabrication phase of manufacture ends with an electrical test. Each die on the wafer is probed to determine whether it functions correctly. The defective dice are marked with an ink spot to indicate that they should be discarded. A computer-controlled testing machine quickly tests each circuit, steps to the next one and performs the inking without human intervention. It can also keep accurate statistics on the number of good circuits per wafer, their location and the relative incidence of various types of failure. Such information is helpful in finding new ways to improve the yield of good circuits.

The completed circuit must undergo one last operation: packaging. It must be placed in some kind of protective housing and have connections with the outside world. There are many types of packages, but all have in common the fact that they are much larger and stronger than the silicon dice themselves. First the wafer is sectioned to separate the individual chips, usually by simply scribing between the chips and breaking the wafer along the scribe lines. The good circuits are bonded into packages, and they are connected to the electrodes leading out of the package by fine wires. The package is then sealed, and the device is ready for final testing. The packaged circuit goes through an exhaustive series of electrical tests to make sure that it functions perfectly and will continue to do so reliably for many years.

After the individual chips are obtained from the wafer the cost per manufacturing step rises enormously. No longer is the cost shared among many circuits. Accordingly automatic handling during packaging and testing must be introduced wherever possible. The traditional cost-saving technique has

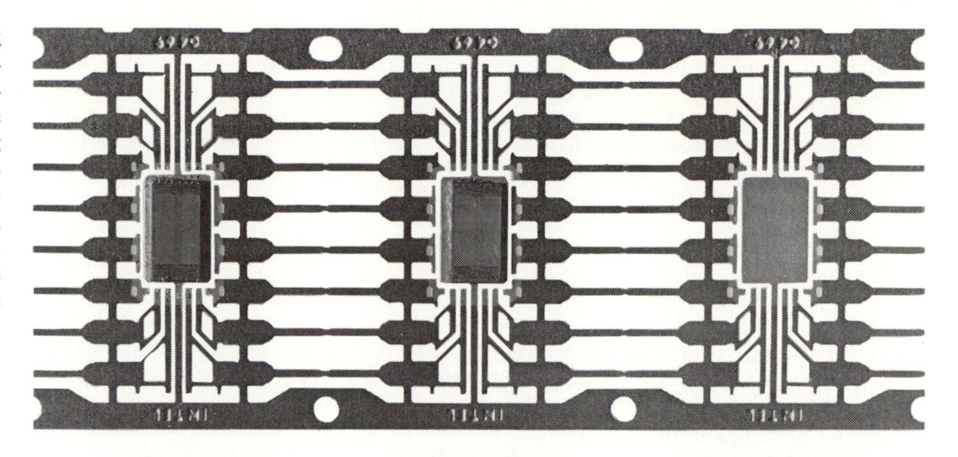

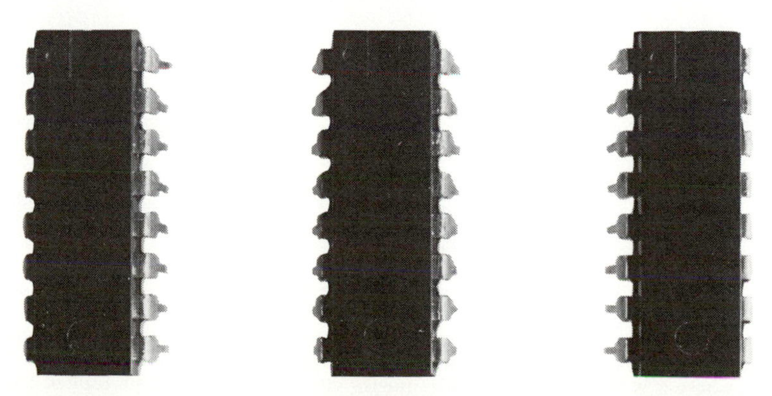

PACKAGING, the final step in the manufacture of a microelectronic circuit, is represented by this sequence of photographs. The dice bearing the finished circuits are first fastened to a metal stamping. Fine wire leads are connected from the bonding pads of the dice to the electrodes of the package; a plastic cover is molded around each die, and the units are separated. The components portrayed in the photographs were supplied by the Intel Corporation.

been to employ less expensive overseas labor for the labor-intensive packaging operation. As the cost of overseas labor rises and improved packaging technology becomes available, overseas hand labor is gradually being supplanted by highly automated domestic assembly.

A number of advanced processing techniques are now under development. For example, the simple wet-etching process in which films of aluminum or polycrystalline silicon are selectively removed is already being supplanted by dry-etching processes. Polycrystalline silicon can be "plasma-etched" in electrically excited gas of carbon tetrafluoride molecules (Freon). A high-frequency electric discharge at low pressure breaks the Freon molecule down into a variety of ions and free radicals (such as atomic fluorine). The free radicals attack the film but do not react with the photoresist mask. In addition to being a more controllable method for the selective removal of silicon, plasma-etching promises to be much less harmful to the environment. Instead of yielding large quantities of corrosive acids, the reaction products are very small quantities of fluorine and fluorides of silicon, which are easily trapped from the output of the system.

The technology of photolithography, which was stable for about 10 years, is also undergoing several changes. First, as projection lithography replaces contact lithography, the number of defects is decreasing. Second, the availability of masks of higher quality is steadily reducing the size and cost of microelectronic devices. Third, new methods of lithography are being developed that could result in a tenfold reduction in the size of individual circuit elements and a hundredfold reduction in circuit area.

The smallest features that can be formed by the conventional photolithographic process are ultimately limited by the wavelength of light. Present technology can routinely reproduce elements a few micrometers across, and it appears possible to reduce the smallest features to about one micrometer. Electron beams and X rays, however, have wavelengths measured in nanometers and smaller; hence they are capable of producing extremely fine features.

X-ray lithography is simply a form of contact photolithography in which soft X rays are substituted for ultraviolet radiation. The X-ray technique does indeed offer high resolution; simple structures less than a tenth of a micrometer across have already been produced. Because the entire wafer is exposed the process is also potentially quick and cheap. There are still many unsolved problems, however. X-ray masks, which consist of a heavy metallic pattern on a thin membrane such as Mylar, are fragile and difficult to make. It is also hard to align the mask with respect to the pattern on the wafer. Because of the attenuation of X rays in air, the wafer must be exposed in a vacuum or in an atmosphere of helium. Present-day X-ray sources are comparatively weak, and long exposures are required. As a result no commercial integrated circuits have yet been manufactured with the aid of X-ray lithography.

DRY PLASMA ETCHING, an advanced processing technique currently under development in the author's laboratory at the University of California at Berkeley, is viewed through a quartz window at one end of the reaction chamber. A gas of carbon tetrafluoride molecules (Freon) at a pressure of about a thousandth of atmospheric pressure flows through the quartz chamber, where it is ionized by a radio-frequency electric discharge. A perforated aluminum shield protects the polysilicon-coated wafers from the direct discharge. Free atomic fluorine, created in the discharge by the breakdown of the Freon molecules, diffuses to the surface of the wafers and reacts with the polycrystalline silicon wherever it is unprotected by a layer of photoresist. The process can yield finer patterns than the conventional wet-etching method.

Electron-beam lithography is an older and maturer technology, having its basis in electron microscopy. Actually a system for electron-beam lithography is much like a scanning electron microscope. A fine beam of electrons scans the wafer to expose an electron-sensitive resist in the desired areas. Although impressive results have been demonstrated, the application of electron-beam lithography is limited by its present high cost. The machines are expensive (roughly $1 million per machine), and because the electron beam must scan the wafer rather than exposing it all at once the time needed to put a pattern on the wafer is quite long. The rate of progress in this area is rapid, however, and more practical systems are clearly on the way.

Although electron-beam lithography is currently too expensive to be part of the wafer-fabrication process, it is already a routine production technique for the making of photolithographic masks. With the aid of the electron-beam method it is possible to eliminate two photographic-reduction steps and write the pattern directly on the mask from the information stored in the computer memory. Masks can thereby be created in a few hours after the design is finished. The advantages of higher resolution, simplicity of manufacture and shorter production time may well result in the complete conversion of the industry to electron-beam mask-making. Gradually, as the cost decreases, electron-beam lithography will be introduced directly into the fabrication of wafers, and a new generation of even more complex microelectronic circuits will be born.

5

MICROELECTRONIC MEMORIES

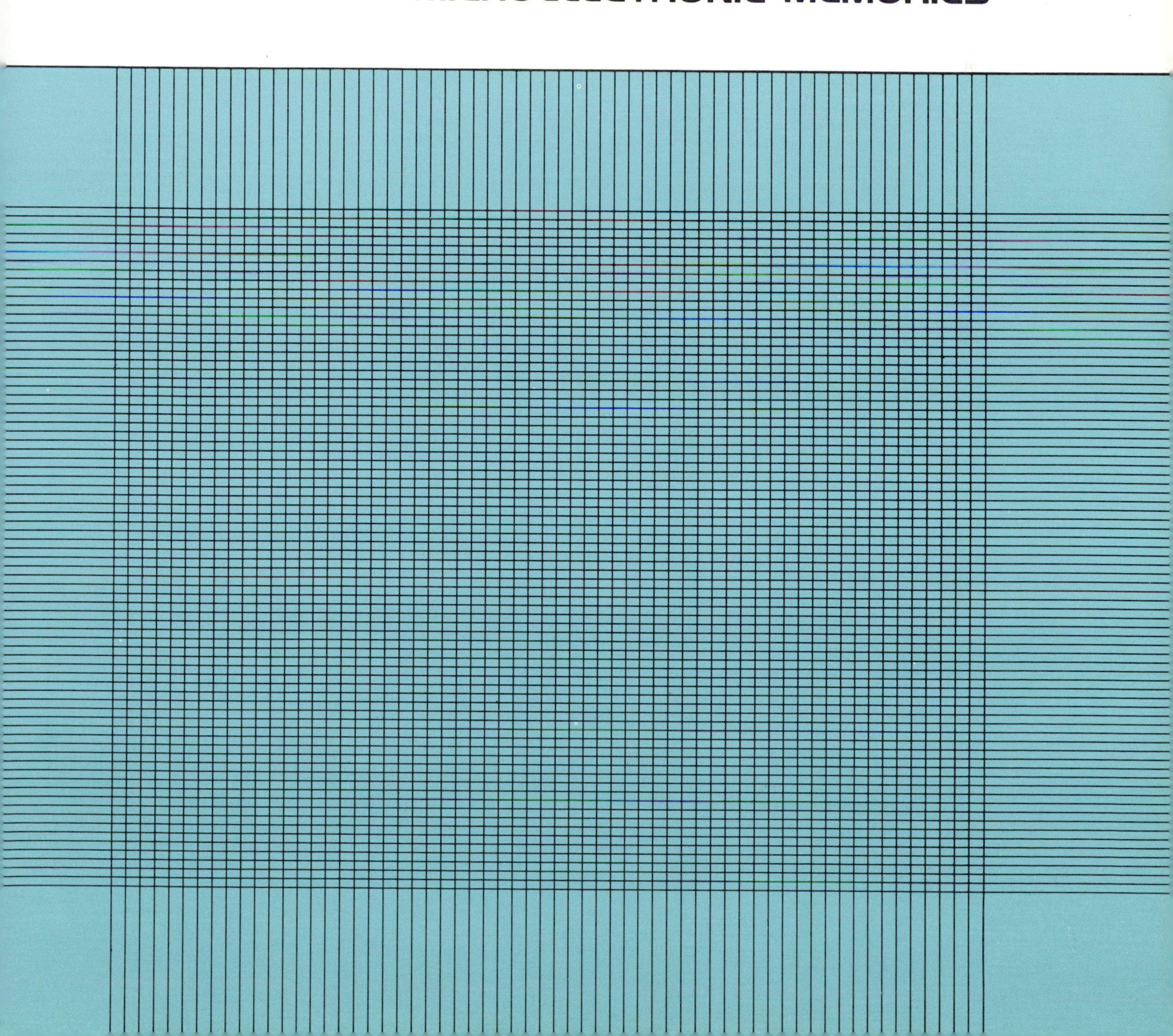

Microelectronic Memories

by DAVID A. HODGES

Present memories based on transistors typically store some 16,000 bits (binary digits) on a chip. Magnetic-bubble and charge-coupled devices are providing an even higher density of information storage

Modern information-processing and control systems call for the rapid storage and retrieval of digital information. The amount of information to be stored ranges from fewer than 100 bits (binary digits) for a simple pocket calculator to somewhere between a billion and a trillion bits for a large computer. The technology of digital storage is perhaps the most rapidly changing sector in all microelectronics. Over the past decade operating speed and reliability have increased by at least an order of magnitude as physical size, power consumption and cost per bit of storage have been reduced by factors ranging from 100 to 1,000. Improvements of comparable magnitude can be envisioned for the decade ahead before fundamental physical limitations enforce a slowdown.

In the context of electronics "memory" (or, in British usage, "store") usually refers to a device for storing digital information. Storage ("write") and retrieval ("read") operations are completely under electronic control. The storage of auditory or visual information in analogue form, which is commonly done on magnetic tape, is usually referred to as recording. In analogue recording the intensity of the "track" on the storage medium directly mimics the varying intensity of the input signal. Although there is some overlap between analogue and digital recording, I shall concentrate here on digital memory.

The most widely used digital memories are read/write memories, the term signifying that they perform read and write operations at an identical or similar rate. It is obvious that storage capacity, cost per bit and reliability are important characteristics for memories. Other important characteristics are speed of operation (defined in terms of access time), cycle time and data-transfer rate. Access time is simply the time it takes to read or write at any storage location. Some memories, such as random-access memories (RAM's), have the same access time to any storage location. Serial-access and block-access memories have access times that depend on the storage location selected.

Random-access memories can complete read or write operations in the specified minimum period known as the cycle time. Serial-access and block-access memories have a variable and relatively large access time, after which the data-transfer rate is constant. The data-transfer rate is the rate at which information is transferred to or from sequential storage positions.

Digital memories can be conveniently classified as either moving-surface devices or entirely electronic devices, with subdivisions in each category. Although moving-surface devices cannot really be considered microelectronic, microelectronic techniques are being fruitfully exploited to improve their performance. For many memory applications either a moving-surface system or an entirely electronic one is clearly superior, but in certain current and imminent applications the two systems are closely competitive. For this reason I shall briefly describe the characteristics of moving-surface devices.

Most commonly moving-surface memory devices provide information storage in localized areas of a thin magnetic film that is coated onto a nonmagnetic supporting surface. The magnetic material must be capable of holding a remnant magnetic flux in the absence of externally applied fields. The supporting surface can be flexible, in the form of a tape or disk made of plastic, or it can be rigid, in the form of a disk or drum made of aluminum or some other material. The magnetic film and the read/write head (a small electromagnet) move in relation to each other in order to bring a storage site into position for writing or reading of information. Information is stored in the form of tiny magnetized spots in the magnetic film. The stored information is sensed in the form of a weak current that is induced in the read/write head as the magnetic film moves under it.

The simplest digital magnetic-tape memories are adapted from audio cassette tape recorders. One cassette, or tape cartridge, is capable of storing 10^6 to 10^7 bits and can provide access time to any chosen storage location ranging from 10 to 100 seconds. The speed of the tape is limited by wear and heat because the tape is in frictional contact with the read/write head (or heads).

More sophisticated digital tape memory systems, and disk and drum memories as well, employ a laminar-flow air bearing to hold the read/write head a few micrometers above the magnetic surface. This stratagem makes it possible to have much higher surface velocities and improves the reliability of data storage.

Large magnetic-disk memories store 10^9 to 10^{10} bits on one or more disk surfaces mounted on a single spindle rotated by a single motor. Access to storage locations is achieved by moving one or more read/write heads radially across the spinning disk or disks. For typical systems the average access time is 20 milliseconds.

Moving-surface memories are block-accessed devices. After an access time determined by rates of mechanical motion and the location being sought, data is transferred at a rate determined by the surface velocity and the density of data storage on the surface. Typical transfer rates range from 10^6 to 10^7 bits per second.

Information stored in magnetic mediums is usually retained when the external power is turned off; such storage is said to be nonvolatile. It is normally feasible to remove the recording medium from tape and disk systems, making it possible to physically transfer information from one location to another and providing access to virtually unlimited quantities of information simply by the interchange of disks or reels of tape. These useful properties are rarely found in electronic memories.

Moving-surface memory devices serve widely as "file" memories in computer systems, where they store information that is changed much less often than it is read. The cost per bit of information stored in moving-surface memories is lower by one to four orders of magnitude than that of information stored in electronic memories (roughly 10^{-5} cent per bit for the least expensive magnetic-tape storage as against .1 cent per bit for the least expensive microelectronic memory). The nonvolatility of moving-surface memories is often an essential feature. On the other hand, the access time to electronic memories can be faster than 10^{-8} second, or six to eight orders of magnitude faster than the access time to moving-surface memories.

Continuing improvements are being made in the cost per bit, the data-transfer rate and the reliability of moving-surface memories. Microelectronic fabrication techniques are helping to shrink the size of read/write heads, yielding higher bit densities on the recording surface. Microelectronic circuits also serve to implement error-correcting codes in moving-surface memory sytems, substantially reducing the rate at which errors are made in the total storage and retrieval process. It is clear that for certain essential tasks of information storage moving-surface systems are not likely to be supplanted by electronic memories in the foreseeable future.

The newest electronic-memory systems have been made possible by modern semiconductor technology: the ability to emplace thousands of electronic elements on a tiny chip of silicon. In the 1950's and 1960's electronic memories were arrays of cores, or rings, of a ferrite material a millimeter or less in diameter, strung by the thousands on a grid of wires. Ferrite-core memories have now been largely succeeded in new designs by semiconductor memories that provide faster data access, smaller physical size and lower power consumption, and all at significantly lower cost. In the 1980's new memory technologies involving magnetic bubbles, superconducting tunnel-junction devices and devices accessed by laser beams or electron beams may come into play. I shall describe some of these devices below. The fact remains that because semiconductor memories are extremely versatile, are highly compatible with other electronic devices in both small and large systems and have much potential for further improvement in performance and cost, they are expected to dominate the electronic-memory market for at least another decade.

The smallest block of information accessible in a memory system can be a single bit (represented by 0 or 1), a larger group of bits such as a byte or character (usually eight or nine bits), or a "word" (12 to 64 bits depending on the particular system). Most memories are location-addressable, which means that a desired bit, byte or word has a speci-

RANDOM-ACCESS MEMORY (RAM) circuit provides storage for 16,384 bits (binary digits). Each bit is held in a single-transistor storage cell *(see illustration on page 57)*. **The time required to write one bit in any arbitrary location or to read it out is about 200 nanoseconds. RAM chip shown here, which is the MK 4116 manufactured by Mostek Corporation, measures 2.8 by 5.1 millimeters. Packaged, each chip sells for $30 in orders of 1,000 or more.**

fied address, or physical location, to which it is assigned.

Content-addressable memories, in which partial knowledge of a stored word is sufficient to find the complete word, would be extremely useful in some applications. Electronic content-addressable memories have never been common because the cost per bit is far higher than it is for location-addressable memories.

The most widely used form of electronic memory is the random-access read/write memory fabricated in the form of a single large-scale-integrated memory chip capable of storing as many as 16,384 bits in an area less than half a centimeter on a side. A number of individual circuits, each storing one binary bit, are organized in a rectangular array. Access to the location of a single bit is provided by a binary-coded address presented as an input to address decoders that select one row and one column for a read or write operation. Only the storage element at the intersection of the selected row and column is the target for the reading or writing of one bit of information. A read/write control signal determines which of the two operations is to be performed. The memory array can be designed with a single input-output line for the transfer of data or with several parallel lines for the simultaneous input or output of four, eight or more bits.

As William G. Oldham points out in the preceding article [see "The Fabrication of Microelectronic Circuits," page 40], the cost of a microelectronic circuit is proportional to the area on a wafer of silicon that it occupies. Accordingly much ingenuity has gone into the development of simple, small-area memory circuits. The individual circuit, or "cell," that stores one bit is a critical design element because it must be repeated thousands of times. Address decoders and other ancillary circuits can be more complex because they are employed only once per row or column or even less frequently.

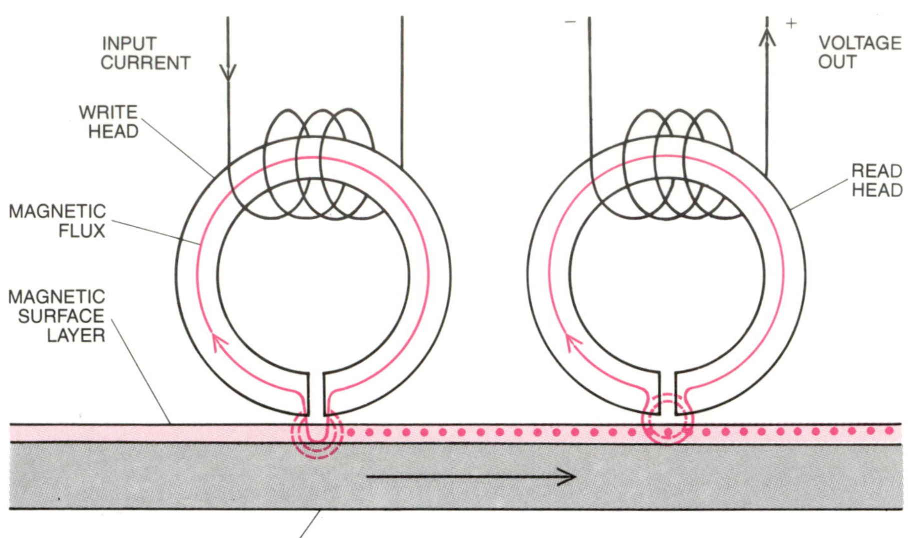

MOVING MAGNETIC SURFACE stores binary information at lowest cost: 100,000 or more bits for a cent. The information is permanently retained unless it is intentionally erased. Binary digits are stored sequentially on a moving magnetic medium by the application of electric pulses to a tiny electromagnet, which serves as the "write" head. When the film passes below a "read" head of similar design, magnetized spots on film generate tiny electric pulses, which are readily amplified. One head can serve for both reading and writing on a time-shared basis.

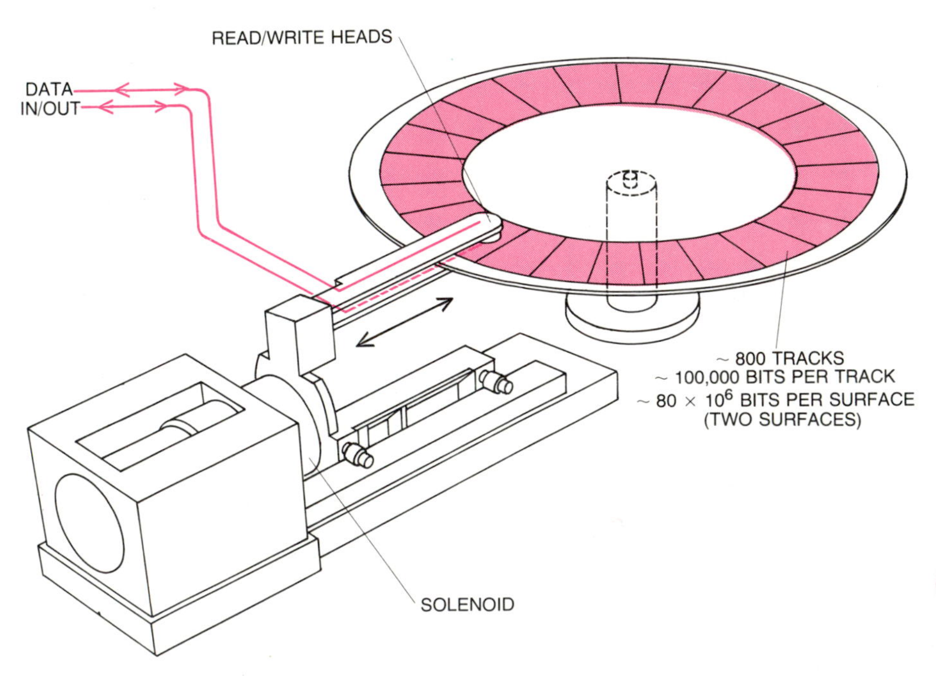

MAGNETIC-DISK MEMORY SYSTEMS are widely used for large, frequently searched files, or information stores. The simple system shown here has only one disk, but multidisk systems are common. Information is stored on both sides of a disk by separate read/write heads. A typical modern system has about 800 tracks per side, with storage for 100,000 bits per track. The total storage capacity of both sides of one disk is therefore 1,600 × 100,000, or 160 million bits. Sequential information can be transferred into and out of system at rates of five million to 10 million bits per second. Average access time to a random location is 25 milliseconds.

A simple one-transistor storage cell is widely used today in the most economical random-access read/write memory devices. Information is stored as an electric charge on a small capacitor. The value of the capacitance is on the order of 50 femtofarad (50×10^{-15} farad). For the representation of binary information two different levels of stored charge are needed. A binary 0 might be represented by zero charge and a binary 1 by a charge of 500 femtocoulombs (equivalent to 10 volts on the storage capacitor). Although this may seem like a tiny amount of charge, it is the charge on three million electrons. Reliable binary storage should ultimately be attainable with a charge 1,000 times smaller.

The transistor in the cell functions only as an on-off switch to connect the storage capacitor to the data line. The data line is shared by many identical cells, but only one cell is connected to it at any one time. The data line corresponds to a column in the memory array. Only one of the many selection lines, corresponding to the rows at right angles to the columns in the array, is activated at any one time. When a selection line is activated, it turns on all the transistor switches connected to it, but only one of these switches is on a simultaneously activated data line.

The capacitor storage cell can lose stored information in two ways. First, the capacitor itself has a significant leakage current. Through this process stored charge can be lost in as little as a few milliseconds. Second, when the cell is selected for a read operation, the charge stored in the cell is shared between the cell capacitor and the larger capacitance of the data line. In typical designs the data-line capacitance is 10 to 20 times larger than the cell capacitance because the data line is connected to many cells. As a result the voltage representing the stored information is attenuated by a factor of 10 to 20. The stored information that is to be retained must therefore be regenerated after every read operation.

It is also necessary to regenerate the stored charge periodically, usually every two milliseconds, in order to compensate for leakage. The necessary regeneration functions are performed by a thresholding amplifier. A sampling switch takes a sample of the signal on the data line at precisely the right time: just after a chosen storage cell has been connected to the data line by a signal on the selection line. The sampled signal is compared with a threshold level that is chosen to be midway between the nominal binary levels for 0 and 1. The amplifier then regenerates the binary level.

There is only one thresholding amplifier on each data line. It is time-shared by all the storage cells connected to that line, typically 64 or 128 in present designs. If each cell requires a regeneration cycle every two milliseconds and a single cycle occupies 400 nanoseconds, the regeneration cycles for 64 cells will occupy 25.6 microseconds of every two-millisecond interval and the cycles for 128 cells will occupy twice that amount of time, or 51.2 microseconds. The requirement for "refresh," or regeneration, operations adds a complication to the design of memory systems. (Although there are ways to avoid refresh operations, as I shall explain, they add to the cost of the device.) The "write" function is performed simply by actuating switches that are under the control of a read/write input signal.

Additional circuits are needed to complete a useful semiconductor memory chip. Most important are the address decoders, which select a unique row and column for each unique address input. Address decoders can be implemented with logic gates that perform the logical "and" function. There is room for much ingenuity in striking a good balance among conflicting demands for high-speed operation, low power consumption and minimum chip area (lowest cost). A clever innovation in some recent designs involves using the address input connections to a component twice in succession on each memory cycle. First the row address is applied; then about halfway through the cycle the column address is presented. The advantage is that the number of external address connections is cut in half. The number of external connections (the "pin count") is a significant factor affecting the cost, reliability and size of microelectronic components, so that cutting it in half is highly advantageous. Other peripheral circuits usually included with a microelectronic memory chip are data buffers (which transform the levels of data signals entering and leaving the chip) and timing generators for the sequencing of functions.

The semiconductor memory chip with only a single transistor in each storage cell is just one of many designs. Storage cells with two, three, four, six and more transistors offer useful additional features but with the penalty of increased silicon area and cost. For example, in metal-oxide-semiconductor (MOS) technology storage cells with three transistors provide amplification

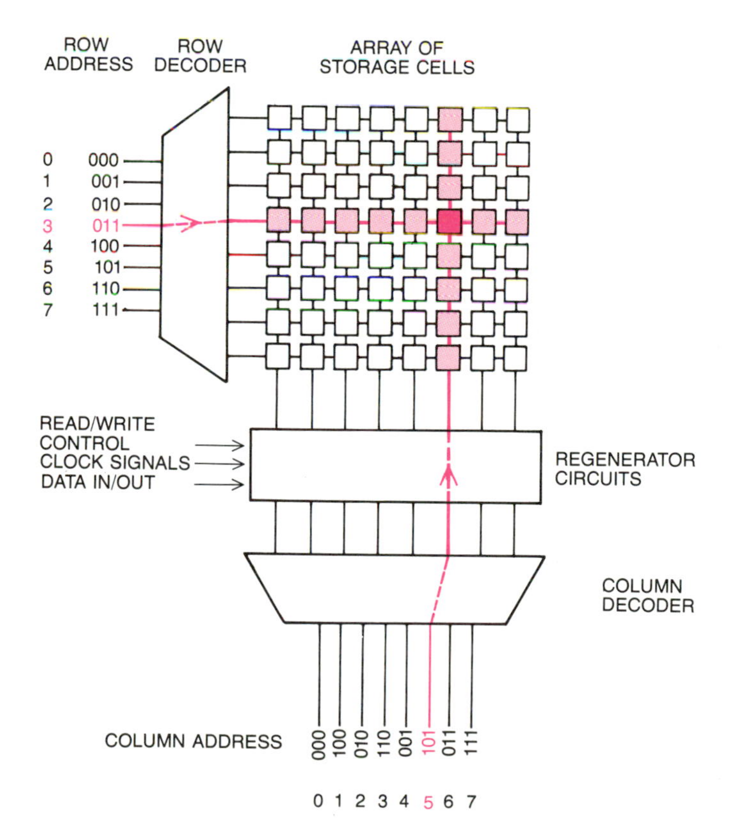

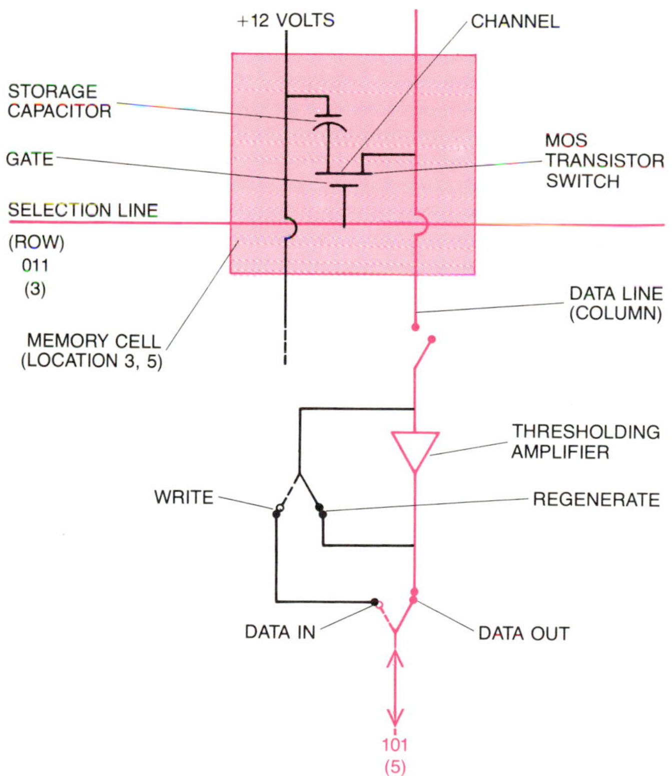

RANDOM-ACCESS MEMORIES are usually organized in rectangular arrays of rows and columns. The diagram at the left shows an eight-by-eight array for storage of 64 bits, one bit being stored in the cell at each intersection. To specify a particular memory location three binary digits are needed to indicate the row and another three to indicate the column. In this example row address 011 (binary for 3) and column address 101 (binary for 5) specify the memory location 3,5. (The locations start with 0,0 at the upper left and end with 7,7 at the lower right, specifying 64 locations in all.) The organization of a single one-transistor memory cell of the array is depicted at the right. Binary information is stored as a charge on a small capacitor. For example, zero charge might represent a binary 0 and a charge of 500 × 10^{-15} coulomb might represent a binary 1. When one of the selection lines, or rows, in an array is activated (here it is row 3), it turns on all the transistor switches connected to it. The transistor functions as an on-off switch to connect the storage capacitor to its particular data line, which corresponds to a column in the array. The simultaneous activation of a row and a column identifies the cell selected for reading or writing (here cell 3,5). Because the storage capacitor loses charge both by being read and by leakage it must be regenerated periodically, usually once every two milliseconds. The regenerated charge, which is supplied by the thresholding amplifier, is returned to the capacitor by the closing of the switch in the data line. All the switching in this type of memory device is accomplished by transistors.

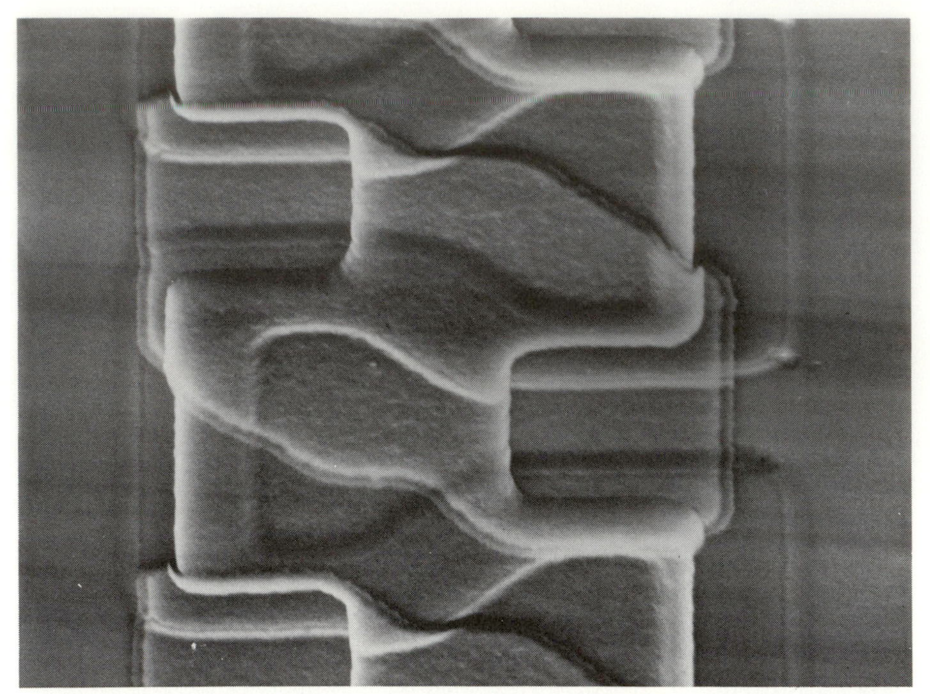

TWO RAM STORAGE CELLS appear in this scanning electron micrograph made by the author. Each cell includes one transistor and one capacitor. For maximum clarity this picture of a circuit manufactured by the Intel Corporation was made after only three of the five major fabrication steps had been completed. The fourth step forms a series of small openings, one for each transistor, in the insulating layer that covers the third level. The last step creates the metallic pattern that is needed for the selection lines (rows) in the finished memory array. The metal makes contact with the transistor gates in third level, as is shown in the exploded view below.

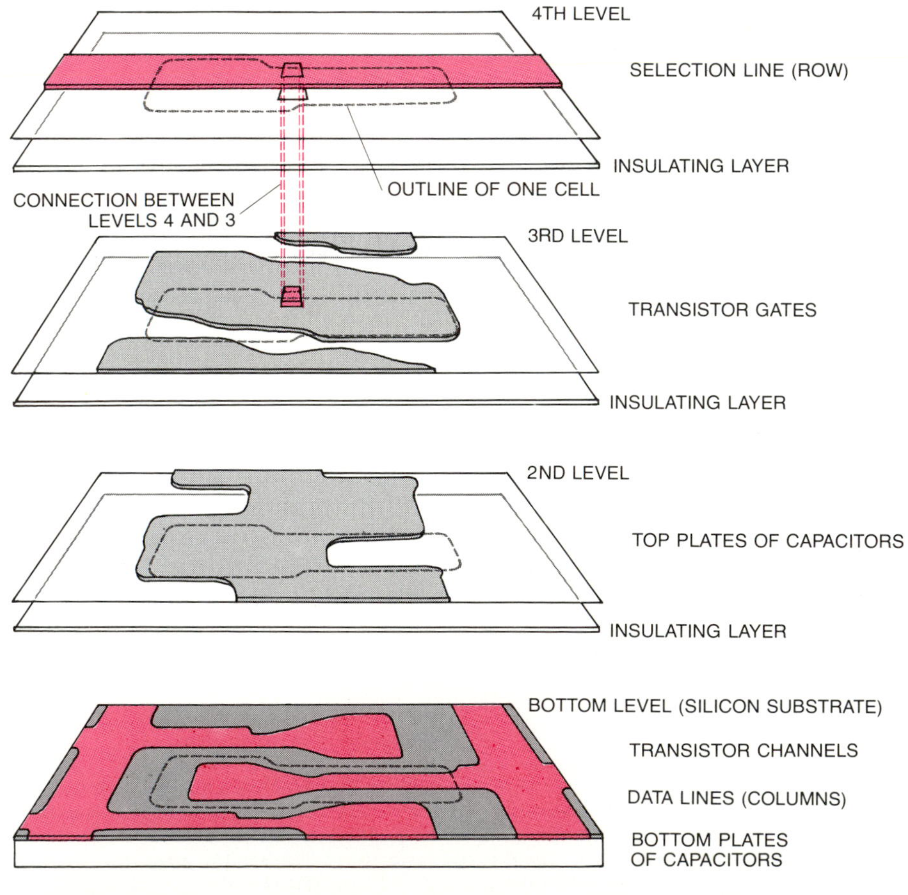

CONSTRUCTION OF A RAM CELL begins with the creation of multipurpose structures (*color*) in the bottom level by "doping" certain areas of the substrate with arsenic, which converts those areas into *n*-type silicon. The undoped area is chemically transformed into silicon nitride, an insulator (*gray*). The second and third levels of the cell are formed by the deposition of polycrystalline silicon. Fourth level is formed by deposition of aluminum. Thin insulating layers of silicon dioxide are deposited between levels. This cell measures 15 by 30 micrometers.

of the stored signal, sharply reducing the problem of detecting the "read" signal on the data lines. Refreshing the stored signal every two milliseconds, however, is still required. Continuous refreshing within each storage cell can be achieved by adding from one transistor to three more transistors per cell, but in addition to taking up more silicon area the design demands that each cell consume power continuously.

Memories that do not require refresh operations are termed static memories. In spite of their higher cost per bit of storage they are favored for small memory systems because they call for a minimum of external support circuitry. At a further premium in cost the power consumption of static memories can be reduced to such a negligible value that small batteries will power them for days or weeks. Such memories exploit the "complementary" MOS (CMOS) technology; they are found in some pocket calculators that hold their data or their program even when the power switch is in the "off" position.

The memories with the highest speed of operation usually employ bipolar-transistor technology rather than the fundamentally somewhat slower MOS or CMOS technologies. The price paid for attaining the higher speed is a more complex manufacturing process and in most instances higher power consumption. A recently developed form of the bipolar-transistor technology known as integrated injection logic (I^2L), or merged transistor logic (MTL), has been widely discussed as a memory technology, but it has not yet proved to have advantages either in performance or cost over memories based on the older bipolar-transistor technologies.

Some applications require random-access memories containing permanently stored or rarely altered information. For example, the control programs in pocket calculators are usually permanently stored. Such storage is provided by read-only memories (ROM's). Information is placed in the storage array when the chip is manufactured. A read-only memory can be obtained by replacing the storage capacitor in a one-transistor memory cell with either an open circuit or a connection to ground, thus representing one or the other of the two binary states. As an alternative to forming this pattern of connections in the initial manufacture of the memory, the cells can be fabricated with a small fusible link from the transistor to the ground in place of the storage capacitor. A pattern of stored information can then be placed in the array simply by applying a pattern of electrical signals strong enough to blow out the unwanted connections. Of course, once a fusible link is blown it cannot be re-formed. Fusible-link memories are one type of programmable read-only memory (PROM). Information stored in ROM's and PROM's is nonvolatile, that is, it is not lost when external power is removed.

Another variation on the read-only memory is the read-mostly memory, which is desired when read operations are far more frequent than write operations but for which nonvolatile storage is required. Read-mostly memories have two forms. The commonest is the optically erasable read-only memory. This memory is read and written by entirely electronic means, but before a write operation all the storage cells must be erased to the same initial state by exposing the packaged chip to ultraviolet radiation.

Each cell in the most modern form of such memory chips contains an MOS transistor with two gate electrodes, one above the other. The lower gate is a "floating" one totally surrounded by silicon dioxide, an excellent insulator, and is not connected electrically to anything. The turn-on threshold voltage of the transistor at the upper gate electrode is controlled by the charge on the floating gate. The insulated floating gate can be charged with electrons by applying a fairly high voltage (about 25 volts) at the gate and one of the drain electrodes while the *p*-type substrate material of the device is held at low voltage. This gives some electrons enough energy (about four electron volts) to cross the silicon dioxide insulating barrier and charge the floating gate. By doing this selectively in an array of cells one can store the desired pattern of information.

Because of the high quality of the silicon dioxide insulator the floating gate will remain reliably charged for years. To discharge the floating gate one exposes the entire memory chip to ultraviolet radiation, which makes the silicon dioxide sufficiently conductive to allow the charge on the gate to leak away. Because of the tiny size of the individual storage cells it is not practical to erase them selectively. Instead all the memory cells are erased simultaneously, after which a new information pattern is imposed.

An alternative form of read-mostly memory is the electrically alterable read-only memory (EAROM), which can be altered without the necessity of erasing the entire array. Its cell structure is roughly similar to the optically erasable cell except that the function of the floating gate is taken over by an interface between two dielectric materials, usually silicon dioxide and silicon nitride, which may be selectively charged and discharged by signals applied to a single overlying gate electrode. These read-mostly memories are called metal-nitride-oxide-semiconductor (MNOS) devices. A variant of the read-mostly memory is made from certain amorphous semiconductor materials, or semiconducting glasses, that hold a pattern of local charges until the charges are erased by a strong pulse of current, after which a new pattern can be electrically imposed. Although the potential usefulness of EAROM devices is large, they have not yet been endowed with the combination of low cost, fast writing, long-term retention and reliability after many erasures that is desired for most applications. As a result EAROM's have so far found only a few applications.

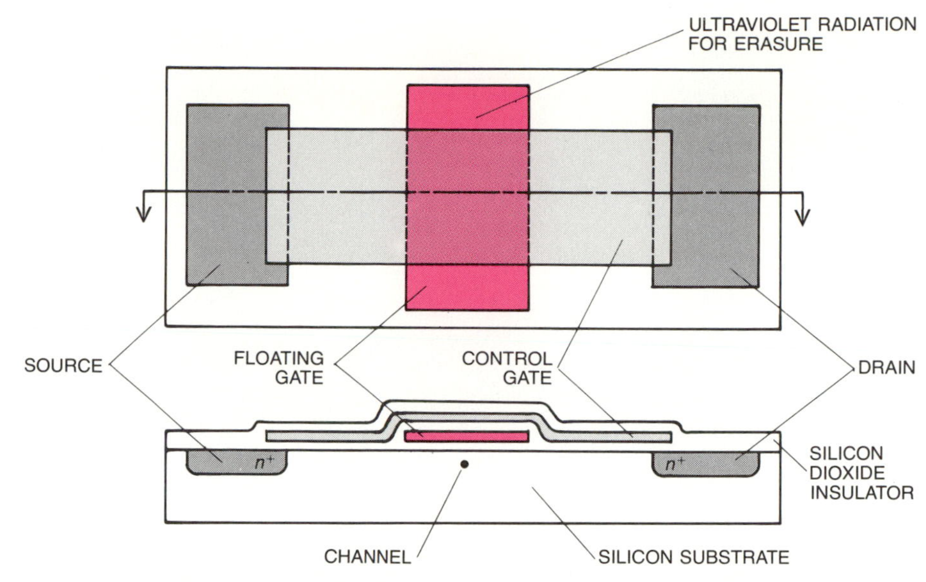

OPTICALLY ERASABLE READ-ONLY MEMORY CELL is shown in plan view and in cross section. The floating gate, which is not electrically connected to anything, holds a binary digit of information in the form of a stored charge that modifies the electrical characteristics of the device. Information is "written" (stored) in the floating gate by applying about 25 volts to the control gate and drain while the source and the substrate are grounded. The resulting high electric fields in the channel accelerate electrons to a considerable velocity. A small fraction of the electrons are able to cross the thin insulator and become trapped on the floating gate. The charge on the gate is not lost during "read" operations or when external power is removed. The stored charge can be erased, however, by exposing the cell to ultraviolet radiation, which temporarily makes the silicon dioxide sufficiently conductive for charge to leak away. All the cells in a memory are erased at one time, after which a new pattern can be written in.

All the electronic memories I have been discussing are designed for random access, that is, the time required for a read or write operation is independent of the physical location within the storage array of the cell being accessed. The access times of random-access memories are on the order of .1 microsecond to one microsecond. Substantial economies can be achieved in the design of memories if random access is not required. One alternative scheme of growing importance is serial access, in which the stored bits circulate as if they were in a closed pipeline.

Leading examples of electronic memory devices organized in this way are charge-coupled devices (CCD's) and magnetic-bubble devices. Each bit that is stored is transferred sequentially through 64 or more storage locations between the time it is written into the memory and the first time it becomes available for reading. The rate at which bits are shifted from one storage site to the next in a CCD memory is about the same as the cycle time for a random-access memory. Thus the longest access time for a serial memory with 64 storage locations is 64 times the cycle time for a random-access memory. There are many applications where serial access is entirely satisfactory. For example, a memory used to refresh the information presented in a conventional video display, which is scanned point by point in a repeating linear pattern, does not require a memory with random access.

There are several reasons why CCD memories can be designed to have a smaller total area per bit (and hence can have a potentially lower cost) than random-access semiconductor memories. First, when the process technology is optimized for CCD purposes, the area required for an individual storage cell is somewhat smaller than it is for a RAM cell. Second, the amount of address decoding required in a serial memory is inherently less than it is for a RAM, since decoding to select individual locations is not needed. Third, the charge representing the stored information is always conveyed and held on tiny nodes, or intersections, in the circuit, which have very little capacitance. As a result the signal voltage in a CCD is not attenuated as it is when the signal is shared between the storage cell and the data line in a random-access memory. Therefore the amplifiers and refreshing circuits for the CCD memory can be somewhat simpler. The net result of these simplifications is that the total silicon area per bit for complete memory components is about a factor of two to

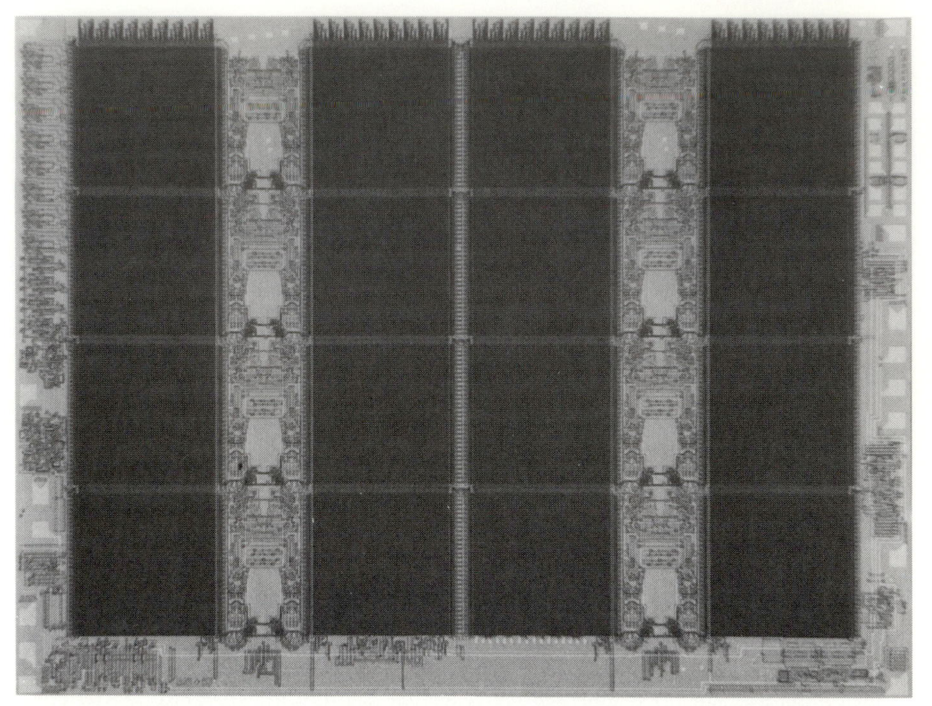

CHARGE-COUPLED DEVICE (CCD) made by the Fairchild Camera and Instrument Corporation provides serial access to 65,536 bits of memory. The total number of bits is divided among the 16 square arrays, each of which therefore holds 4,096 bits. Since the bits in all the arrays are circulating continuously and synchronously, the time it takes to access any of the 65,536 bits is set by the circulation time of the 4,096 bits in any one array (*see top illustration on page 61*). The average access time is .5 millisecond. Chip measures 4.4 by 5.8 millimeters.

MAGNETIC-BUBBLE MEMORY made by Texas Instruments Incorporated provides serial access to 100,637 bits, with an average access time of four milliseconds. Magnetic bubbles are domains, or islands, of magnetic polarization embedded in thin magnetic films of the opposite polarization. The bubbles can be circulated along prescribed tracks at high velocity (*see bottom illustration on page 61*) and do not disappear when power is turned off. The density in present devices is about 150,000 bits per square centimeter, or somewhat less than in CCD's.

three smaller for a CCD memory than it is for a random-access memory. One of the latest designs of a CCD serial-access memory has storage for 65,536 bits on a chip measuring about 3.5 by five millimeters.

The other principal form of microelectronic serial-access memory exploits the mobility of magnetic bubbles, or microscopic domains of magnetic polarization, in a thin magnetic film of orthoferrite or garnet. In the presence of a steady magnetic field of the appropriate strength, with its lines of force perpendicular to the plane of the film, domains of one polarization (say upward) are stable within a larger area of the opposite polarization (downward). The domains can be moved in the plane of the film by applying weaker magnetic fields at right angles to the principal field. When the domains are appropriately organized, they can constitute a magnetic-bubble memory.

If a small domain with an upward polarization is used to designate a binary 1, the absence of such a domain can be considered as designating a binary 0. Domains can be formed or eliminated with single-turn electromagnetic loops on the surface of the magnetic film. Arrays of electrodes in the shape of chevrons, *T's* or disks, made of another magnetic material, are used to create local variations in the magnetic field in order to localize and separate the individual bubbles. A rotating magnetic field in the plane of the magnetic film can be used to shift the domains along a line of fixed chevrons (or some other pattern).

A key feature of bubble memories is that stored information is retained when external power is interrupted, a valuable property that is exploited in most present applications. The polarization of the bubbles is preserved by using a permanent magnet to maintain the steady perpendicular magnetic field. Electronic circuits are required to drive the bubble-forming and -steering loops and to amplify the small output signal that is obtained by electromagnetic induction or a magnetoresistive effect when a bubble is passed under a sensing device. These circuits are powered during the read and write operations.

Bubble memories are inherently serial in organization, so that access time depends on the number of storage locations in a serial path and on the maximum shifting rate. In present devices serial paths range in length from about 10 locations to 1,000 or more; shifting rates range from a fraction of a microsecond to several microseconds. Like CCD memories, bubble memories cannot compete with random-access electronic memories in speed. Their most attractive potential application is the replacement of tape and disk memories with a capacity of between one million and 10 million bits. In such applications semiconductor memories may be ruled out because their storage of information is volatile. At present the cost per bit of bubble memories, if they are produced in quantity, should be roughly comparable to that of moving-surface memories for storage capacities of up to about 10 million bits. For the storage of larger quantities of information moving-surface memories still have a major cost advantage.

A promising variation of the bubble memory, the bubble-lattice file or structureless bubble circuit, eliminates the requirement for one or two physically patterned surface electrodes per bit of information stored. The bubbles are packed tightly together; the distinction between binary states is determined by changes in the magnetization within the wall of a single domain rather than by the presence or absence of a domain. The successful development of bubble-lattice memories would overcome the limitations imposed by the techniques for defining patterns, which now determine the maximum storage density.

The device that offers perhaps the most dramatic potential for improving microelectronic memories, indeed a potential for creating a new generation of superfast computers, is the superconductive tunnel junction. Such junctions are multiterminal switching devices that harness the transition between a superconducting tunneling state and a normal tunneling state in response to a small change in a local magnetic field. "Tunneling" refers to the ability of electrons under certain conditions to penetrate energy barriers they would ordinarily lack the energy to surmount. In order for the tunnel junction to go into the superconducting state the device must be cooled to cryogenic temperatures by immersion in liquid helium.

Since tunnel junctions operate at temperatures close to absolute zero and at the very low voltages characteristic of the superconducting state, they are capable of performing switching operations one or two orders of magnitude faster than semiconductor circuits can perform, and they consume two or three orders of magnitude less power. The reduction in power consumption makes it possible to pack circuits very densely without creating problems arising from the necessity to dissipate heat. The signal-propagation times in a large system are in turn reduced because the interconnecting lines can be shorter. In view of these considerations one can foresee the development of cryoelectronic memories with extremely high component densities, operating at speeds 10 to 100 times faster than today's fastest electronic memories.

A number of difficult problems remain to be solved before cryoelectronic memories and computers become practical. The fabrication of tunnel junctions

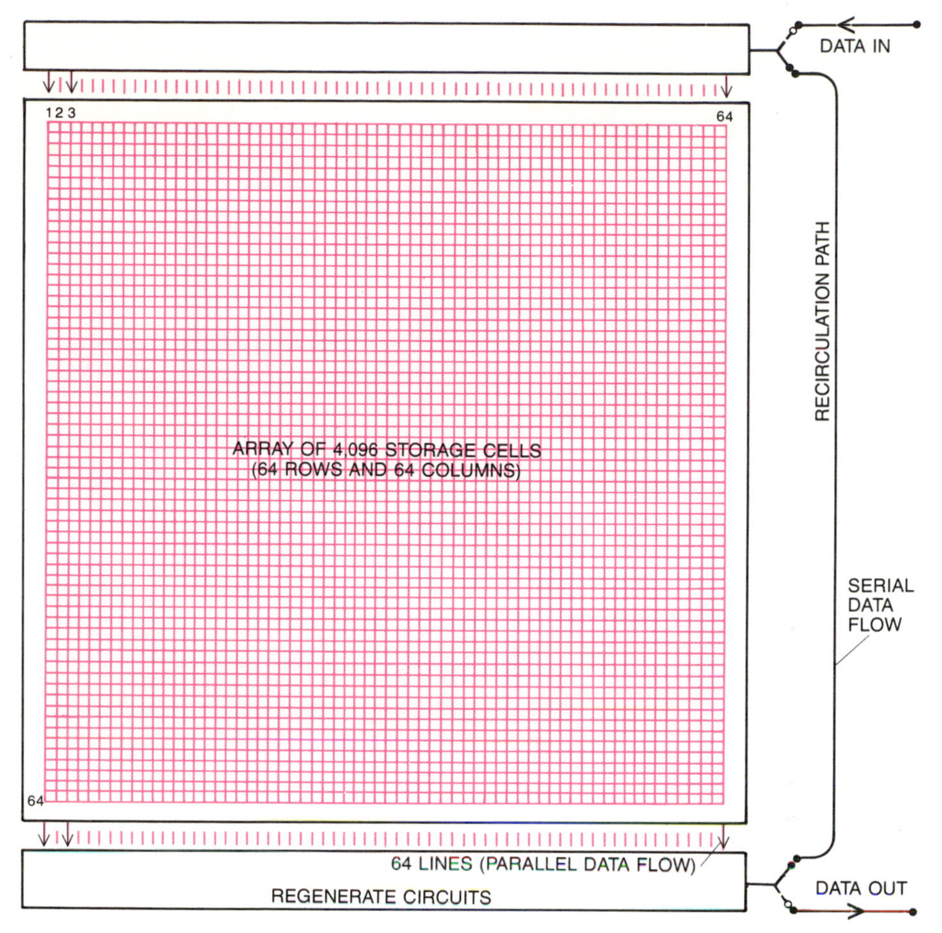

DIAGRAM OF CCD MEMORY depicts the circulation pattern of the 4,096 bits stored in each array of the 16-array, 65,536-bit memory device at the top of page 60. As in most random-access memory devices, the pattern of stored bits must be continuously regenerated, or refreshed. The regeneration cycle also establishes the access time for reading and writing. Here the 64 bits in each of the 64 columns shift downward synchronously into the regenerator along parallel lines at a rate of some 80,000 bits per second. The 4,096 bits (64 × 64) flow serially out of the regenerator at a rate 64 times higher (that is, at about five million bits per second) and reenter the top of the array, distributed along 64 parallel lines into their original columns. Since all 16 arrays operate in a similar fashion, the 65,536-bit device is externally indistinguishable from a single 4,096-bit serial shift register operating at five million cycles per second.

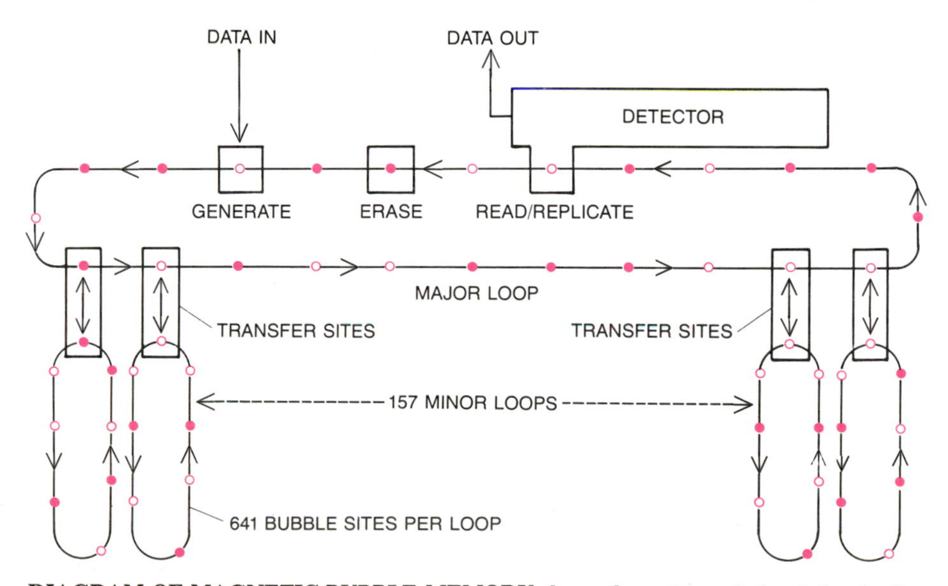

DIAGRAM OF MAGNETIC-BUBBLE MEMORY shows the pattern of circulation in the 100,637-bit memory. The major loop holds a single data block, consisting of 1's and 0's (bubbles or no bubbles) that are being written into the memory, read out, replicated or erased. In this particular device the data block contains 157 bits. In the writing cycle the 157 bits first enter the major loop, whence they are transferred simultaneously, at a signal, to the 157 minor loops, one bit per loop. Each minor loop in turn provides sites for 641 bubbles. Thus total capacity of device is 157 × 641, or 100,637, bits. In read cycle 157 bits are transferred simultaneously, at a signal, from minor loops into major loop, which carries them past read head.

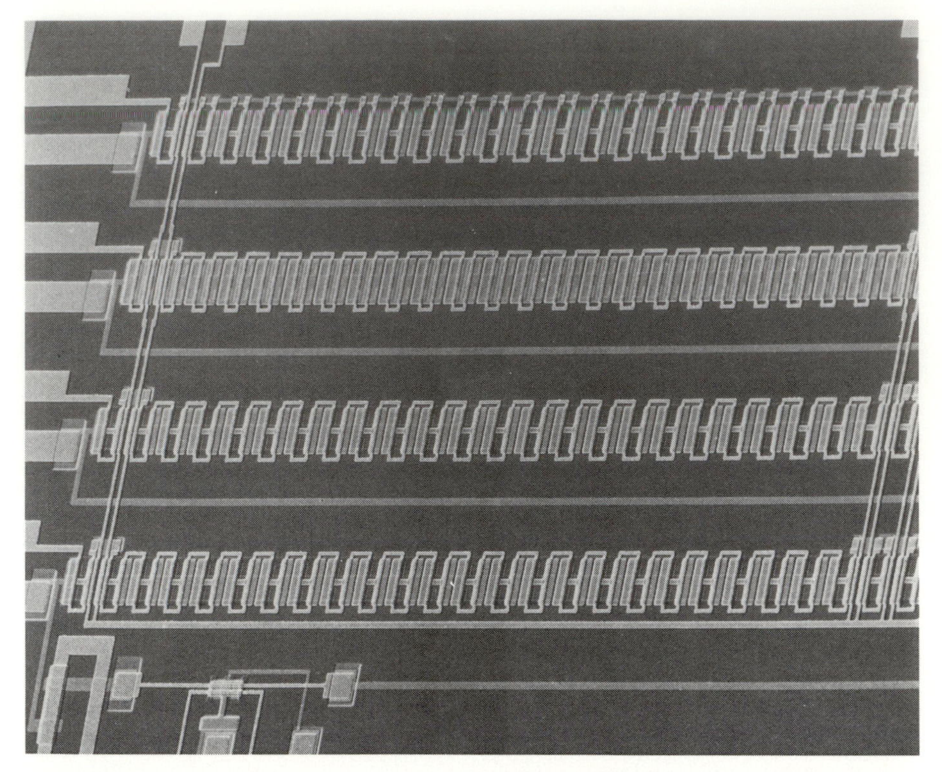

JOSEPHSON MEMORY CELLS, which operate close to absolute zero, can store information as a single quantum of magnetic flux, thereby achieving the ultimate energy economy attainable in a magnetic storage device. This experimental device, built by the Zurich Research Laboratory of the International Business Machines Corporation, has 392 Josephson memory cells arranged in four strings of 98 cells each. Test chip, only a section of which is seen here, was fabricated in part to see whether entering or removing information would disturb information stored in other cells. Results confirm storage of single quanta in closely spaced cells is feasible.

requires perfect insulating films (usually amorphous oxides grown in situ) with a thickness of about a two-hundredth the wavelength of blue light. New packaging and interconnection techniques are needed to achieve the high packing density required for minimum interconnection delays. Mechanical stresses generated by differing thermal-expansion coefficients must be dealt with in order to avoid damage when the temperature is reduced from room temperature to nearly absolute zero. Such considerations suggest that some years of work will be needed before cryoelectronic systems become available.

So far I have been describing mostly the functional features and high-speed performance of alternative memory technologies. At this point I should like to comment briefly on matters that may be the most important in the selection of a memory technology: the price per bit of storage capacity and reliability. Finally I shall touch on the prospects for further advances in the technology of moving-surface memories and all-electronic memories.

The price per bit of a memory system is roughly correlated with the complexity of the structure or system on a per-bit basis. For example, the highest-speed electronic memories require several transistors, resistors and interconnecting lines for each cell. In contrast, the simplest electronic memories require only one transistor and two interconnecting lines per bit. The fabrication of semiconductor memories calls for four to eight sequential pattern-transfer processes. Bubble memories are potentially cheaper than semiconductor memories because they require only one to three pattern-transfer processes in the fabrication of the storage-cell array.

Moving-surface memories are cheaper than electronic memories because there is no need to define individual physical patterns or structures for each individual storage cell. The minimum price of a moving-surface memory system, however, is relatively high because of the need for precision mechanical components to transport the magnetic storage medium. The economic attractiveness of the bubble-lattice memory is based on the fact that it requires neither moving parts nor the definition of features at each cell.

One way to consider the prices of various forms of memories is to compare them as a function of average random-access time. In the currently available systems the access time ranges from about 10 nanoseconds for bipolar-transistor memories to 10 seconds or more for magnetic-tape memories. The corresponding cost per bit varies from about one cent for the bipolar memories to less than 10^{-5} cent for tape. Across a million-to-one range in price per bit the pattern is uniform: one must accept a two orders of magnitude increase in access time in order to obtain a one order of magnitude saving in price.

The reliability of memory systems is a function of both fundamental and practical problems. The fundamental problems have to do with phenomena such as corrosion. For the well-established memory technologies these problems are now quite well understood and typically are not the cause of many failures. The practical problems have to do with defective manufacturing, packaging or testing and with mistakes in the use and

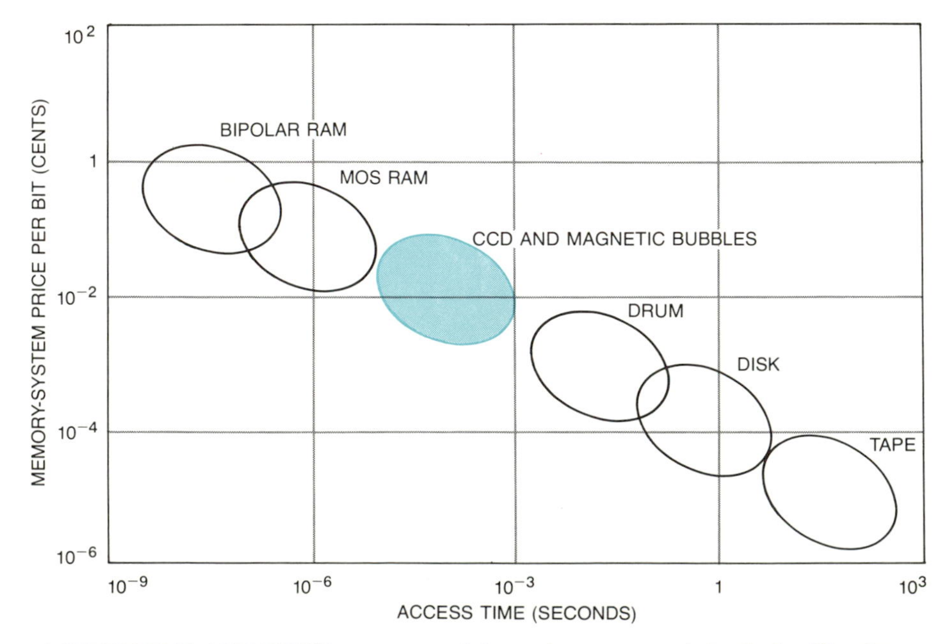

ACCESS TIME AND PRICE are compared for various memory technologies. The charge-coupled devices and magnetic-bubble memories are filling an important gap between ultra-high-speed, random-access memories and the much slower moving-surface magnetic memories. Cost for all memory systems should drop by a factor of 10 or more in coming decade.

maintenance of components and systems. Most of the failures in today's memory systems result from such practical problems. For example, in semiconductor memories failure rates in field service are better correlated with the number of separately packaged chips in the system than with the total number of bits the system stores.

For many years it has been empirically observed that the manufacturing cost of technologically complex products tends to decrease as accumulated production experience increases. A similar pattern is believed to exist in failure rates, which for a given electronic product seem to decrease with increasing manufacturing experience. This phenomenon, which is known in graphical representation as the learning curve or the progress function, creates a barrier to the introduction of new memory technologies. Over the past 20 years dozens of new memory technologies have fallen by the wayside in the face of continued improvements in established memory technologies. Moving-surface magnetic memories have improved steadily in this period. The success of semiconductor memories in overtaking ferrite-core memories as the dominant electronic storage technique is the only example that breaks the pattern. Many observers think this success resulted only from the fact that semiconductor microelectronic technology was already proceeding along a steep learning curve before it was employed for memory.

A number of promising memory technologies have not achieved commercial success in spite of intensive research and development and in some instances even limited production. Examples are the magnetic-film memories (such as plated-wire and planar-film memories) and the electron-beam and optical-beam memories. Often the investment required to bring a new memory technology to the stage of being manufacturable at low cost and with high reliability has been considered too great in relation to the risks and the potential rewards.

The next decade is likely to bring substantial improvements in the performance of both moving-surface and electronic memories, together with reductions in cost. There are no fundamental barriers to increasing the bit-storage density on moving magnetic surfaces a hundredfold, with little accompanying increase in the price of the system. The anticipated introduction of electron-beam and X-ray techniques in the fabrication of microelectronic circuits should also make it possible to increase the bit density of these devices by a factor of 100, again with only minor increases in price per component. Thus the expectation is that over the next 10 years there will be a reduction of more than an order of magnitude in the price per bit of all forms of digital memory.

6

MICROPROCESSORS

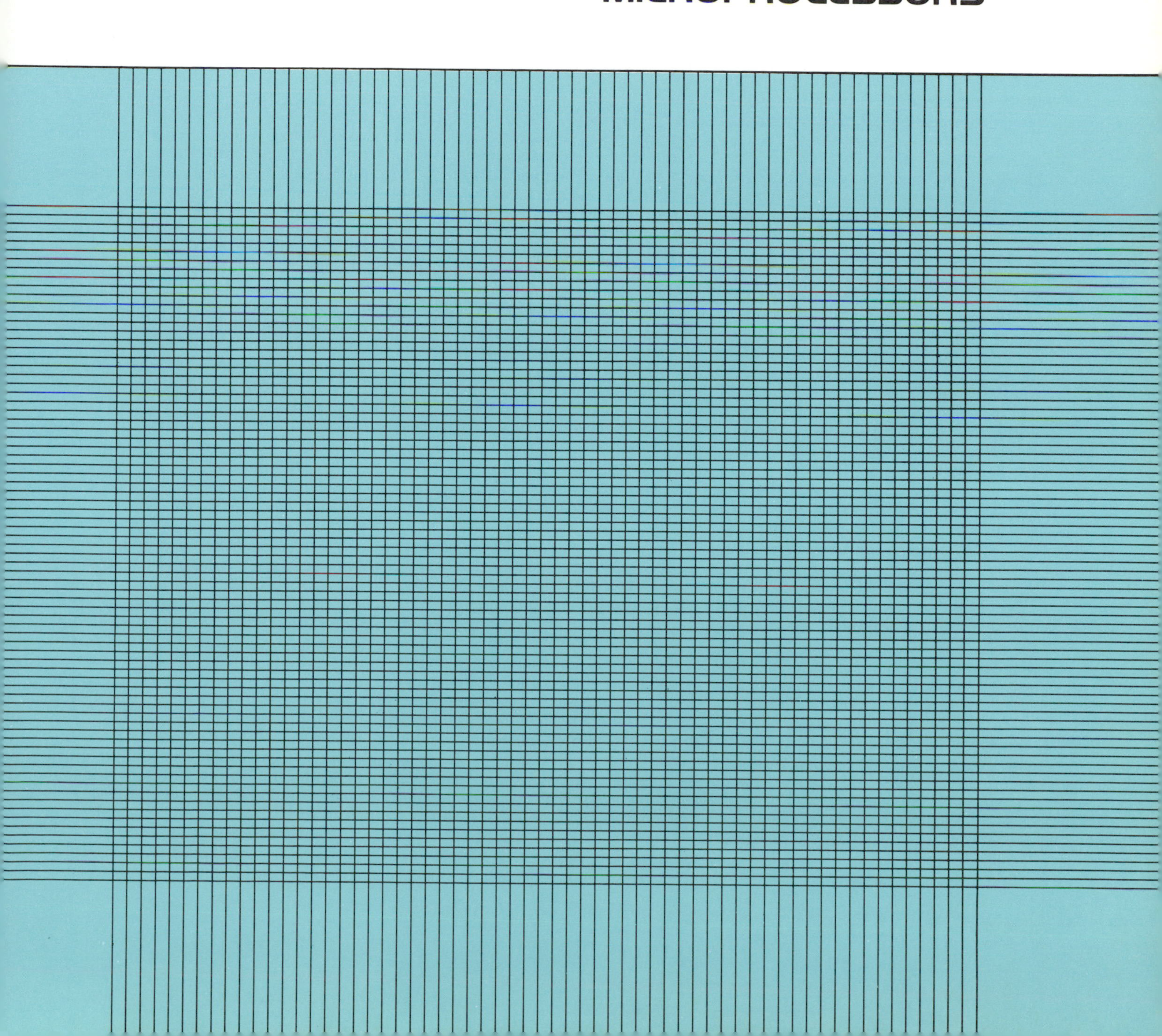

Microprocessors

by HOO-MIN D. TOONG

A microprocessor is a computer central processing unit on a single chip. Currently it is associated with other chips in a microcomputing system. Now emerging, however, are complete computer systems on a single chip

A microprocessor is the central arithmetic and logic unit of a computer, together with its associated circuitry, scaled down so that it fits on a single silicon chip (sometimes several chips) holding tens of thousands of transistors, resistors and similar circuit elements. It is a member of the family of large-scale integrated circuits that reflect the present state of evolution of a miniaturization process that began with the development of the transistor in the late 1940's. A typical microprocessor chip measures half a centimeter on a side. By adding anywhere from 10 to 80 chips to provide timing, program memory, random-access memory, interfaces for input and output signals and other ancillary functions one can assemble a complete computer system on a board whose area does not exceed the size of this page. Such an assembly is a microcomputer, in which the microprocessor serves as the master component. About 20 U.S. companies are now manufacturing some 30 different designs of microprocessor chips, ranging in price from $10 to $300. More than 120 companies are incorporating these chips in microcomputer systems selling for $100 and up. The number of applications for microprocessors is proliferating daily in industry, in banking, in power generation and distribution, in telecommunications and in scores of consumer products ranging from automobiles to electronic games.

As in the central processing unit, or CPU, of a larger computer, the task of the microprocessor is to receive data in the form of strings of binary digits (0's and 1's), to store the data for later processing, to perform arithmetic and logic operations on the data in accordance with previously stored instructions and to deliver the results to the user through an output mechanism such as an electric typewriter, a cathode-ray-tube display or a two-dimensional plotter. The block diagram of a typical microprocessor would show the following units: a decode and control unit (to interpret instructions from the stored program), the arithmetic and logic unit, or ALU (to perform arithmetic and logic operations), registers (to serve as an easily accessible memory for data frequently manipulated), an accumulator (a special register closely associated with the ALU), address buffers (to supply the control memory with the address from which to fetch the next instruction) and input-output buffers (to read instructions or data into the microprocessor or to send them out).

Present microprocessors vary in their detailed architecture depending on their manufacture and in some cases on the particular semiconductor technology adopted. One of the major distinctions is whether all the elements of the microprocessor are embodied in one chip or are divided among several identical modular chips that can be linked in parallel, the total number of chips depending on the length of the "word" the user wants to process: four bits (binary digits), eight bits, 16 bits or more. Such a multichip arrangement is known as a bit-sliced organization. A feature of bit-sliced chips made by the bipolar technology is that they are "microprogrammable": they allow the user to create specific sets of instructions, a definite advantage for many applications.

The flood of microprocessors and microcomputers reaching the market, combined with the rapid rate of innovation, guarantees that any attempt to catalogue them will be instantly obsolete. A more fruitful introduction to the "micro" marketplace is to classify systems hierarchically according to their capability and function. Along these two dimensions there is a well-defined upward progression in both hardware and software. In hardware the levels are chips, modules, "breadboard" systems, small computer systems, full-development systems and multiprocessor systems.

This hierarchy is not absolute because the evolving technology creates ever more powerful chips, some of which can bridge two or three hierarchic levels. Chips are used to construct a module, modules to construct a small computer system (SCS) and small computers to construct a full development system (FDS). Multiprocessor systems can incorporate modules, SCS's or FDS's, depending on the application and complexity.

At the first level of the hierarchy are the microprocessor chips, representing the large-scale integration of tens of thousands of individual electronic devices: transistors, diodes, resistors and capacitors. At this level there are also more specialized chips: random-access memories (RAM's), read-only memories (ROM's), programmable read-only memories (PROM's), input-output (I/O) interfaces and others. The cutting edge of the technology works most directly at the chip level, providing, for example, RAM's of ever higher storage capacity. (Currently the most advanced commercially available RAM can store 16,384 bits; within a year or two the maximum storage capacity will be 64,536 bits.)

Generally the various kinds of chips are grouped into families that are compatible with particular microprocessors. The families will include a series of RAM, ROM and PROM chips to create a memory system, a series of interface chips capable of handling both parallel and serial input-output functions and miscellaneous chips to enhance system capabilities, such as high-speed arithmetic operations. Master-control chips are needed to establish priorities and to keep signals flowing smoothly through the complex maze of interconnections. The compatibility of chips and chip families made by different manufacturers varies widely. For example, the microprocessors of different builders are generally not physically interchangeable, whereas several types of memory chips often are.

The second level of the hierarchy, the module and breadboard systems, represents the simplest true computer systems. They can be created by combining a microprocessor with a limited array of memory chips (RAM's and ROM's) and input-output chips. In order to communicate with such a minimal system the user will also need a simple device such as a numeric keyboard and a device capable of displaying or recording the computer output. Such single-board systems are useful for introductory teaching purposes or can serve as breadboard prototypes for more sophisticated systems. For a modest investment (usually under $300) a beginner can learn the fundamentals of microprocessor programming. Because of the system's limited memory, its lack of software-development tools and its crude interface with the user, however, even a novice is apt to outgrow a single-board system quickly.

At the next level in the hierarchy of capability and function are the small computer systems that are prepackaged as stand-alone units. Unlike the single-board modules, they have a self-contained power supply, the capability for memory expansion and room for a series of plug-in interface modules. Some of the more powerful single-board development modules can be expanded with the appropriate hardware to create such a single-box computer system. All the small computer systems have software capabilities that approach in sophistication those found in much larger conventional systems. They also provide an interface for a cathode-ray tube or a keyboard-display console. In addition many of the small systems can be interfaced to such peripheral devices as "floppy disk" memories, tape cassettes, paper tapes and line printers. With such enhancements a small computer system could serve as a full development system. For the most part, however, the single-box computers find their principal market today among computer hobbyists, who employ them for small programming tasks, word processing, general computation and game playing.

At the next level in the hierarchy we come to the FDS, or full development system. Perhaps its most important role is to provide a quick and efficient means for developing a low-cost microprocessor module that will later be manufac-

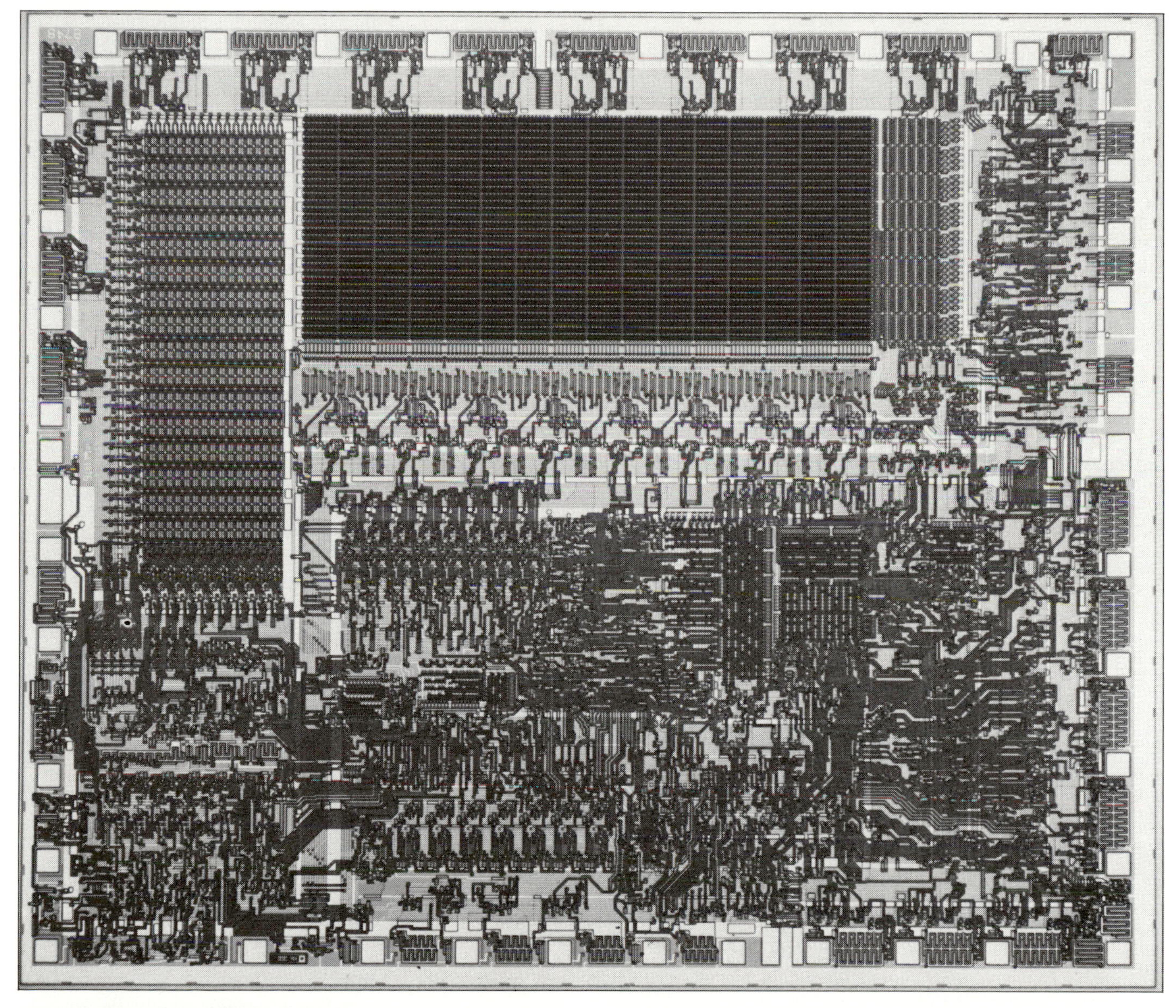

SINGLE-CHIP MICROCOMPUTER is a complete general-purpose digital processing and control system in one large-scale integrated circuit. The device combines a microprocessor, which would ordinarily occupy an entire chip, with a variety of supplementary functions such as program memory, data memory, multiple input-output (I/O) interfaces and timing circuits. The device shown, the 8748 made by the Intel Corporation, measures 5.6 by 6.6 millimeters. The program is stored in an erasable and reprogrammable read-only memory (EPROM), which has a capacity of one kilobyte, or 8,192 bits (binary digits). The program is erased by exposing the circuit to ultraviolet radiation, which causes the electric charges stored in the EPROM to leak away, after which a new program can be entered electrically. In volumes of 25 or more a packaged 8748 sells for $210. A functional map of chip appears in the illustration at bottom of page 68.

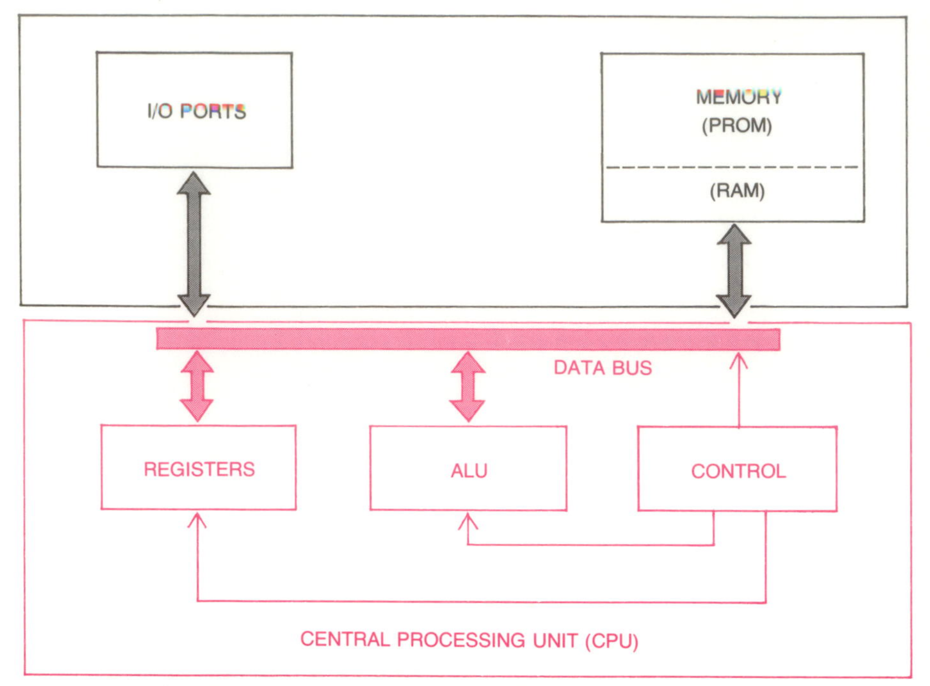

BASIC COMPONENTS OF COMPUTER SYSTEM can now be compressed onto a single chip, as in the Intel 8748. In this block diagram "control" includes control logic and instructions for decoding and executing the program stored in "memory." "Registers" provide control with temporary storage in the form of random-access memories (RAM's) and their associated functions. "ALU" (for arithmetic and logic unit) carries out arithmetic and logic operations under supervision of control. "I/O ports" provide access to peripheral devices such as a keyboard, a cathode-ray-tube display terminal, "floppy disk" information storage and a line printer. The functions that are in black convert a microprocessor (*color*) into a complete microcomputer.

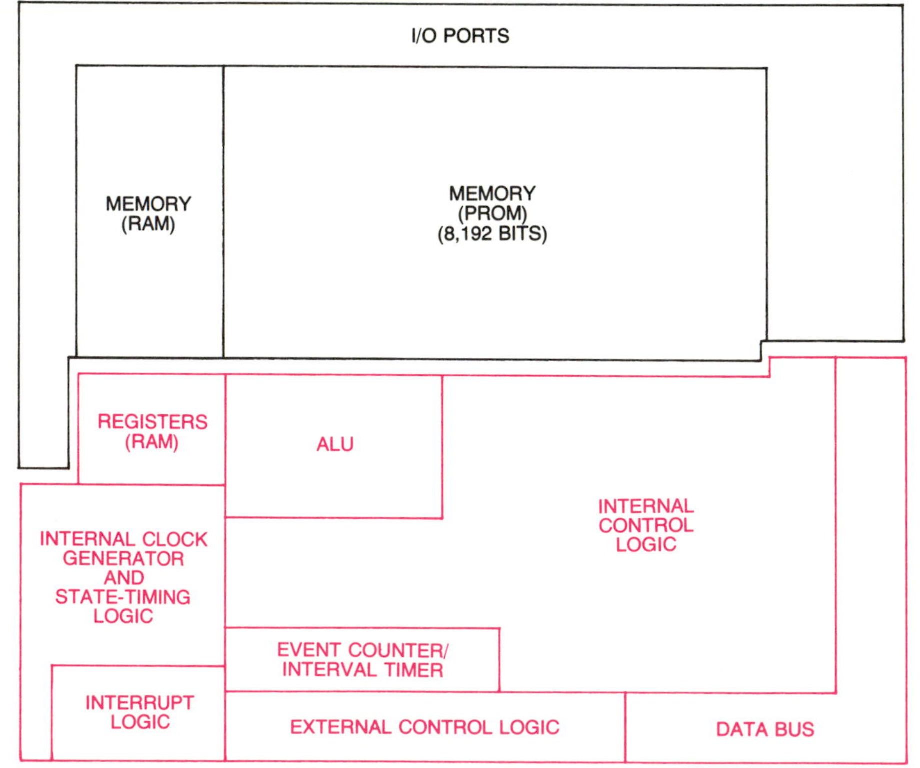

MAP OF 8748 MICROCOMPUTER identifies the location of the various computer functions. The color scheme used in the preceding illustration is repeated here. Each function can be assigned to one of the five basic functional blocks: control, memory, registers, ALU and I/O ports. The portions of the chip outlined in black represent the functions that transform the 8748 from a simple microprocessor into a microcomputer. Device holds some 20,000 transistors fabricated by *n*-channel silicon-gate metal-oxide-semiconductor (*n*-MOS) technology. Eight-bit central processor responds to 96 instructions in average time of 2.5 microseconds.

tured in volume to solve a manufacturing, telecommunications or business problem. In other words, the FDS is a full-capability microelectronic system that can serve as a vehicle for helping to develop a smaller target system.

Whereas the FDS may represent an investment of some $15,000, counting both hardware and software, the target system will be a microprocessor or microcomputer costing perhaps $500, or even less when it is manufactured in volume. For example, if one wanted to develop a microprocessor to optimize the performance of an automobile engine (by continuously adjusting the amount of fuel, the ignition timing and the fuel mixture), the final unit might be a small integrated-circuit module affixed to each engine and costing well under $100. One would use an FDS to develop the programs that would ultimately be placed in the ROM and PROM memories of the target system. A full development system normally includes a small computer system, a dual floppy-disk drive with a controller, a line printer, a cathode-ray-tube display terminal or teleprinter console, a ROM or PROM programmer and possibly a few other specialized pieces of hardware.

The final level of microcomputer usage is the multiprocessor system. The microprocessor represents truly low-cost computing. Its economics are so compelling that microcomputers are serving not only in many applications where computing power was previously too costly but also in applications where several dozen dedicated microprocessor modules can now be teamed to monitor and control parts of existing industrial or commercial systems where computer control was formerly unthinkable. Such an assembly of microprocessors or microcomputers can be organized in two functionally distinct ways.

In the first type of organization a tightly coupled group of microprocessors is designed to exchange data at high rates over short distances with a high degree of parallelism to achieve a maximum of computational power. Such a system could be used to emulate a large computer, to provide high reliability or to handle a specific problem that can take advantage of several processors operating in parallel.

The second organization, with by far the greatest application potential, is a loosely coupled system in which several widely distributed microcomputers communicate at low data rates with little or no parallelism. Examples of such distributed systems would be applications to factory automation, the control of oil refineries and chemical plants and the control of electrical devices in a home. Distributed systems are approaching reality in sizable numbers.

As can be imagined, the design and

development tools for multiprocessor systems are much more primitive than those for single processor systems. Software problems that are difficult enough to solve for one microcomputer expand almost geometrically in complexity as more units are added to the system. The problems include the organization of distributed files, or information-storage systems, process scheduling and achieving what programmers like to call "graceful" degradation, which means that the system should not merely "fail safe" but fail in gentle stages—gracefully. On the hardware side of the problem manufacturers have so far given scant attention to the configuration of microcomputers that would lend themselves efficiently to distributed installations.

So far I have used the term software without being very specific about its meaning. Since an understanding and a manipulation of software are fundamental to all computer usage, let me be somewhat more explicit. Like the hardware described above, software has its hierarchies. In the broadest sense software provides the means for telling a computer explicitly what to do through a step-by-step sequence of instructions that form a program. Each computer is provided with an "instruction set": a list of all the basic operations the computer is capable of performing. Each instruction is written in binary machine code: a sequence of 0's and 1's, typically eight or 16 bits long. Although a complete program could be written in this low-level language, the task is so tedious that an intermediate representation known as assembly language was developed and is currently the commonest language employed for programming microprocessors. Usually each symbolic instruction written in assembly language represents a single instruction in machine language. The translation is done by the computer itself with an "assembler" program.

To make programming still easier "higher level" languages were developed in which the instructions more nearly approximate ordinary English and the notations of mathematics. Examples are FORTRAN, ALGOL, COBOL and PL/1. One statement in such languages usually corresponds to many statements in machine language. The translation is done by the computer with the aid of a program called a compiler.

Even with such simplifications writing a program is arduous. Discovering the inevitable mistakes and correcting them is known as debugging. To simplify making changes in a program, the user

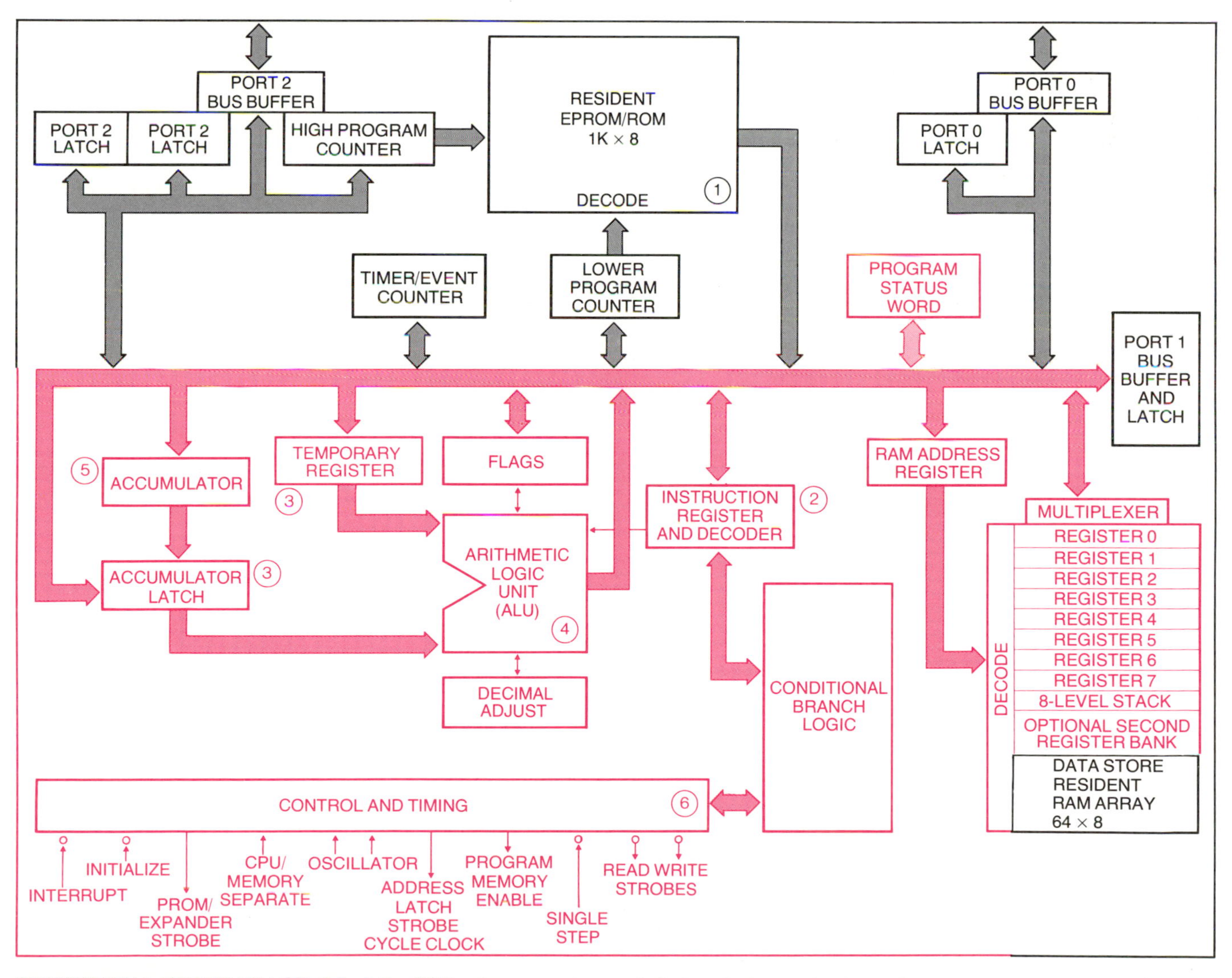

FUNCTIONAL BLOCK DIAGRAM of the 8748 microcomputer can be used to follow the sequence of steps involved in a simple operation, for example the addition of the contents of two registers, *A* and *R*, where *A* is the accumulator and *R* is any of the registers in the array at the lower right. The computer's first step (*1*) is to fetch the instruction from memory: "ADD A, R." The next step is to place the instruction in the instruction register and decoder (*2*), where the decoder finds that the instruction is to add *R* to *A* and to leave the result in *A*. In the next step the contents of register *R* are sent to the temporary register (*3*) and the contents of the accumulator to the accumulator latch (*3*). The ALU (*4*) then adds the contents of the two registers and the result is returned to the accumulator (*5*). Instruction ends and a signal is generated (*6*) to fetch next instruction. The 8748 microcomputer is capable of performing some 400,000 such additions per second.

SINGLE-BOARD COMPUTER MODULES provide a capability for the design and development of systems for special tasks. The photograph shows a briefcase microcomputer built by the author's group at the Massachusetts Institute of Technology. A commercial single-board module, an Intel SBC 80/10 (*nearest briefcase*), is combined with a custom-designed module (*lower right*) and two peripheral units: a keyboard and a display unit. The arrangement demonstrates how standard modules can be integrated with custom ones to solve a particular problem.

MICROPROCESSOR SYSTEM DEVELOPMENT is studied in the author's laboratory at M.I.T. The development engineer is working at a terminal tied into a full development system of the kind described in the illustration shown on pages 72 and 73. The microcomputer itself, an Intellec MDS-80, is in the cabinet at the rear immediately to the left of the engineer's head. Directly under the microcomputer is a paper-tape reader. To the left of the microcomputer is a dual floppy disk, and above it is a PROM programmer. The software facilities available to the engineer can be extended beyond full development system to include complete software capabilities of a minicomputer in "Triad" system depicted in upper illustration on page 71.

can employ a special editing program, which facilitates the changing of individual instructions. Once the program is debugged it is usually stored in some nonvolatile memory device, such as a magnetic tape or disk. When the program is ready for the computer, it is rapidly transferred from the tape or disk into the computer's high-speed random-access memory. As we have seen, however, the usual goal of a full development system is to create a program that can be stored in the permanent (ROM or PROM) memory of a microcomputer targeted to solve a specific problem repetitively.

Over the years there has been a proliferation of symbolic and higher-level languages for special purposes, each with its own assemblers or compilers for making it intelligible to particular models of computers. As a result "cross software" systems have been developed to facilitate communication between computers. Thus users of large computer systems and time-sharing services have access to cross-software assemblers, compilers and simulators (programs that enable a computer of one make or model to duplicate the actions of another). At present this represents an expensive alternative to a full development system for creating microprocessor software. In addition software simulators do not readily duplicate real-time input-output and execution speeds of the target microprocessor.

In order to project the rate of introduction of future microprocessor applications it is helpful to understand the process by which applications are typically developed. Given an identifiable use for a microprocessor in a product or system, how does a manufacturing company go about developing a suitable device equipped with a suitable program? As I have indicated, the prospective user does not as a rule try to design a special microprocessor chip for his particular task. He starts with one of the chips already on the market and selects from the wide assortment of other available chips: RAM's, ROM's, PROM's, I/O interfaces or whatever may be needed to construct a module capable of carrying out the task he hopes to perform. In many cases commercially available general-purpose, single-board microcomputer modules will be satisfactory for his task. Microprocessors have become so inexpensive that it is usually cheaper to exploit as little as 10 or even 5 percent of the computing power of an existing chip or module than to invest in the design and programming of a special unit that would do the job with only the minimum number of electronic components. Given a sufficient volume of production, however, development of a special unit may be justified.

The process of developing a microprocessor application begins with the identification of a need. Often the person in an organization who perceives the need is unfamiliar with the details of the new microelectronic technology. As a result in most cases the need is communicated to an engineering project manager, who makes an evaluation to determine whether the use of a microprocessor is justified. Such an evaluation would include an analysis to determine which, if any, of the available microprocessors have suitable capabilities and to estimate the time and manpower needed to develop the necessary software, by far the most time-consuming and costly part of the job.

If a microprocessor appears to be justified, the task is broken down into two distinct paths: the hardware requirement is turned over to a design engineer and the computation and control requirements are given to a software programmer. In the typical case hardware and software efforts are carried out in parallel by two separate groups. The key to a successful application is close communication between the two as the system evolves. Unfortunately there is no standard methodology for achieving a good design. As in all engineering, much depends on intuition, a good working knowledge of available products and past experience.

With "ad hoc" design procedures for both hardware and software a prototype system is developed. Hardware prototyping mechanisms commonly include wire-wrap breadboard models, plugboard setups or printed-circuit prototypes. The corresponding prototyping system in software is a target, or object, program that can be derived from any of several translation mechanisms: hand assembly (by means of a code book), the use of a resident assembler on a full development system or the use of a cross assembler on a time-sharing system.

When prototypes of both hardware and software are sufficiently advanced, the two must be mated, a task usually performed by a systems engineer. The target program is loaded into the hardware prototype to see if the resulting system meets the original specifications of the program manager. Deficiencies in either hardware or software at this level must be fed back through the hardware and software "loops" in iterative fashion until satisfactory performance is achieved.

Now a major decision must be made: whether or not to put the microprocessor module into production. The length of time from the start of development to a successful prototype is often critical. Many prototypes are shelved at this stage because development has taken so long (it can take as long as two years) that the original problem requirements have changed, the perception of the problem has changed or a competitor may have reached the market first with an equivalent product, so that it is now necessary to come out with something better. Since time is often critical, software- and hardware-design groups have an urgent need for a development tool that is more powerful than those I have outlined and that will make it possible to compress the development time of prototype systems into a few weeks.

Let me describe such a tool. As we have seen, the development of a prototype microprocessor application obliges hardware designers and software programmers to work at nearly all levels of

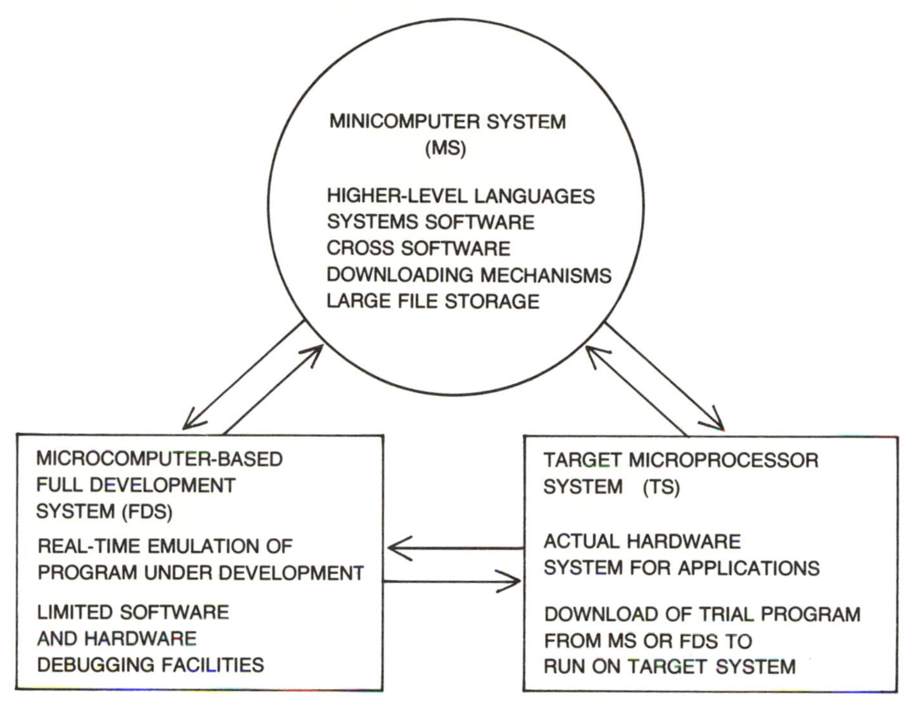

"TRIAD" STRUCTURE has as its goal reducing the time needed to develop an applications program for a microprocessor module or a microcomputer. The use of the centralized minicomputer system greatly facilitates the task of the programmer in developing software for the target system. In addition microprocessor of target system is tied into effort from the outset.

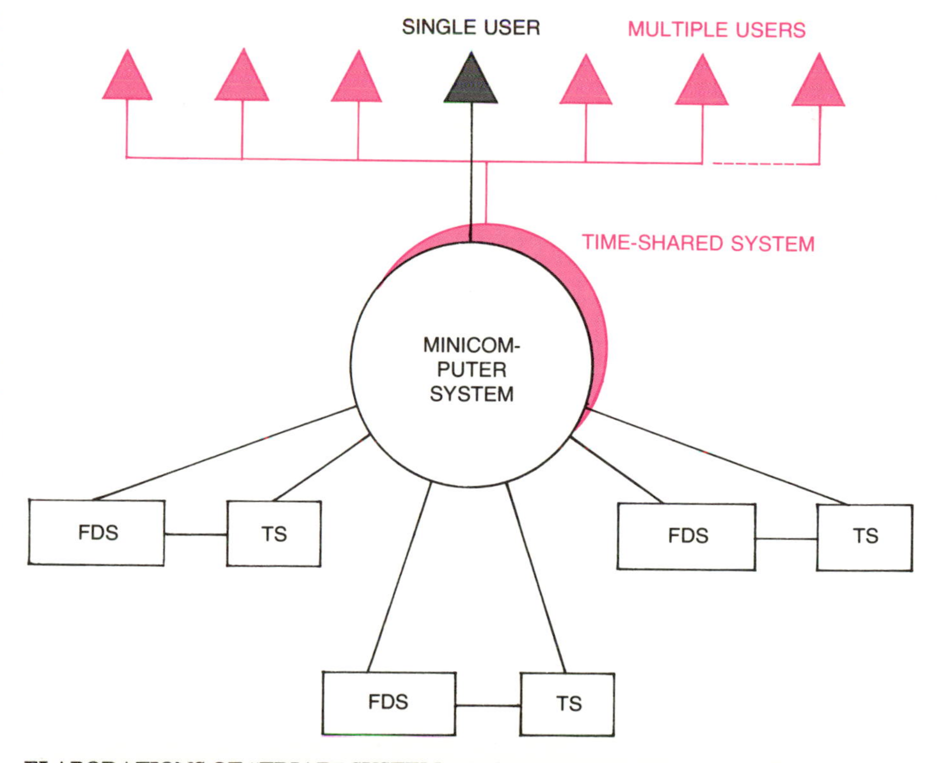

ELABORATIONS OF "TRIAD" SYSTEM may be warranted in large organizations where a number of microprocessor applications are under development. The system entirely in black is designed to give one programmer access to several different Triads, each dedicated to a different microprocessor. In the more powerful time-sharing system (*color*) engineers and programmers can work simultaneously on development efforts involving different microprocessors.

the hierarchies that have been evolving in their respective disciplines. The hardware designers must work from the level of chips up to full development systems, and the programmers must work from the level of the computer's instruction set up to floppy-disk operating systems with the various compilers, assemblers and editors associated with a full development system. Moving from level to level in the two hierarchies is inevitably inefficient because of incompatibilities between levels and the difficulty of moving results upward or downward between levels.

It turns out that even the FDS does not provide all the development features that many system designers would like, such as large file-handling capabilities, quick response to system commands, powerful editing programs and a library of software for the microprocessors of different makers. Designers who require such features are finding it desirable to use minicomputers (the small but high-performance computing systems that first reached the market in the early 1960's) as a supplementary development tool.

Although some cross-software packages exist for enabling minicomputers to develop microprocessor programs, there are very few systems for closely coupling minicomputers, FDS's and the microcomputer target system. Such

CAPABILITIES	TYPICAL USES AND USERS	MICRO HARDWARE HIERARCHY	
		LEVEL	REPRESENTATION
DISTRIBUTED COMPUTING TIGHTLY COUPLED PARALLEL PROCESSING	PROCESS AUTOMATION COORDINATION AND CONTROL ON A DISTRIBUTED AND LOCAL BASIS	MULTIPROCESSOR SYSTEM	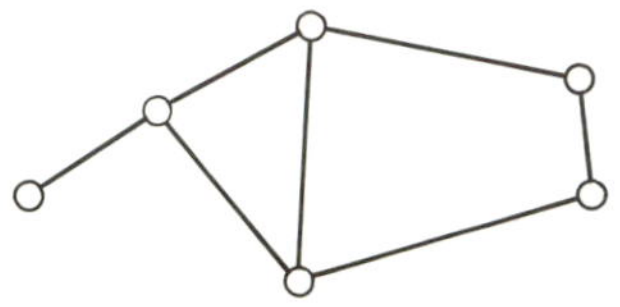 FDS, SCS OR MODULE
FULL SOFTWARE DEVELOPMENT HARDWARE DEBUGGING HIGHER-LEVEL LANGUAGE PROGRAMMING	SOFTWARE-APPLICATIONS PROGRAMMING DEBUGGING OF HARDWARE TARGET SYSTEM	FULL DEVELOPMENT SYSTEM (FDS)	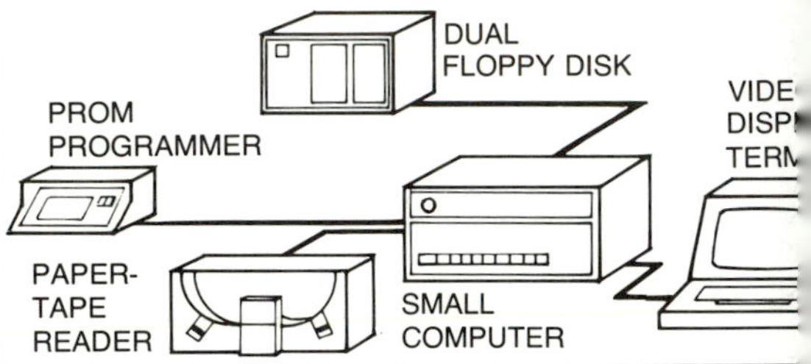 DUAL FLOPPY DISK PROM PROGRAMMER VIDE DISP TERM PAPER-TAPE READER SMALL COMPUTER
INTERMEDIATE-COMPLEXITY APPLICATIONS PROGRAMS (1–10 K) SOME HIGHER-LEVEL LANGUAGE CAPABILITY (E.G., BASIC)	PERSONAL COMPUTER HOBBYIST	SMALL COMPUTER SYSTEM (SCS)	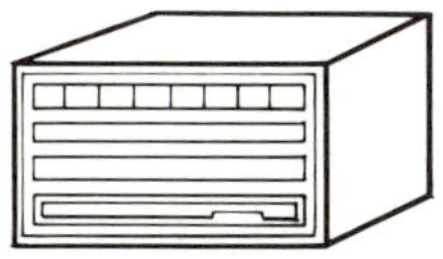
SMALL DEVELOPMENT SYSTEM FOR LEARNING MICROPROCESSOR CHARACTERISTICS SMALL-USER PROGRAMS (UNDER 1 K)	BEGINNING USERS OF MICROPROCESSORS ELEMENTARY PROTOTYPING USER EVALUATION OF MICROPROCESSOR	MODULES	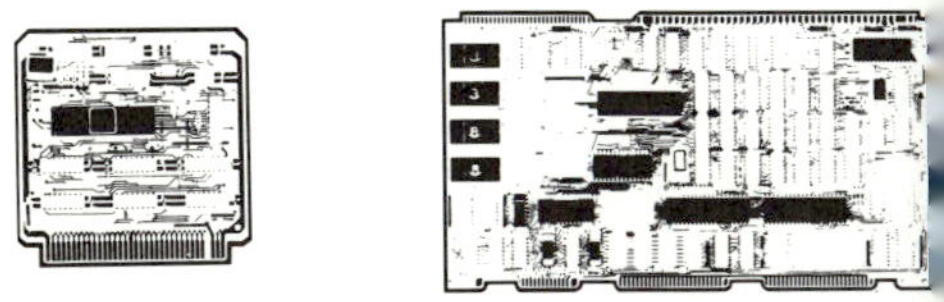
CUSTOM DESIGN OF A HARDWARE SYSTEM FOR PARTICULAR NEED	HARDWARE DESIGNERS	CHIPS	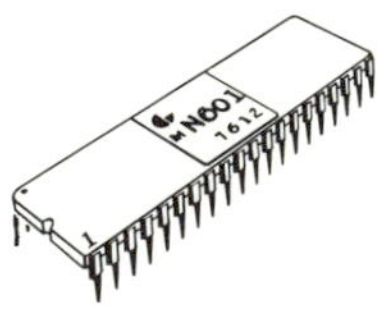

MICROPROCESSOR SYSTEMS can be arranged in an ascending hierarchy of hardware and software in which smaller components are assembled into successively larger systems with more powerful capabilities. The building blocks are the families of chips designed for various functions. To solve an application problem, for example the control system of an airplane, designers usually assemble modules or small computer systems and provide them with a suitable program for the task. Chips and modules have become so cheap (less than $30

coupling is desirable because it enables a development engineer to move easily within this hierarchy as well as to exploit the distinctive features of each system: the minicomputer to provide efficient editing, mass storage, documentation tools and shared data bases; the FDS to emulate the real-time performance of the microcomputer as programs are evolving and to debug the hardware, and the microcomputer target system itself to evaluate the final programs and control routines under the actual environmental and electrical constraints of the application setting.

The cleanest integration of these three systems is the "Triad," a recently developed research tool that enables the programmer or engineer to work at any level in achieving a desired program. The Triad closely couples three systems: the minicomputer, with its powerful software capabilities; the FDS, with its real-time emulation and hardware-debugging facilities, and the target system, with its application-defined construction. The Triad provides fast and direct access to all levels of the hierarchy.

The hardware designer and the software programmer would use the Triad as follows. The hardware engineer develops the target system, which might consist of a commercial general-purpose, single-board computer module in conjunction with an additional module of his own design that provides an inter-

	MICRO SOFTWARE HIERARCHY		
COMPONENTS	LEVEL	REPRESENTATION	COMPONENTS
CROCOMPUTER SYSTEMS)MMUNICATION SUB- STEMS AL-TIME CONTROL TERFACES WITH NSORS AND ACTUATORS	DISTRIBUTED-SYSTEMS SOFTWARE	DISTRIBUTED NETWORK MODULES FOR FILE MANAGEMENT, DEVICE CONTROL, COMMUNICATION APPLICATIONS, ETC.	DISTRIBUTED OPERATING SYSTEM
CROCOMPUTER,)EO DISPLAY TERMINAL, OPPY DISK, PROM OGRAMMER, ETC. NICOMPUTER-BASED OSS-SOFTWARE STEM	DEVELOPMENTAL SOFTWARE	DISK-BASED OPERATING SYSTEM HIGHER-LEVEL LANGUAGES (E.G., FORTRAN, BASIC, PL/M) EXAMPLE: X = X + 1 WHERE X IS AN 8-BIT NUMBER AND Y IS ITS ADDRESS	FLOPPY-DISK/CASSETTE OPERATING SYSTEM COMPILERS IN-CIRCUIT DEBUGGERS

CKAGED VERSION
MODULES
TH EXPANSION
PABILITY FOR
MORY AND
ERFACES
	STAND-ALONE SOFTWARE (NO PERIPHERALS)	SIMPLE HIGHER-LEVEL LANGUAGES (E.G., BASIC)
		ASSEMBLY LANGUAGE (E.G., 8080)
		LXI H,Y
		MOV A,M
		ADI 1
		MOV M,A
			STAND-ALONE MONITOR
			ASSEMBLER
			EDITOR

RE PLUG-IN CIRCUIT ARD INCORPORATING P FAMILY WITH PANSION CAPABILITY R MULTIPLE BOARDS	ELEMENTARY SOFTWARE	ASSEMBLY LANGUAGES (E.G., HEXADECIMAL) 21 Y 7E C6 01 77	SIMPLE MONITOR LIMITED DEBUGGER
NTRAL PROCESSING UNITS (CPU'S) NDOM-ACCESS MEMORIES (RAM'S) AD-ONLY MEMORIES (ROM'S))GRAMMABLE ROM'S (PROM'S) UT-OUTPUT (I/O) INTERFACES HER SPECIAL-PURPOSE CHIPS	BARE INSTRUCTION SET	BINARY MACHINE CODE 0010 0001 Y 0111 1110 1100 0110 0000 0001 0111 0111	OPERATIONS HARD-WIRED OR MICROPROGRAMMED AT CHIP LEVEL

for a microprocessor and less than $300 for a single-board module) that a major cost in engineering an application is the cost of developing the software to create the final program for the "target" system. Improvements in semiconductor technology are steadily making it possible for systems at each level to include more of capabilities once assigned to level above. Symbols and characters in color show how the same instruction looks when it is written in different languages of an increasingly higher level, beginning with binary machine code.

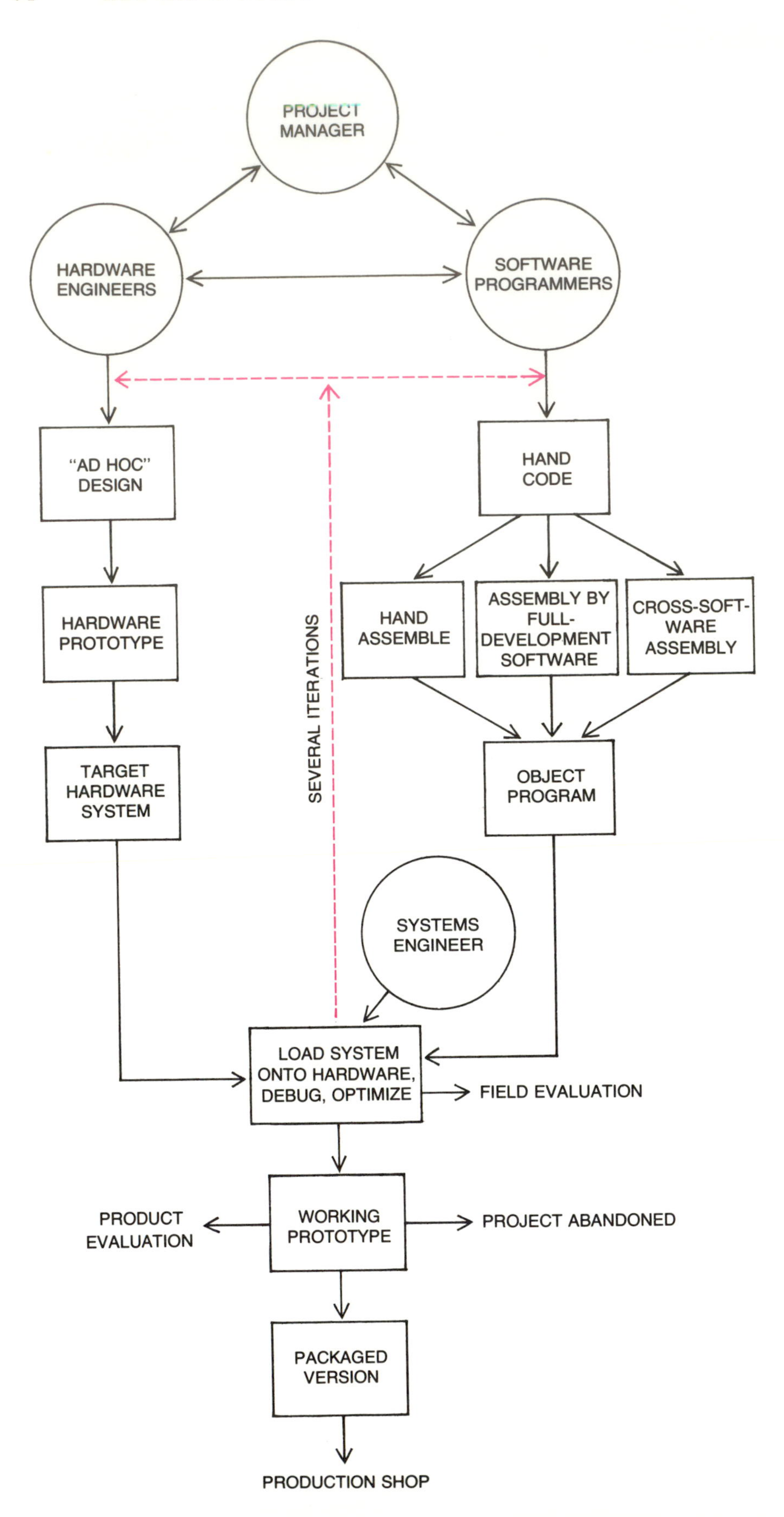

TASK OF DEVELOPING APPLICATIONS FOR MICROPROCESSORS requires close cooperation between hardware engineers, concerned with the selection and interfacing of chips, and software programmers, whose job is to provide a "debugged" program that will meet the project objectives. Because the time needed for programming is often underestimated the original objectives are sometimes obsolete by the time prototype is debugged and ready for production. As a result many projects are abandoned at this stage. It is not unusual for development programs to take two years. Author believes the time can be greatly compressed with Triad.

face to the larger system to which the microprocessor is being applied. In order to debug the target system he will need some test programs. Such programs can be quickly prepared with the editor on the minicomputer, cross-assembled into the machine code of the microprocessor and loaded directly into the undebugged target hardware. As the engineer works out the errors in his hardware he will continually update his test programs. It takes only a few minutes to reassemble and load each change.

Concurrently with the hardware development the programmer can be editing, assembling and simulating his programs on the minicomputer and on the FDS. He would maintain all his programs in the central file system of the minicomputer and share a library of applications software with other users of the Triad. System integration can proceed smoothly, because the same facilities the hardware engineer has been using to load test programs into the target-system hardware can also be used by the programmer to load the final applications program downward into the hardware. Changes at this stage are readily achieved by a simple process of re-edit, assemble and download. In this way the Triad system can sharply reduce the time needed for development of a microprocessor application.

Several Triad systems can be supervised simultaneously by a single central minicomputer system. Such an arrangement enables a user to develop application packages utilizing several different microprocessors while he is still maintaining all his programs, documentation and engineering reports within the file system of the central minicomputer. Moreover, other users can share his programs, thus making the development cycle more efficient. As the next step the minicomputer operating system could be rewritten to allow time-sharing among simultaneous users. This gives rise to the most powerful variation of the Triad concept.

The potential applications of microprocessor technology are so numerous that it is hard to visualize any aspect of contemporary life that will escape its impact. Here are some of the areas that will feel the effect soonest.

In automobiles a few 1977 models already employ microprocessors to meet Federal emission standards by controlling exhaust-gas recirculation systems. Many more 1978 models will include microprocessors both for emission control and for optimizing engine adjustments to improve gasoline mileage. In the near future microprocessors will also be connected to safety devices, such as sensors to prevent skidding on wet or icy surfaces.

In business offices among the first applications of microprocessors will involve the distribution and control of information. Desk-sized computers will become nearly as common as typewriters. They will handle small, specialized data bases appropriate to each person's job as well as accounting information and personnel data. The transfer of typewritten documents between offices will be largely replaced by electronic memorandums, relayed through the office computer system. Complete office systems are under development by major companies such as the International Business Machines Corporation and the Xerox Corporation.

In industry microprocessors are now used for such diverse tasks as machine-tool control and remote monitoring of oil fields. Microcomputers will also make possible a new generation of "intelligent" robot arms and hands capable of factory assembly operations heretofore too complex for mechanization.

In the home microprocessors have already appeared in a host of video games and such household appliances as microwave ovens and food blenders. They will rapidly be extended to temperature controls, refrigerators, telephones, solar-energy systems and to fire- and burglary-alarm systems. In time microcomputers linked directly to one's bank, and provided with multimillion-bit files for personal records, will be as commonplace as high-fidelity sound systems are today.

Devising a successful microprocessor application, however, is only the first step toward achieving its adoption and acceptance. The advent of the microprocessor presages more than just a technical revolution. It will probably touch more aspects of daily life than have been affected by all previous computer technology. In many instances society is neither aware of nor prepared for the microprocessor's nontechnical impact. For example, the introduction of microprocessors in automobiles to improve fuel economy and reduce exhaust emissions will have a profound effect on tens of thousands of small automobile-repair shops and hundreds of thousands of gasoline service stations. Literally millions of maintenance workers who at this moment may never have heard of microprocessors, much less seen one, must quickly become acquainted with them and become reasonably expert in their testing and replacement. Otherwise the entire network of automobile service will have to be drastically revamped.

In other areas microprocessor technology is likely to move more slowly than one might expect. For example, it would not be difficult to equip every gasoline service station with a microprocessor terminal that would record the details of every credit-card transaction. At the end of each day the recorded information could be transmitted rapidly by code over an ordinary telephone circuit to computers in the credit-card company's central office, speeding up by several weeks the billings on millions of dollars' worth of sales. Although such systems are eminently practicable, they have not been adopted. One can infer that the innovation is unwelcome because it threatens the existence of entire divisions of credit-card organizations: the optical-character-recognition (OCR) divisions that now transcribe into computer-usable form the transaction data from millions of separate pieces of paper filled out and mailed in by service-station attendants. Entire divisions of companies do not willingly disappear overnight.

These are just two examples of how thousands of business organizations and millions of individuals may be affected by what appears to be a straightforward engineering decision either to apply or to not apply the new microprocessor technology. Within the business organization itself the microprocessor and microcomputer are making more acute the already difficult question of how to distribute computing and information resources for maximum effectiveness. Although many companies are rapidly consolidating and centralizing all aspects of their computing resources, other companies are decentralizing with equal aggressiveness. The elusive "optimal" strategy involves delicate considerations of management control, strategies of system development and operational procedures. Business managers who had to penetrate the mysteries of the $250,000 room-sized computer barely 15 years ago and of the $25,000 minicomputer six or seven years ago must now try to weigh the costs and benefits of the $250 microcomputer and the $25 microprocessor.

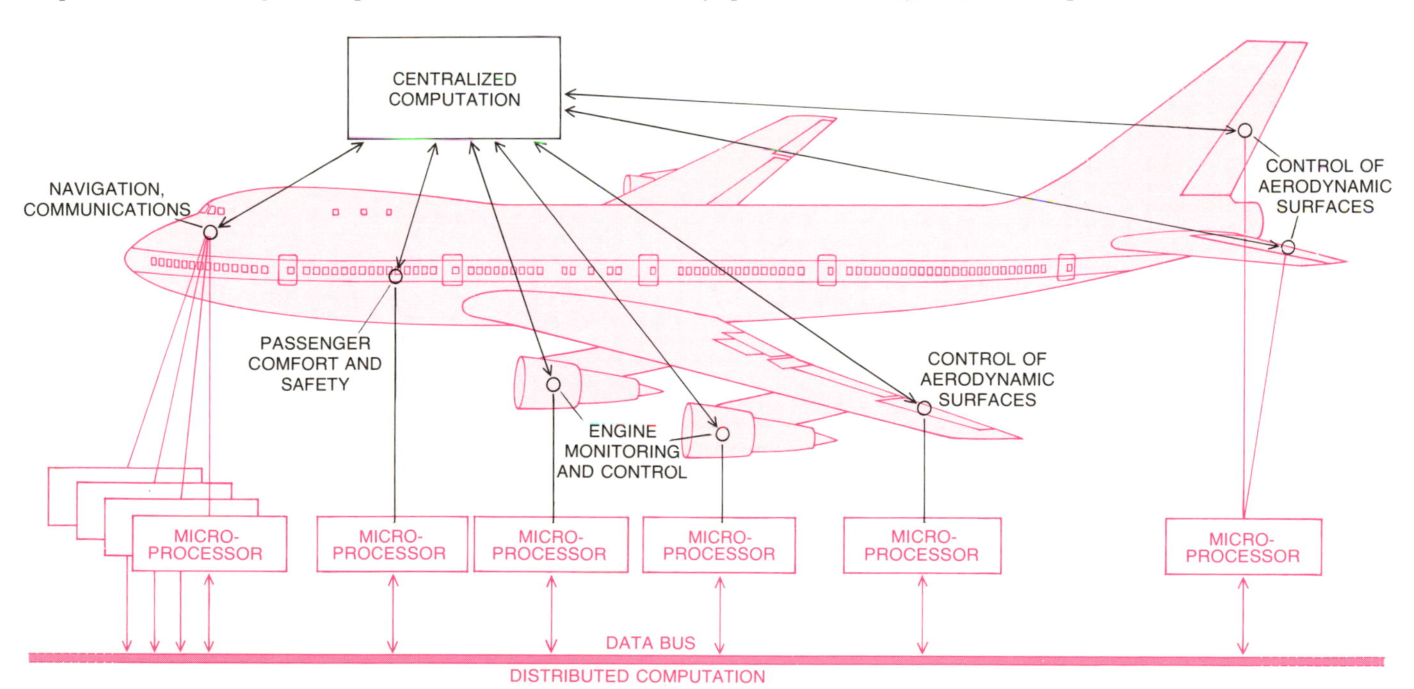

MODERN JET AIRCRAFT depend on a variety of sophisticated systems for navigation, communication, passenger comfort and safety, engine control and the control of aerodynamic surfaces. At present the sensors that monitor these various systems transmit their data to a central computer, which generates the control signals needed to keep the systems working properly. The miles of cables required for such centralized systems have become a significant fraction of the total cost of modern aircraft. In principle various systems of aircraft could be controlled locally by microprocessors with a great saving in cable costs, increased reliability, increased computing power and lower maintenance costs. Such distributed computing networks are under active examination for a wide variety of similar applications.

7

THE ROLE OF MICROELECTRONICS IN DATA PROCESSING

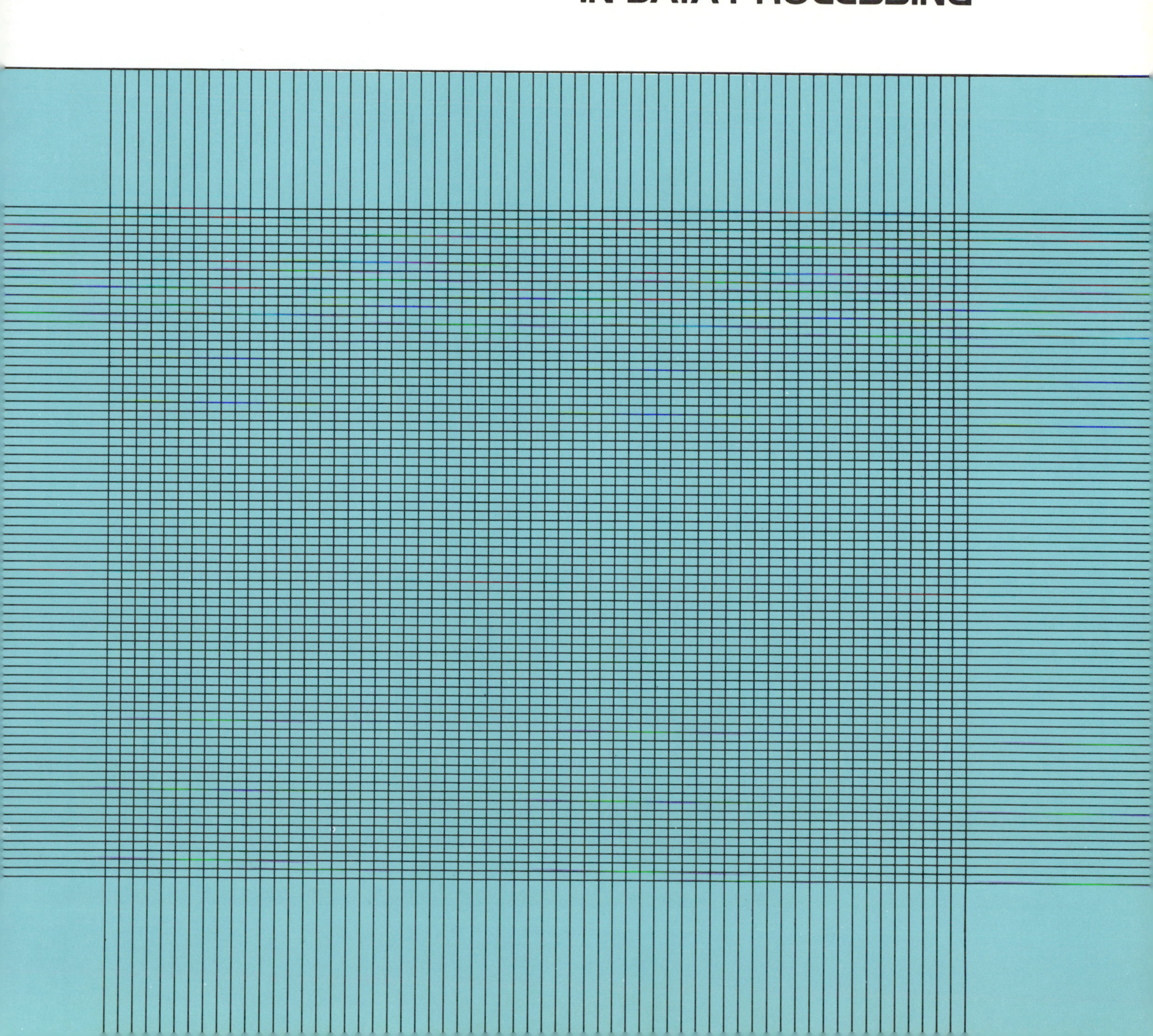

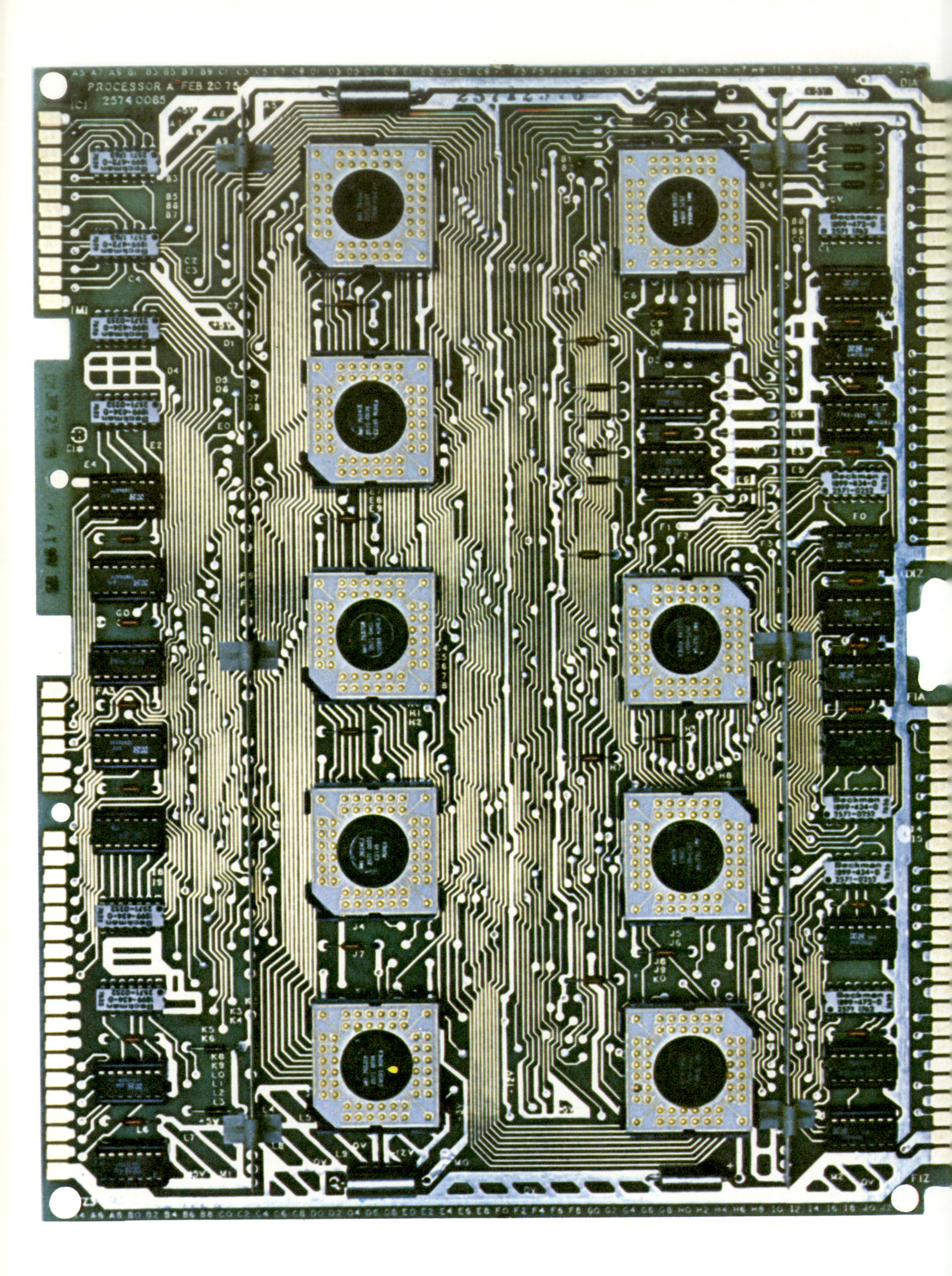
PROCESSOR A FEB 20 75
2574 0085

The Role of Microelectronics in Data Processing

by LEWIS M. TERMAN

Large modern computers could not exist without microelectronic components. Microelectronics has also led to the rise of smaller computers able to perform tasks that do not call for larger ones

Modern computers and microelectronic devices have interacted so closely in their evolution that they can be regarded as virtually symbiotic. The first large stored-program digital computers were based on vacuum tubes; their successors were based on discrete transistors. Such elements could not be used to build computers with the capabilities of those we have today. They would be too costly and too unreliable, and they would require too much power. Above all, they would be too large and too slow. The number of operations per second a computer can perform depends on the speed with which the computer's active elements communicate with one another, and that speed depends on how fast the active elements operate and how close together they are. Computer technology has demanded microelectronics, and microelectronics has evolved largely in response to that demand.

Computers themselves have evolved largely in response to the demand for the processing of data: large quantities of characters, such as numbers or letters, representing information of almost every conceivable kind. In science computers digest and analyze masses of measurements, such as the sequential positions and velocities of a spacecraft, and solve extraordinarily long and complex mathematical problems, such as the trajectory of the spacecraft. In commerce they record and process inventories, purchases, bills, payrolls, bank deposits and the like and keep track of ongoing business transactions. In industry they monitor and control manufacturing processes. In government they keep statistics, analyze economic information and bill people for taxes. Today it would be difficult to find any task that calls for the processing of large amounts of information that is not performed by a computer. Moreover, many lesser tasks, requiring comparatively small amounts of data processing, are also being taken over by computers.

Although a large computing system is expensive, it can perform millions of operations a second; even a small computing system can perform 100,000 operations a second. Therein lies the fundamental reason for the growth of data processing by computer: a computer can perform so many operations in so little time that the cost of each operation is very low.

Perhaps the key element in the growth of data processing by computer is the fact that with the advent of microelectronics the cost of computer hardware has steadily decreased as capacity and performance have steadily increased. Since 1960 the cost of a computer divided by its computing power (measured in millions of operations per second) has dropped by a factor of more than 100. In the same period the cycle time (the time it takes to do an operation) of the largest machines has decreased by a factor of 10. The increase in the performance of computer hardware has been important for the development of "high end" or "mainframe" computers: the largest and fastest machines. It has been even more important for the development of low-end computers: the smaller and slower machines. Many tasks do not require the great processing capability, flexibility and speed of even a small mainframe computer system. In the 1960's advances in microelectronic components led to the development of the minicomputer, followed more recently by the even smaller microcomputer. Both have filled a need for small but relatively flexible processing systems able to execute comparatively simple computing functions at lower cost. Moreover, with the development of the microprocessor there has been an almost Darwinian speciation of computers into machines of different sizes and organizations, each tailored to a different range of functions. The progress toward smaller computers is likely to continue; there is already talk of nanocomputers and picocomputers.

Today a large mainframe computer system may consist of 100,000 logic circuits and between four million and eight million bytes of memory. (One byte is either eight or nine bits; a bit is a binary

PROCESSOR BOARD of a Burroughs B 80 small "mainframe" computer contains the decision logic for the entire computer: it is responsible for executing the programs entered into the machine. The actual processor board measures 10 inches by 12½ inches. The logic components are nine metal-oxide-semiconductor (MOS) large-scale integrated-circuit chips, each housed in one of the nine white-and-gold packages. On this board the logic functions, instead of being executed by logic circuits, are stored as permanent bit patterns in small read-only memories, which are contained in the top, middle and bottom white-and-gold packages in the left column. The second and fourth packages in that column house working-register chips, which store data for execution in the processor. The top chip in the column at the right controls the operation of the input-output devices of the computer. The second chip in that column is the main-memory address register, which holds the address of the bit word currently being accessed from the computer's main memory. The chip next to the bottom in the right column is a microstack, another type of memory-address register. At the bottom of that column is the "timing-machine-state" chip, which controls the timing and the sequence of operations executed by the entire processor board. The remaining components on the board have no logic functions. The black rectangles are amplifiers. The white rectangles and the tiny striped cylinders are resistors. The red sausage-shaped components, the black cylinders and the large silver cylinders are capacitors. The two white bars running the entire length of the board are busbars, which distribute power and ground across the board. All the components are interconnected by soldered tracks. Board is inserted into computer by way of gold connectors along the edges.

digit.) A small mainframe system might have 20,000 logic circuits and 128,000 bytes of memory. A typical minicomputer system might have between 5,000 and 10,000 logic circuits and between 16,000 and 32,000 bytes of memory. A microcomputer system might have some 1,000 to 2,000 logic circuits and a few hundred to a few thousand bytes of memory. There is a corresponding range in the cost of such systems: the largest mainframe computers cost several million dollars, and a complete microcomputer system costs as little as a few hundred dollars.

Most computer systems, regardless of their size, consist of three basic elements: the input-output ports, the memory hierarchy and the central processing unit. The input-output ports are the paths whereby information (instructions and data) is fed into the computer or taken out of it by such means as punch cards, magnetic tapes and terminals. The memory hierarchy stores the instructions (the program) and the data in the system so that they can be retrieved quickly on demand by the central processing unit. The central processing unit, which consists of a control logic unit and an arithmetic and logic unit, controls the operation of the entire system by issuing commands to other parts of the system and by acting on the responses. It reads information from the memory, interprets instructions, performs operations on the data according to the instructions, writes the results back into the memory and moves information between memory levels or through the input-output ports. The operations it performs on the data can be either arithmetic, such as addition or subtraction, or logical, such as the comparison of two fields of numbers to determine whether or not they are equivalent.

The memory organization of a computer is hierarchical in terms both of speed and of cost. Interacting directly with the central processing unit is a buffer memory, which transfers information into the central processing unit at high speed. Speed is expensive, however, and the cost per bit of storing information in the buffer memory is high. Below the buffer memory in the hierarchy is the computer's main memory, which is larger and slower than the buffer and costs much less per bit of information stored. Below the main memory are storage systems such as magnetic disks and tapes, which are larger and slower still and cost even less per bit. Hence the memory levels further removed from the central processing unit are progressively larger, slower and cheaper.

The information that is currently in active use is moved to the fastest memory (the buffer) for ready access; the less active or inactive information is placed in the low-cost-per-bit storage. The central processing unit thus finds in the buffer the information it needs for immediate processing, and only occasionally does it have to retrieve information in the lower levels of the hierarchy. The performance of the entire computer system is determined by the speed of the buffer memory, whereas the system's capacity and cost per bit is determined by the number and the size of the storage levels.

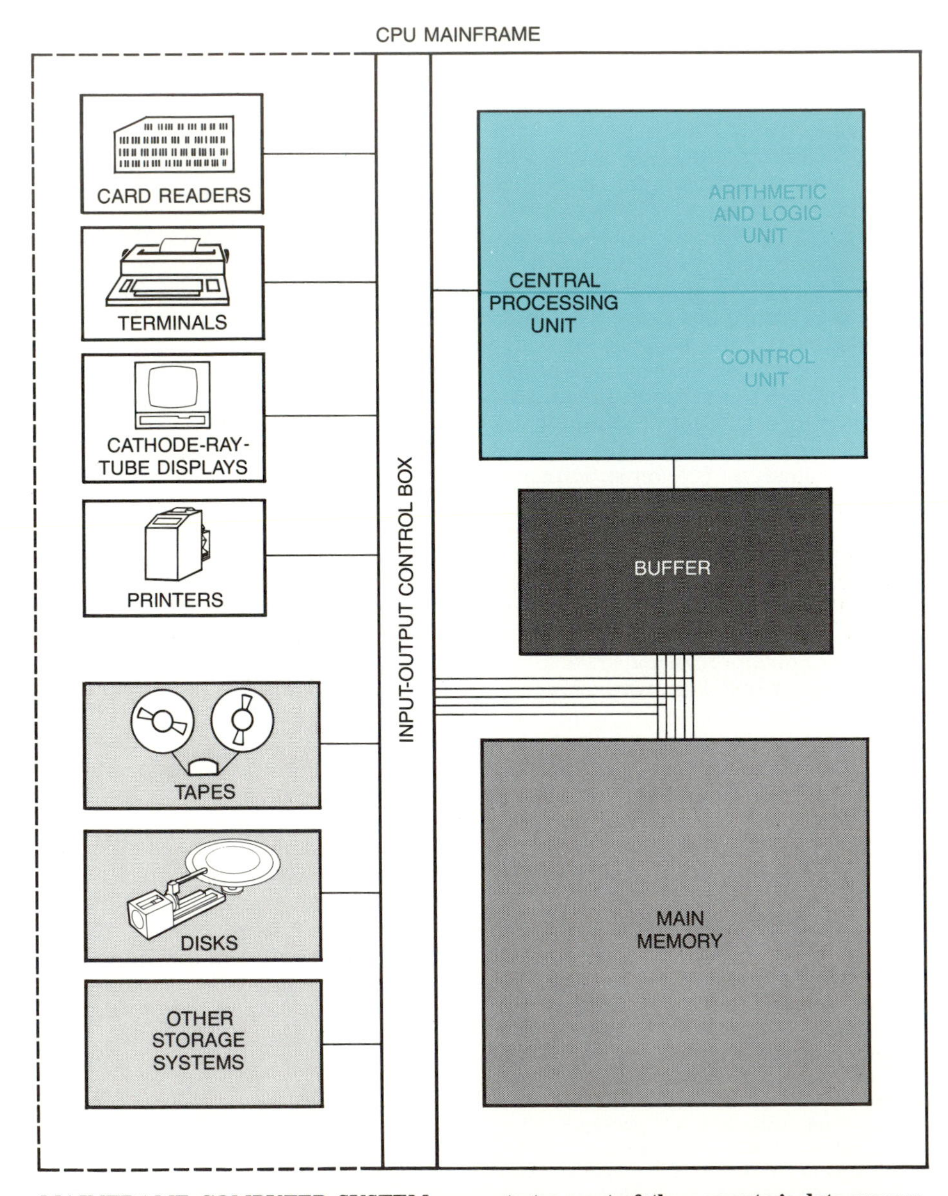

MAINFRAME COMPUTER SYSTEM concentrates most of the computer's data-processing capability in the computer's central processing unit, or CPU (*color*). Physically the CPU mainframe of a large computer system holds the central processing unit itself, the high-speed buffer memory and the main memory, and it may also hold several of the input-output ports. The central processing unit itself consists of one unit that executes the arithmetic and logic operations in the computer program, and another unit that controls the operation of the entire computer system. The computer program and the data in active use are stored in the buffer memory (*dark gray*), where they can be accessed by the central processing unit at high speed. Behind the buffer is a larger and slower main memory (*medium gray*). Away from the CPU mainframe itself are disks, tapes and other memory-storage systems (*light gray*), which are the largest and slowest of all. The input-output devices are tape readers, card readers, keyboards, cathode-ray-tube displays, disks, tape drives, teletypewriters, high-speed printers and so forth.

The impact of microelectronics on computer memories has been different at the different levels in the hierarchy. From the first the buffer memory has consisted of purely electronic circuits. With the advent of microelectronics the circuits have become smaller and faster than their predecessors of five or 10 years ago and have contributed to the increase in the speed of high-end computers.

The structure of the main memory has been transformed to an even greater degree by microelectronic devices. From the mid-1950's through the 1960's the individual information-storage element in the main memory of a computer was the magnetic "core," a small toroid of ferrite material in which one bit of information was stored as the direction of magnetization around the toroid. Each core was fabricated individually, and millions of them were strung together in large arrays by wires passing through the hole in each of them. Information was written into the cores and read out of them by electronic support circuits. Because of the characteristics of the cores and the weak magnetic coupling between the cores and the wires on which they were strung, the support circuitry required for writing the information into the cores and reading it out was complex, and neither it nor the cores were amenable to being fabricated as integrated circuits.

In the early 1970's semiconductor memory cells that served the same purpose as cores were developed, and integrated memory circuits began to be installed as the main computer memory. In an integrated memory circuit the information-storage mechanism (the memory cell) is not a passive piece of material but is itself a semiconductor circuit. Moreover, the support circuits required for writing information into the memory cell and reading information out of it are much simpler than what was required for writing and reading with cores. Thus both the memory cell and the support circuits could be combined on the same silicon chip. Microelectronic semiconductor devices are found in the main memory of the bulk of computers being built today, although magnetic cores are still being manufactured for some purposes.

One particularly valuable characteristic of a main computer memory consisting of microelectronic devices is that the cost per bit of the memory is more or less independent of the memory's size. With core memories the support circuitry was complex and expensive no matter how large the system was, making the cost per bit of a computer with a small core memory far higher than the cost per bit of a computer with a large core memory. With microelectronic memories the cost of the system is primarily in the memory chips and their packaging, not in the support circuits. Hence the cost of a computer system is proportional only to the size of the memory, and the memory's cost per bit remains the same. In other words, the user of a low-end minicomputer or microcomputer system who requires only a small amount of memory no longer has to pay a substantial penalty in the system's cost per bit. This factor has been quite important for the growth of low-end computer systems.

Microelectronics has not penetrated very far into the construction of storage devices such as magnetic disks and tapes and probably will not penetrate much further in the foreseeable future [see "Microelectronic Memories," by David A. Hodges, page 54]. There are several reasons. First, the disks and tapes themselves are inexpensive. The electrical

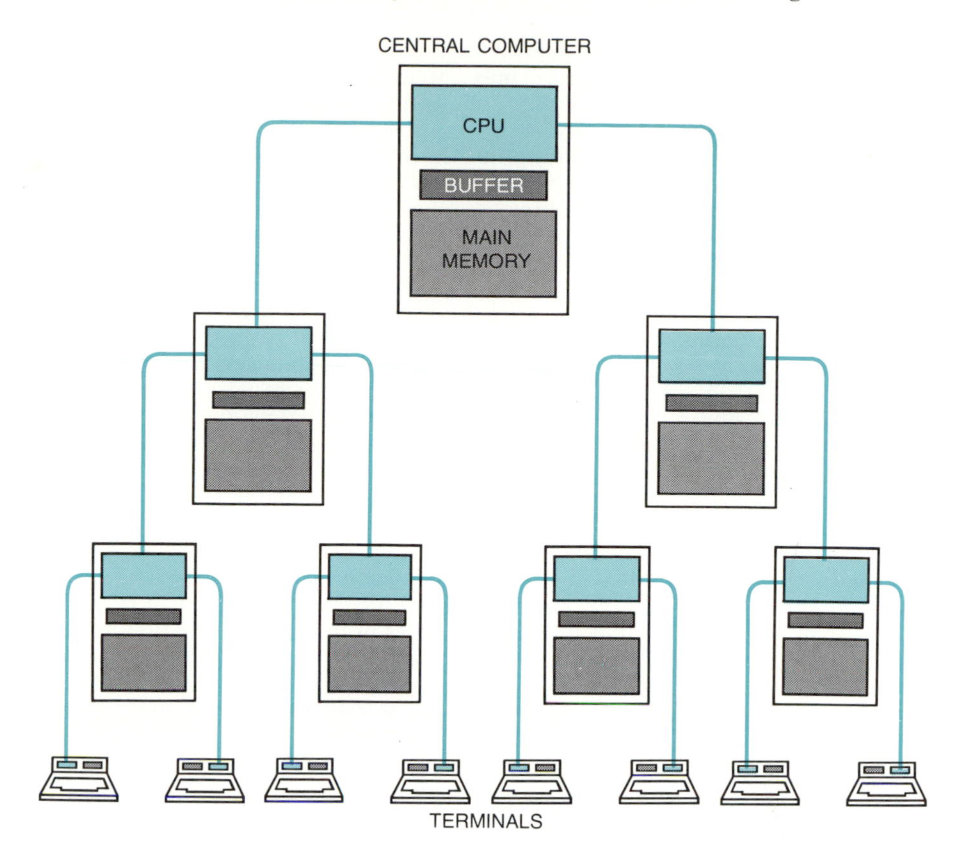

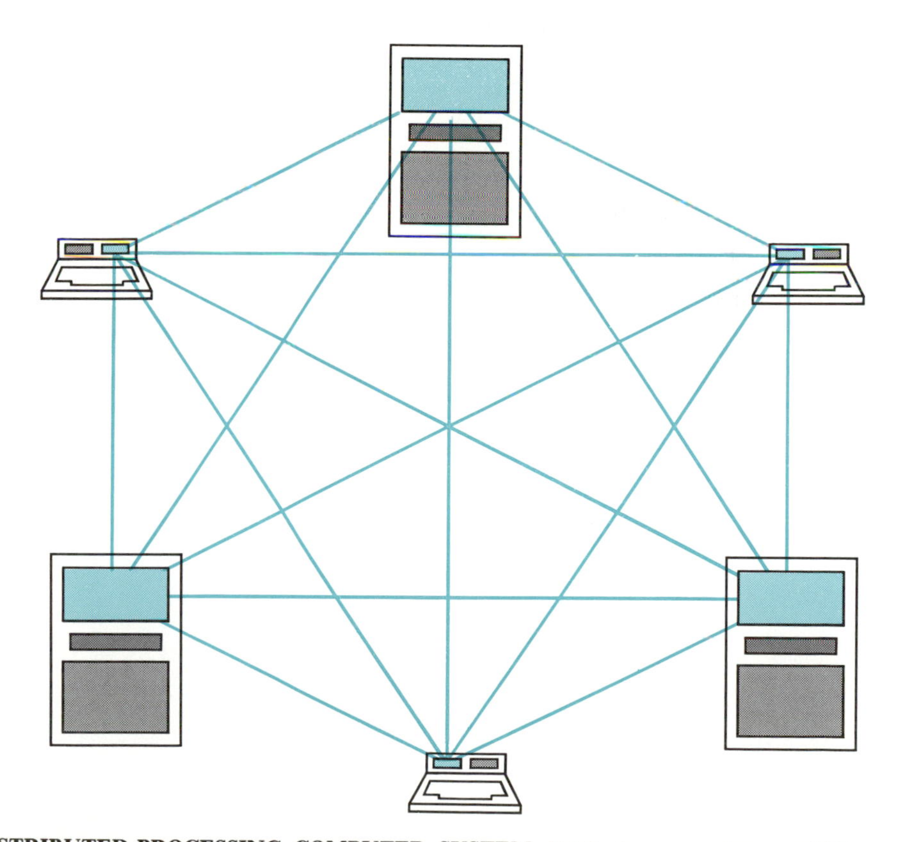

DISTRIBUTED-PROCESSING COMPUTER SYSTEM distributes processing capability throughout a network of computers instead of concentrating it in one central processing unit. Physically that allows the network to be spread over a wide geographic area and enables each local processor (*small boxes in color*) to handle information independently before transmitting it to one or more other elements in the network. The two basic forms of a distributed-processing system are the hierarchical structure (*top*) and the peer structure (*bottom*). In the hierarchical structure there is one central computer, which communicates with smaller computers on a lower level, each of which communicates with computers on a next lower level, and so forth. In the peer structure all the elements are on the same level and each element can communicate with every other element. In practice the two forms are often combined into a hybrid system.

electrical and mechanical machinery required for writing information into them and reading it out, however, is quite costly. Therefore the cost per bit of such storage systems is a strong inverse function of the size of the system, and the lowest cost per bit is obtained with the very largest systems. For example, a disk-storage system capable of retaining 100 million bytes of memory costs only a few thousandths of a cent per bit. For a large computing system the cost per bit of microelectronic storage methods could not compete with the cost per bit of storage by disk or tape.

Second, disks and tapes are designed to be nonvolatile: if the power fails in the computer system, the information stored on the disk or tape remains intact. Information stored in a microelectronic memory would be lost if the power failed. Third, disks and tapes are relatively compact, and for some purposes it is important to have the physical convenience of being able to transport them or to store them apart from the computer.

For a small computer system the requirements are quite different. For example, the user of a minicomputer does not need anywhere near the full capacity of a large disk-storage unit, and he certainly would not want to pay the cost of it. He is quite happy to get a smaller system at a lower overall cost even though the cost per bit may be higher. Thus microelectronics may penetrate into the area of storage devices with a smaller capacity. Semiconductor storage mechanisms such as charge-coupled devices will become cheaper than main semiconductor memory technologies, and their cost should become competitive with, or even lower than, the cost of small magnetic disk or tape storage systems. Since a semiconductor storage system would be accessed electronically instead of physically, as tapes or disks must be, the time required to enter or retrieve information in a semiconductor storage system would be much shorter. Even if the semiconductor storage system cost more per unit, the user might feel that it would be desirable to accept higher cost for higher performance. It is still too early to determine how successful microelectronic storage devices may become.

In the memory and storage hierarchy of a computer information is stored passively and is retrieved unchanged. In the central processing unit of the computer information is operated on actively. The elements out of which the processing portions of the computer system are made are logic circuits. Logic circuits are distinctly different from memory circuits. In a microelectronic memory each memory cell retains one bit of information. Only a small fraction of the bits in the memory system are being accessed at any particular moment, so that the standby power needed to keep the memory functioning properly can be low. The energy required by the support circuits needed to write information into the memory or to read information out of it is supplied on demand by circuits

SUPERCOMPUTER CRAY-1, made by Cray Research, Inc., of Minneapolis, Minn., is a powerful high-speed scientific digital computer for executing highly complex tasks. The CPU mainframe of the CRAY-1, the cylindrical structure in the center of the room, is composed of more than 1,000 modular logic boards inserted horizontally into 24 vertical chassis. The CPU mainframe holds the central processing unit and the computer's memory. Benches around the base of the CPU mainframe are cabinets housing the computer's 12 power supplies. At left an operator is seated at the keyboard of a terminal of a minicomputer that functions as a maintenance control unit for the CRAY-1 computer. Beyond him another operator is loading a magnetic tape onto a tape drive of the maintenance control unit. Cabinets in foreground are disk-storage devices. This CRAY-1 computer is at research laboratory of Cray Research in Chippewa Falls, Wis.

outside the array of memory cells. In a computer processor, however, every logic circuit is in continuous operation. Only a few circuits may be in the process of switching at any particular instant, but every remaining circuit is giving an output corresponding to its input and must be ready to switch at any moment if the input should change.

A logic circuit essentially consists of a comparatively high-impedance load device (such as a resistor) connected in series with one or more switching devices that are connected in parallel. Each switching device has an impedance that depends on the strength of the input voltage. When the input voltage is high, the impedance of the switching device is comparatively low; current flows through the switch, and so it is "on." When the input voltage is low, the impedance of the switching device is essentially infinite; no current flows through the switch, and so it is "off."

A typical kind of logic circuit is the negative-or (or "nor") circuit. A common "nor" logic circuit has three inputs, three switching devices (each of which is associated with an input) and one output, which is located between the load device and the three switching devices connected in parallel. One "nor" circuit is incorporated into an assembly of circuits (as in a processor) by connecting its output to an input of one or more similar circuits, and by connecting each input to the output of one or more other circuits. If the voltage on any one of the inputs to the logic circuit is high (corresponding to a binary 1), its switch will close, and the output will be low (corresponding to a binary 0). If all the inputs of the circuit are low (0), all the switching devices are essentially open circuits, and the output will be high (1). It is quite possible to design the logic of an entire computer with nothing other than "nor" circuits.

A logic circuit dissipates power and has a small but significant switching delay, caused by the fact that the output cannot respond immediately to a change in the inputs. Both the amount of power dissipated and the length of the switching delay depend on the impedance of the load device. A high-impedance load yields a smaller current and so will dissipate less power, but a smaller current requires more time to open or close the switch. In fact, for any particular design of circuit, the product of the power dissipation and the switching delay (the power-delay product) is a constant over a fairly broad range, and one disadvantage may be traded off in favor of the other.

The power-delay product of logic circuits has been considerably improved with the advent of microelectronic devices. Smaller devices are inherently faster than larger ones. A logic circuit that has smaller dimensions presents a lighter load to the switching device, and the current can switch the lighter load faster. It is interesting to note exactly how much the switching delay has improved as circuit technology has progressed. In the mid-1950's a large vacuum-tube logic circuit had a switching delay of one microsecond (one millionth of a second). In the early 1960's a printed-circuit card several inches on a side with discrete transistors and other components mounted on it had a typical switching delay of 100 nanoseconds (100 billionths of a second). A typical contemporary logic integrated-circuit chip for a mainframe system is less than a tenth of an inch on a side, has between five and 10 circuits and has an average switching delay of less than five nanoseconds. Even faster logic circuits are possible.

In the semiconductor logic circuits of today two basic types of devices are in use: bipolar devices and metal-oxide-semiconductor field-effect transistor (MOSFET) devices [see "Microelectronic Circuit Elements," by James D. Meindl, page 12]. Circuits made with bipolar devices are more complex to fabricate and require a larger area on a

LOGIC BOARD FROM CRAY-1 supercomputer is one of 1,056 such boards in the central processing unit. Each board measures six by eight inches. On it can be mounted up to 288 integrated-circuit chips, 144 on each side; on the side of the board shown here there are 137 of them. Entire CPU of the CRAY-1 is built up of only six different types of elements: four types of integrated circuits (*large white packages*) and two types of resistors (*small dark components*).

			CRAY-1
CENTRAL PROCESSING UNIT	NUMBER OF LOGIC CHIPS		278,000
	CPU CYCLE TIME		12.5 NANOSECONDS
MEMORY	BUFFER	TECHNOLOGY	—
		CAPACITY	—
		CYCLE TIME	—
	MAIN	TECHNOLOGY	BIPOLAR SEMICONDUCTOR
		CAPACITY	UP TO 1,048,576 64-BIT WORDS
		CYCLE TIME	50 NANOSECONDS
OTHER	SIZE OF CPU		9-FOOT DIAMETER BASE 4.5-FOOT DIAMETER CENTRAL PART 6.5 FEET HIGH
	WEIGHT OF CPU		5.25 TONS
	BASIC PURCHASE PRICE		$8,000,000

TABLE OF INFORMATION gives a few pertinent statistics about the CRAY-1 computer. CPU cycle time is the pulse rate of the clock in the CPU. The CRAY-1 has only one memory. The memory is divided into 16 independently operating banks that can be accessed one after another in an interleaved fashion at intervals of 12.5 nanoseconds. Cycle time of 50 nanoseconds for the memory is the time required to access one word of information from one bank; because consecutively located words can be stored and retrieved from successive banks in a fourth of the time, memory's effective cycle time is 12.5 nanoseconds. Number of chips in the CPU is only a rough approximation of size of a computer as each chip may have many or few logic circuits.

chip. Bipolar devices have a low impedance and are inherently fast. They are installed in the logic of mainframe computers and in the buffer memory, where the emphasis is on high speed and good performance. Circuits made with MOSFET devices are simpler to fabricate and more compact. They are relatively slow, but they cost less because they can be packaged quite densely. MOSFET devices are generally found in the main memory of computers and in microprocessors, where the emphasis is on low cost and lower performance is acceptable. They are also widely used for low-cost logic in a variety of applications.

The reduction in switching delay with the introduction of microelectronic devices is reflected in the improved performance of data-processing machines of all sizes. For example, the cycle time (the time required to execute an operation) of a high-end computer system has dropped from 150 nanoseconds in the early 1960's to 10 or 20 nanoseconds today, and the cycle time of a microprocessor has fallen from about two microseconds in 1971 to half a microsecond or less today.

Not only do smaller circuit elements have a better power-delay product but also more of them can be placed in a given area on a chip. Increasing the number of circuits on a single chip is a more efficient way of fabricating the computer; fewer chips are required, and the connections between the circuits on a chip are shorter and less of a load on the system than the interconnections between chips. When the circuits on one chip must drive circuits on other chips, the switching delays are longer.

What limits the size of individual circuit elements and chips is the "yield" of the fabrication process. First, if the chip is to be a good one, all the elements on it must function properly. A single defect in a critical area will spoil a circuit and make the entire chip worthless. A larger chip will have more critical area, and the probability is higher that it will have a defect that will spoil a circuit.

Second, in logic chips there is a practical limit to the number of random logic circuits it is worthwhile to put on a single chip. A memory chip is a regular array of cells and support circuits. The basic function of all memory chips is the same. The input-output connections are simple. A specific type of memory chip is repeated many times in a single memory system, and it may be found in the memories of other computer systems as well. The result is that one type of memory chip is widely utilized, and the cost per chip is low.

The situation is quite different with logic circuits. Logic is much more random. A particular configuration of logic circuits may appear only once in a machine, and it may not be repeated in that machine or any other. The situation gets worse as the number of circuits on a chip increases. The result is that particular types of random logic chips are not widely utilized, and the cost per chip is high.

In addition, as more logic circuits are put on a chip it becomes more difficult to design the chip. This is a problem not only in initially designing the chip but also in redesigning it to modify it or to correct errors in the initial design. Redesigning a random-logic chip can be quite expensive if the number of circuits on

LARGE MAINFRAME COMPUTER, the IBM System/370 Model 168, made by the International Business Machines Corporation, is the biggest member of the family of IBM 370 computers. It is designed to handle high-speed large-scale tasks for either business or science. The big set of cabinets in the rear holds the central processing unit, the main memory and power supplies. The desk with cathode-ray-tube screen, keyboard and switches is the operator's console. Machine shown is at IBM headquarters in White Plains, N.Y.

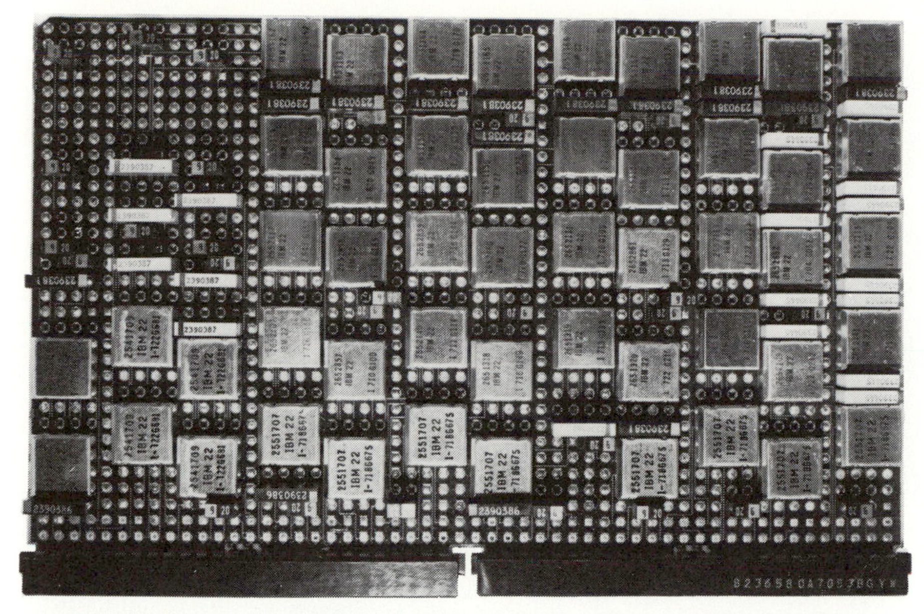

LOGIC CARD FROM IBM 370/168 computer is one of several hundred such boards in the central processing unit. Each board measures $4\frac{1}{2}$ inches wide by $6\frac{7}{8}$ inches long. On it are mounted integrated-circuit modules (*square silver packages*) and other electronic components.

the chip is very large. Moreover, as more logic circuits are put on a chip the number of input-output connections increases significantly, making it more difficult to package the chip in the computer. Thus increasing the number of logic circuits on a chip is not an unmixed blessing.

A designer developing the logic circuitry for a computer can choose among several alternative approaches. The first is that each circuit can be custom-designed and individually placed on a chip as needed in order to achieve optimum performance. Custom-building computer logic has all the drawbacks mentioned above, but it provides the best performance and the highest density of circuits. It also can have the lowest cost if the individual chips are in sufficiently high demand to amortize the cost of designing them.

The second approach is that each circuit can be custom-designed and placed on a chip in a regular array, like squares on a chessboard, with wiring tracks provided for connections between the circuits. Such an arrangement eases the difficulty of designing the entire chip but still retains a degree of versatility for different functions. It is less efficient in density and in performance because the circuits can no longer be organized in the most effective way for each individual function. Moreover, in general only a fraction of the circuits on a chip are actually used.

The third approach is that a number of small functional circuit units known as "macros" can be custom-designed. Such a unit might be an adder, a Boolean-logic unit or a register stack. In this way a "library" of macros can be built up. When a chip is designed, it is constructed of a number of macros connected together by a minimum of custom-designed logic.

The fourth approach has to do with the fact that it is possible to facilitate the design of logic chips by making the logic resemble memory. The commonest way is to use a read-only memory. A read-only memory differs from a read/write memory in that the information is written into it when it is fabricated, and it cannot be changed. In such a memory the sequence of logic operations is stored as bit patterns in a sequence of bit "words"; when the words are accessed, the bit patterns are decoded and the logic operations are carried out. Only a relatively small amount of logic is needed to decode and utilize the bit patterns. The read-only memory cell is smaller than a read/write memory cell, and it is nonvolatile. Between 10 and 20 bits of read-only memory is generally equivalent to a single logic circuit. The logic of the chip can be changed simply by altering the read-only memory bit pattern; no new circuit design or layout is required.

The fifth and last approach to designing logic circuitry is to implement the logic functions by a microprocessor. The microprocessor is the least efficient alternative in terms of performance and density, but it is the most flexible and the easiest to use. Its functions can be changed simply by reprogramming it. The same chip can perform many different functions, so that it is in high demand and is relatively inexpensive. A single microprocessor chip today costs as little as $10 to $20.

The versatility and convenience of the microprocessor has altered the entire architecture of modern computer systems. No longer is the processing of information carried out only in the computer's central processing unit. Today there is a trend toward distributing more processing capability throughout a computer system, with various areas having small local processors for handling operations in those areas.

For example, an input-output port may have a controller to regulate the flow of information through it. At times the controller may accept commands from the central processing unit and send signals back in order to coordinate its operations with those of the rest of the system; at other times the controller may operate independently of the central processing unit. Another example of a local processor is what is usually called an intelligent terminal. At a terminal an operator is connected to the computer and time-shares the computer with other operators. Originally terminals were "dumb": they simply sent inputs from the operator to the central

			IBM-370/115	IBM-370/168
CENTRAL PROCESSING UNIT	NUMBER OF LOGIC CHIPS		1,800	20,000
	CPU CYCLE TIME		480 NANOSECONDS	80 NANOSECONDS
MEMORY	BUFFER	TECHNOLOGY	—	BIPOLAR SEMICONDUCTOR
		CAPACITY	—	32,768 BYTES
		CYCLE TIME	—	80 NANOSECONDS
	MAIN	TECHNOLOGY	MOSFET	MOSFET
		CAPACITY	UP TO 393,216 BYTES	UP TO 8,388,608 BYTES
		CYCLE TIME	480 NANOSECONDS	320 NANOSECONDS
OTHER	SIZE OF CPU		2.5 FEET LONG 5 FEET DEEP 5 FEET HIGH	13 FEET LONG 10 FEET DEEP 6.5 FEET HIGH
	WEIGHT OF CPU		1,800 POUNDS	5,100 POUNDS
	BASIC PURCHASE PRICE		$175,000	$4,500,000

TABLE OF INFORMATION gives a few pertinent statistics about the smallest member (the Model 115) and the largest member (the Model 168) of the IBM 370 family, showing the range that is available today in the computing power of both small and large mainframe computers.

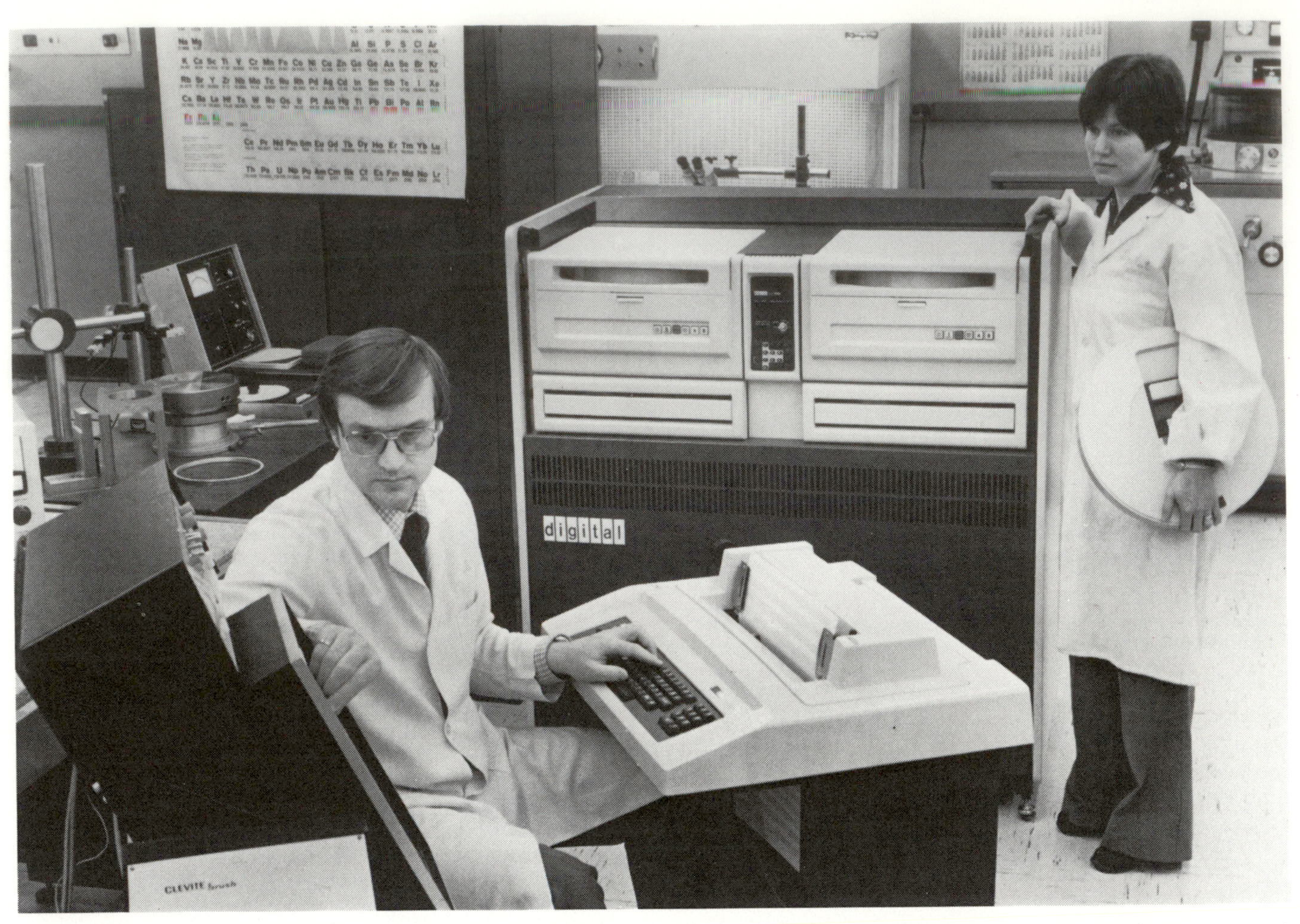

MINICOMPUTER PDP-11 Model 60, made by the Digital Equipment Corporation and photographed in one of their testing laboratories in Marlborough, Mass., is one of the middle-range members of the family of PDP-11 computers. The PDP-11/60 shown here consists of a central processing unit (*middle control panel on large cabinet*) with two disk units flanking it and an operator's console (*keyboard in foreground*). Operator at right is holding a disk cartridge for the computer. Operator at left is testing computer with an oscilloscope.

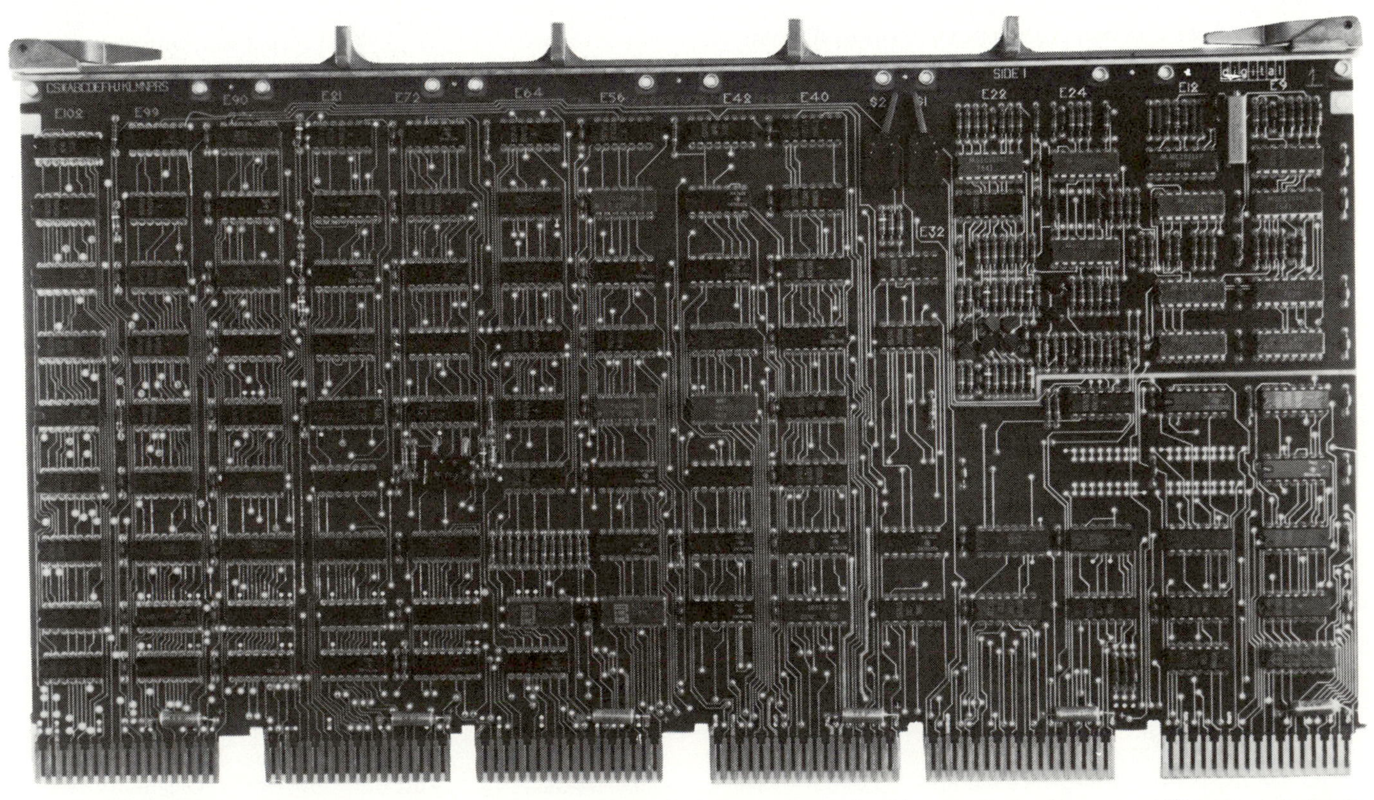

LOGIC BOARD FROM PDP-11/60 minicomputer is one of six similar boards in the central processing unit. The board measures $8\frac{1}{2}$ inches wide by $15\frac{1}{2}$ inches long. It contains 107 integrated circuits (including two read-only memories) and some discrete components.

processing unit of the computer and displayed outputs on a screen. Today increasing numbers of terminals are "smart": they are capable of doing some preliminary and independent processing on the operator's input before communicating with the computer's central processing unit.

In many computer systems today a number of processors are connected together to form a distributed-processing network. Most commonly the network consists of a number of minicomputers, but mainframe computers and microcomputers can also be incorporated into it. Input-output ports and data-transmission hardware are considered an active part of the network only if they are able to process information. Parts of a task are distributed among the elements of the network. Each element works independently for some period of time, communicating as necessary with other elements.

There are a number of advantages to distributed processing. First, since many elements of the computer can be working on different portions of the same task, the work may be done faster. Second, if one element in the network malfunctions, its work load can be shifted to another element or shared among several elements, so that the entire network is relatively immune to failure. Third, the network can be small enough to be contained within a single laboratory or building, or it can be spread out over a wide area, as in various branches of a bank. The ease or difficulty with which each element can communicate with another will affect how much the data are manipulated before they are transmitted through the network. A major obstacle to designing an effective distributed-processing system is the difficulty involved in writing the system's software, which must enable the various elements of the network to operate and interact efficiently.

			PDP-11/03	PDP-11/70
CENTRAL PROCESSING UNIT		NUMBER OF LOGIC CHIPS	4	600
		CPU CYCLE TIME	3.5 MICROSECONDS	300 NANOSECONDS
MEMORY	BUFFER	TECHNOLOGY	—	BIPOLAR SEMICONDUCTOR
		CAPACITY	—	2,048 BYTES
		CYCLE TIME	—	240 NANOSECONDS
	MAIN	TECHNOLOGY	CORE OR MOSFET	CORE
		CAPACITY	UP TO 57,344 BYTES	UP TO 4,096,000 BYTES
		CYCLE TIME	CORE: 1.15 MICROSECONDS MOSFET: 750 NANOSECONDS	1.26 MICROSECONDS
OTHER		SIZE OF CPU	19 INCHES LONG 13.5 INCHES DEEP 3.5 INCHES HIGH	21 INCHES LONG 31 INCHES DEEP 6 FEET HIGH
		WEIGHT OF CPU	35 POUNDS	500 POUNDS
		BASIC PURCHASE PRICE	$2,000	$63,000

TABLE OF INFORMATION gives pertinent statistics about both the smallest member (the Model 03) and the largest member (the Model 70) of the family of PDP-11 minicomputers. The PDP-11/70 can be considered a small mainframe computer. It has a high-speed buffer memory. It and PDP-11/60 are only members of PDP-11 family that have two memories.

Distributed-processing systems can be organized in several ways. A large distributed-processing system can be organized into a hierarchical structure. At the top of the hierarchy is a single mainframe computer that communicates with processors in the network at a secondary level, which in turn can communicate with other processors on a tertiary level and so forth. In a pure hierarchy the processors on any particular level cannot communicate directly with one another. Instead communications must be routed through the next higher level.

Alternatively a distributed-processing system can be organized into a peer structure. All the computers are on the same level and can communicate with one another on an equal footing. Except for very small networks, however, it seldom happens that every element in the network can communicate with every other element. Instead the hierarchical structure and the peer structure can be combined into a hybrid system in which the processors on a particular level can communicate with one another and with processors on the next higher level.

Microelectronics and data processing are inextricably linked. The hardware in data-processing machines, from the largest to the smallest, is built out of microelectronic devices. Advances in microelectronic devices have both increased the performance and lowered the cost of data-processing machinery. These advances will continue into the foreseeable future, although there is some indication that as the industry matures the rate of advance will decrease. And as computing continues to become faster and cheaper, data processing will continue to penetrate further into the activities of everyday life.

8

THE ROLE OF MICROELECTRONICS IN INSTRUMENTATION AND CONTROL

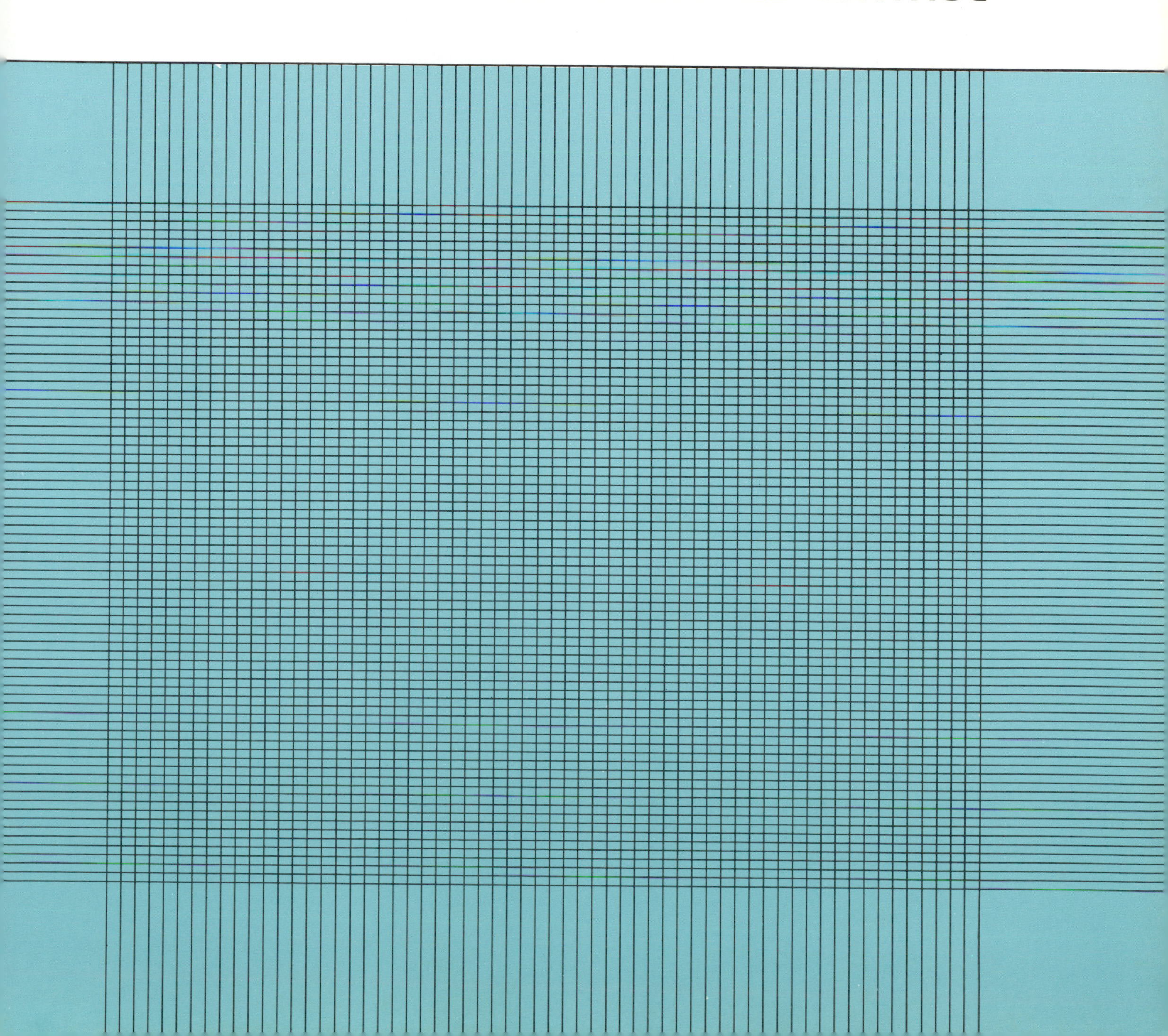

The Role of Microelectronics in Instrumentation and Control

by BERNARD M. OLIVER

Microelectronics enables measuring instruments not only to make measurements but also to analyze them. It has also brought closer the fully automatic control of industrial processes and machinery

Science and technology alike depend on the ability to measure an enormous variety of phenomena the unaided human senses cannot measure with precision and in many instances cannot even detect. Without instruments to make such measurements there can be no efficient analysis and prediction; without devices responding to the readings of instruments there can be no automatic control. Today a new generation of powerful measuring instruments and control devices is beginning to emerge. The high performance of its members is largely the result of incorporating, as an integral part of each instrument or control device, a microprocessor and some digital memory.

The underlying principle of any control system is the feedback loop. A classic example is the household thermostat, where the loop consists of a furnace, the air in the house, an air-temperature sensor and a furnace regulator. When the temperature of the air falls below or rises above predetermined points, the sensor feeds the information back to the regulator, which turns the furnace on or off. The sensor is in effect a measuring instrument. Indeed, most measuring instruments can be said to be part of a feedback loop, since the measurements they make are a guide for action modifying the quantities measured. Many measurements are not simple to make or to integrate into a feedback control loop. It is here that new measuring instruments based on microelectronic devices such as the microprocessor have come into their own.

To appreciate how profoundly measurement technology has changed over the past four decades let us look into a large communications development laboratory in the mid-1930's. A new repeater, or amplifier, for radio-network telephone lines is being tested. Its amplification is being measured over the frequency range of music and speech.

The test signal is generated by a tunable oscillator, and the output of the repeater is measured by a thermocouple meter. The output power of the oscillator varies with frequency and must be manually reset for each new frequency to some desired value, as read on the thermocouple meter. After each resetting the output of the oscillator is fed to the repeater through a calibrated attenuator, and the repeater output is read on the same thermocouple meter. Next the attenuator is adjusted, reducing the input from the oscillator until the attenuator loss cancels the repeater gain and the thermocouple reading is the same as before. The attenuator reading is equal to the amplification the repeater provides at that specific frequency.

The test continues. Each new oscillator frequency setting must be read off a calibration chart. This setting and the corresponding attenuator reading are entered in a notebook. After lunch the results will be plotted and examined to see if there are problems with the frequency response. If not, the next day can be devoted to a further test of the repeater, this time its harmonic distortion with respect to frequency.

How would we test the same repeater today? The venerable device is dusted off and cables from an automatic test set are connected to its input and output. The test conditions are typed in at a console, and a "RUN" key is pressed. In 10 seconds a video terminal displays a graph plotting the gain of the repeater across the entire range of frequencies. Pressing another key changes the display to a similar plot of distortion with respect to frequency. No dials have been turned and no meters have been read, but two days' worth of data has been obtained in a few seconds. In four decades instrumentation has progressed from the performance of the Conestoga wagon to that of the jet airplane.

It is worth noting a few of the major milestones in the development of instrumentation from the 1930's to the present, because without them the microprocessor would not be able to play an effective role in an instrument. The first milestone was the stabilization of the analogue circuits in instruments. By applying automatic gain-control circuits to oscillators and negative feedback to amplifiers it was possible to construct signal generators and voltmeters with a constant output and sensitivity over a wide range of frequencies. It was not necessary to use a substitution method and to adjust attenuators for each reading; instead a dial could be set to each new frequency and the variations in amplification could be read directly.

Digital technology, which had been evolving in computers, was first applied to instrumentation in the digital frequency counter. Earlier methods of measuring frequencies were quite time-consuming; a digital frequency counter displayed the answer in a fraction of a second and in an easily readable numerical form. Frequency measurement became almost overnight the fastest and most accurate of all electrical measurements. An immediate demand for digital readouts in other instruments followed, and soon the analogue readout—the pointer—of the traditional voltmeter was replaced by a digital display.

The replacement of analogue readouts by digital ones led in general to fewer human reading errors. More important, it led to instruments that could be read quickly, tirelessly and totally without error by computers. Computers were already in service for recording the results of expensive, one-of-a-kind tests: a nuclear explosion, say, or a rocket firing. With the aid of devices that converted analogue signals into digital ones computers stored the output of hundreds of instruments simultaneously, enabling the testers to review and re-review the test measurements at their convenience. These early computerized facilities were the precursors of today's automatic measuring systems.

The advent of digital-output instruments brought automatic measuring systems one step closer. The next step was to develop digital-input instruments: instruments that are set digitally and can therefore be controlled by computer. The frequency counter was matched by the frequency synthesizer, the digital voltmeter by the digitally adjustable power supply. One by one signal sources became computer-controllable, and together with the computer-readable instruments they formed the basis for a wide variety of automatic measuring systems.

Such systems speed up electronic measurements by one or two orders of magnitude and improve the accuracy of measurement to a like degree. They are particularly advantageous for the testing of manufactured products, where routine but often complex tests must be repeated over and over again, and where

DIGITAL METER of unusual simplicity instantaneously presents in numerical display the value of any analogue signal input it receives in the form of a voltage or a current, such as the signal from a sensing transducer. Preceding generations of digital meters incorporated 10 times as many components as the 15 in this AD2026 panel meter, made by Analog Devices Inc. of Norwood, Mass. Most of the components drive the light-emitting diodes of the display (*top*). Two others are input controls and one is an integration capacitor. Some 120 components in earlier meters are replaced here by one large-scale-integrated chip (*bottom*) of the integrated-injection-logic (I^2L) type.

there is infrequent need for reprogramming.

Automatic test systems have many advantages in addition to relieving workers of monotonous tasks. The automatic system can compensate for its own systematic measuring errors by making measurements on a reference standard and storing in its memory the difference between these measurements and the known values. The differences, which represent errors, can then be subtracted when a device on the production line is measured. The result is a tenfold to hundredfold reduction in the level of measuring error.

To get a new reading an automatic measuring system need wait no longer than the time it takes for the device being tested and the associated instruments to settle down. This time may be only milliseconds or microseconds and so is orders of magnitude faster than it would be for a human tester. The computer exercises the system thousands of times harder than a human being could and gets correspondingly more data per hour from the test equipment.

The automatic system has inherent data-processing capability. Out of the thousands of measurements that are made every minute only those that show some anomaly need to be presented to the human monitor, and they can be displayed complete with error information. Test data on different production runs can be stored for later reference, reviewed item by item and analyzed statistically to determine if the manufacturing processes are under control.

Automatic test systems do not fudge the data or make mistakes in recording it or get tired or omit tests or do any of the dozens of troublesome things human beings are apt to do. Whatever tests the program specifies will be made regardless of the time of day or the day of the week; no front-office pressure to ship goods by a certain date can compromise the computer's inspection.

What happens after the next step is taken, when each instrument incorporates its own computer in the form of a microprocessor? A host of new capabilities and modes of performance become possible; what we now have is "smart" instrumentation. To gain an understanding of what smart instruments can do let us consider some specific examples.

Compare a traditional spectrum analyzer with a smart one. A spectrum analyzer measures the power of a signal as a function of frequency over some specified portion of the frequency spectrum. Essentially it is a narrow-band receiver that repeatedly sweeps a selected part of the spectrum and displays any signals it receives as peaks with heights proportional to the power of the signal. If the sweep rate is too high, however, the receiver cannot respond fully to the signals it detects; if the sweep rate is too low, the display may be intermittent and may flicker, and changes in the signal may be missed.

To get the highest repetition rate possible without sacrificing either detail or accuracy in the display three interrelated factors must be optimized: the width of the spectrum selected, the resolution within the spectrum and the sweep rate. In the days before microelectronics the task of optimization was left to the operator, and only a careful and knowledgeable one could get consistently accurate and complete results.

In today's analyzers the microprocessor automatically adjusts the spectral resolution to present as much detail as the display can show and then selects the highest sweep rate that will still allow an accurate, full response to each signal detected. Even when the sweep rate is very low, the display will not flicker; it is stored in digital memory and presented to the operator 60 times per second. The microprocessor puts on the display not only the signal peaks themselves but also all other pertinent data: the sweep time, the bandwidth of the receiver, the span of the spectrum, the strength of the signal and the minimum, maximum and center frequencies of the signal. Monitoring photographs of the display make it possible to record the entire story at any time. Instead of having to contend with the shortcomings of the conventional spectrum analyzer the operator finds himself actively helped. The controls are simplified and entire settings

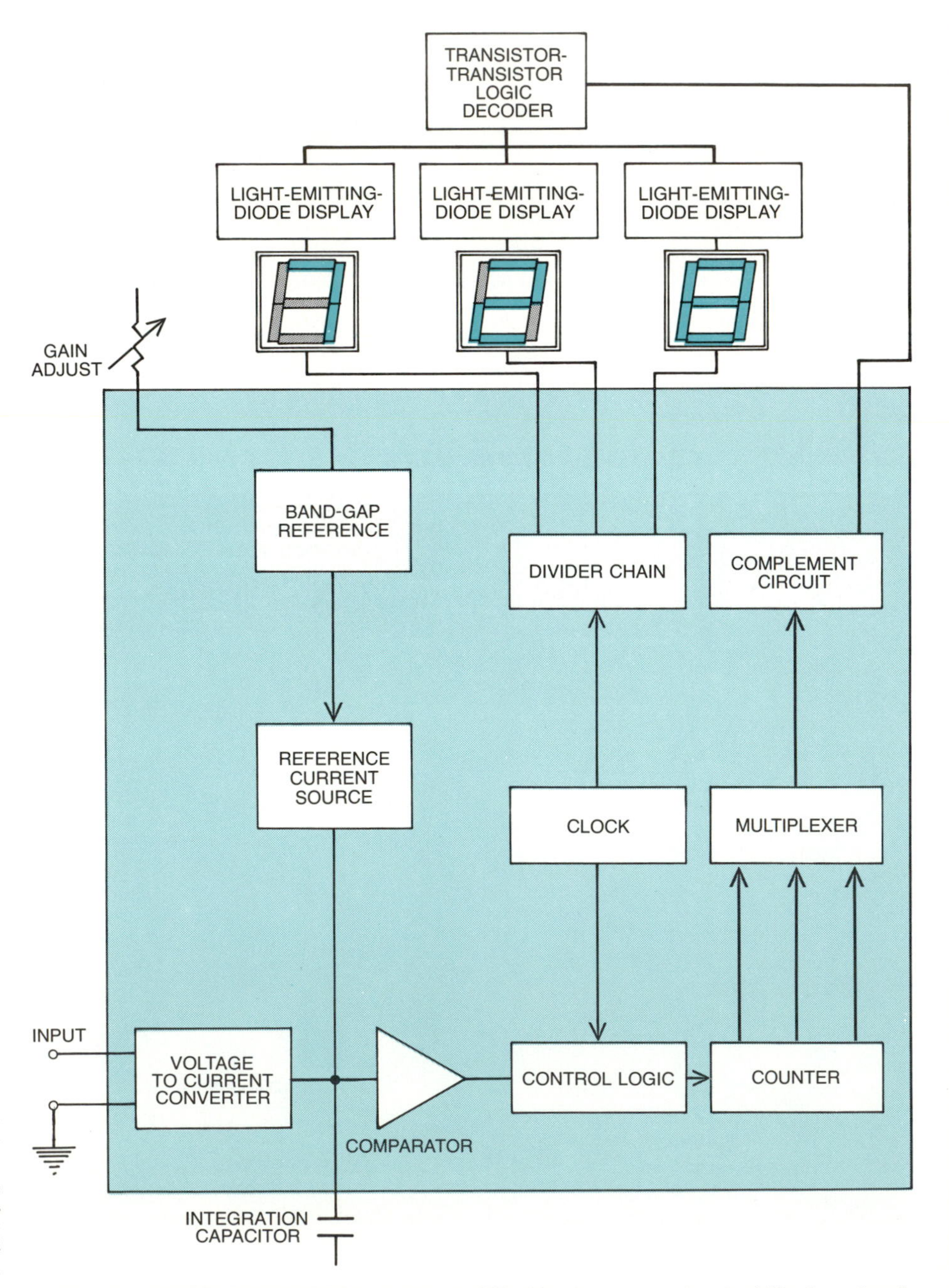

DIGITAL-METER CIRCUITRY includes the I²L chip elements enclosed within the colored box and 14 additional components, most of them related to the numerical display seen at top.

can be stored in memory to be recalled later if the measurement needs to be repeated.

A second example has to do with the combination of two instruments of chemical analysis: the gas chromatograph and the mass spectrometer. When a sample of unknown composition passes through the column of a gas chromatograph, each component of the sample travels at its own characteristic rate, so that two components rarely reach the end of the column at the same time. The arrival of each component at the end of the column is sensed in any one of several ways and is quantitatively recorded as a peak on a strip chart. Such a chart is the chromatogram. For example, a drop of gasoline will produce a chromatogram with a different peak for each hydrocarbon present. The chemist can then determine the percentage of each hydrocarbon in the sample by integrating the area under each peak, a laborious process.

Procedures for gas chromatography were greatly accelerated a few years ago by converting the output of the detector from analogue to digital form and storing the chromatogram in computer memory: the area of each peak could then be integrated numerically with the aid of a suitable computer algorithm to mark the beginning and end of each peak. Today, with microprocessors, each chromatograph can be a self-contained quantitative instrument with its own internal calibration and self-testing routines and its own internal integrator.

In the mass spectrograph a sample of unknown composition is injected into a vacuum and is ionized by a stream of electrons. Ions of differing mass separate in accordance with their ratio of mass to electric charge, and the result is recorded as a series of separate peaks analogous to the peaks of a chromatogram. These multipeaked spectra represent a series of "signatures" uniquely characteristic of individual elements and compounds. Libraries of signatures have been compiled to aid in the identification of samples of unknown composition. The flaw in the system is that with a mixture of compounds the spectra overlap and the identification of the component spectra becomes very difficult and uncertain.

In contrast, each peak in a chromatogram almost always represents a single substance, and so the notion of harnessing the two instruments in tandem arose. The chromatograph would separate the unknown sample into its constituents, and the mass spectrometer could then identify each constituent singly.

The combination is a powerful analytic tool. It is also a complicated system requiring daily correction and recalibration to compensate for drift resulting from contamination by past samples.

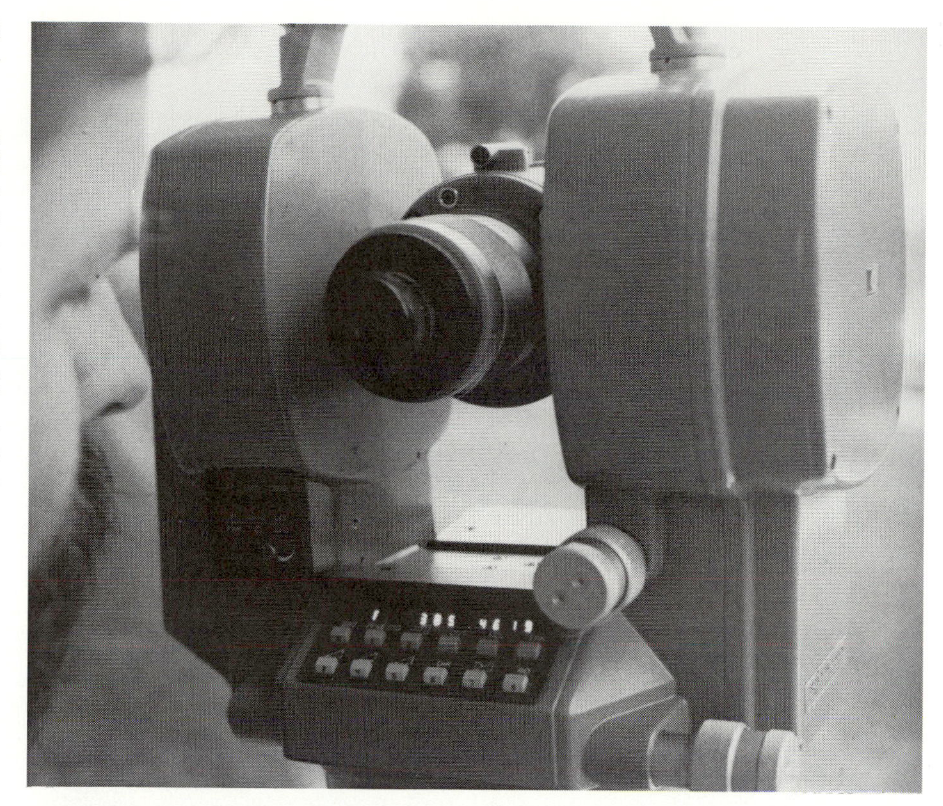

"SMART" SURVEYING INSTRUMENT combines the angle measurements of a theodolite with distance measurements obtained by microelectronically processing reflected waves of laser light. A microprocessor in the instrument presents azimuth and elevation angle readings in degrees, minutes and seconds of arc. An internal sensor detects any leveling errors and corrects the angle readings accordingly. With distance measurements of up to five kilometers the microprocessor converts slant-range readings into their horizontal and vertical components while correcting for the curvature of the earth. The instrument is the Hewlett-Packard 3820A.

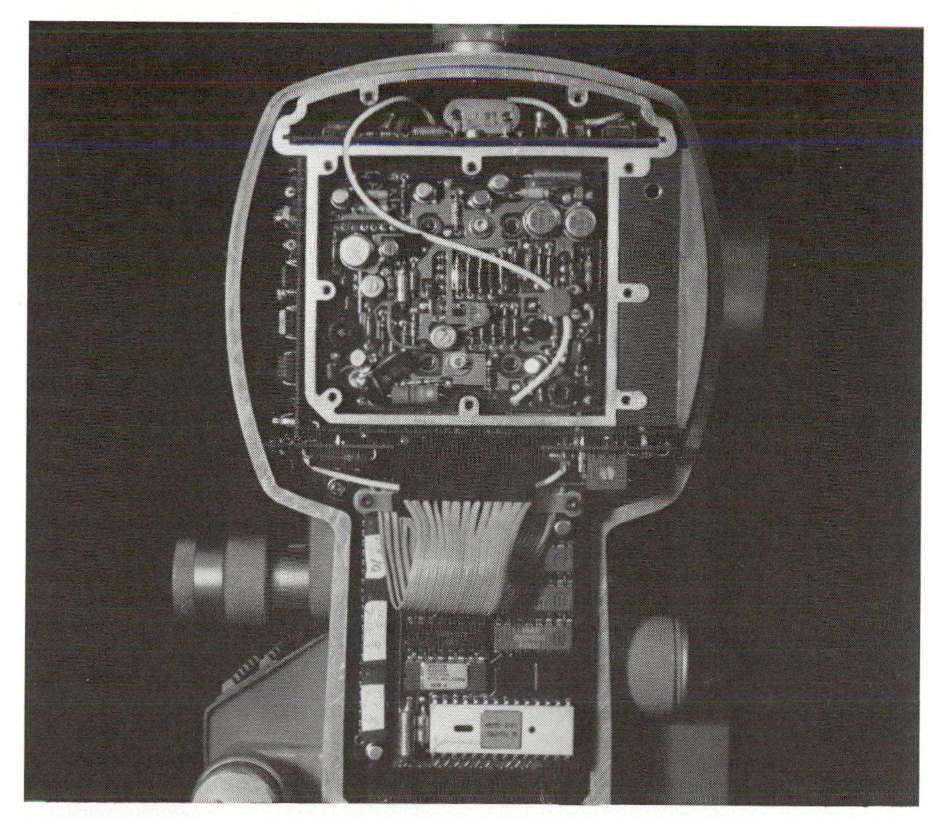

OPENED CASE exposes to view part of the circuitry of the 3820A. By plugging in a solid-state memory the surveyor can record a day's field observations for transfer to a computer at headquarters directly or by telephone. The computer-held data in turn can be fed to a plotter that will automatically convert days of field observations into a standard surveyor's plat.

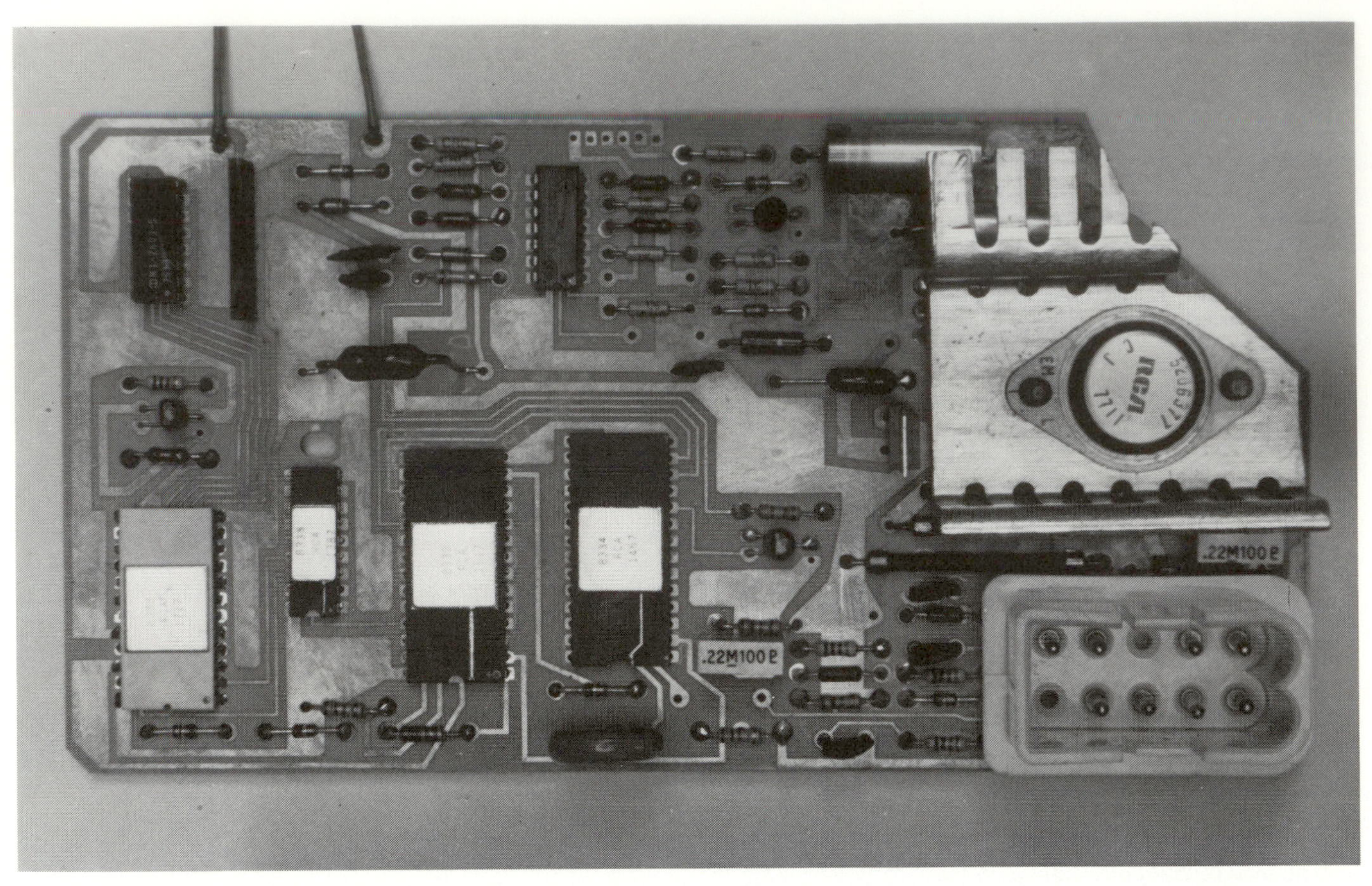

AUTOMOBILE-ENGINE CONTROL incorporates a microprocessor to integrate readings from six sensing instruments and determine the optimum time of ignition at any of a wide range of engine speeds and loads. The control, which is under development by the Chrysler Corporation, uses a microprocessor chip produced by RCA Corporation. Details of the control are shown in illustration below.

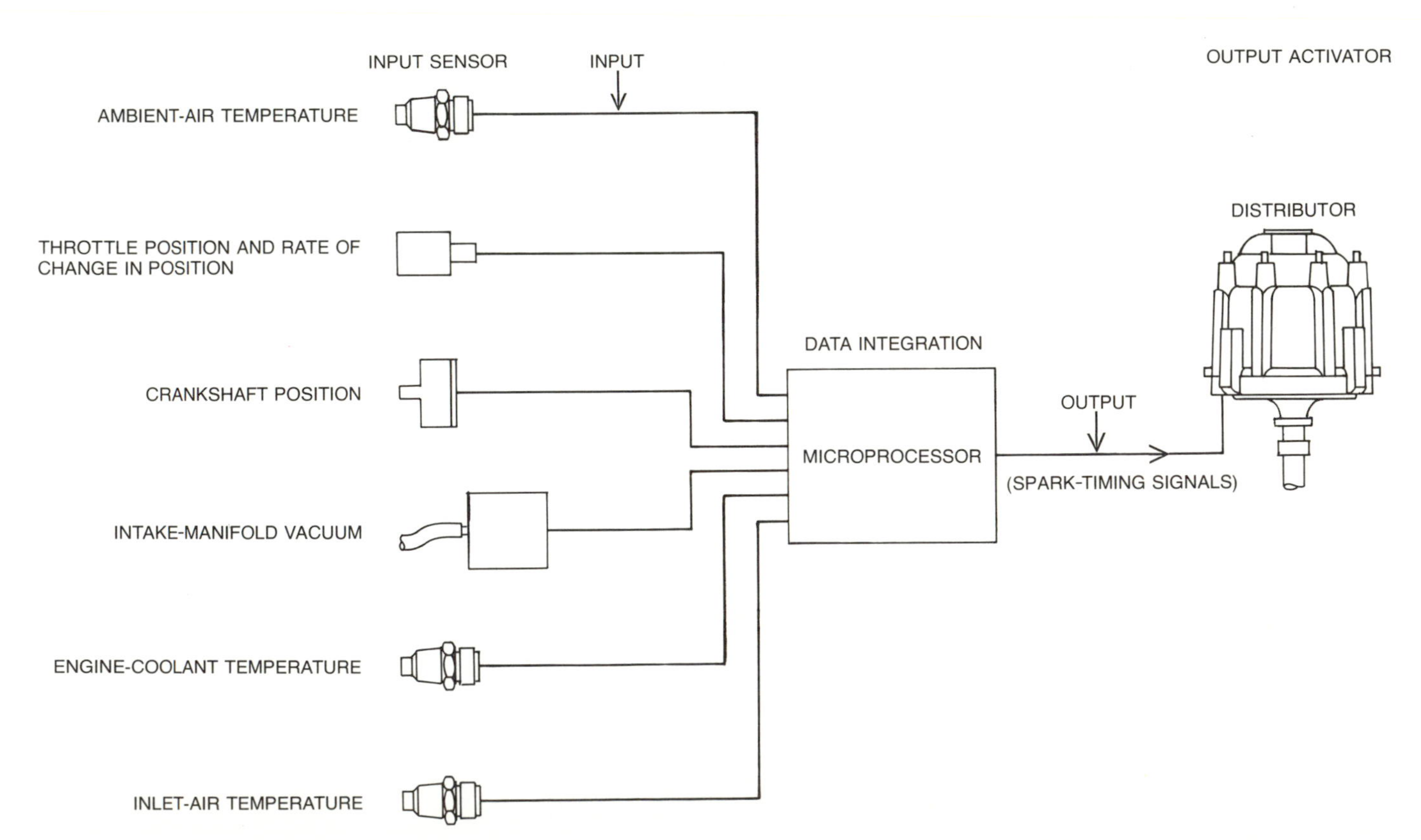

SEVEN INPUTS FROM SENSORS in the Chrysler engine-control monitoring network include the ambient-air temperature, the inlet-air temperature, the engine-coolant temperature, the intake-manifold vacuum, the crankshaft position, the throttle position and the rate of change of the throttle position. The best engine performance with the fuel-to-air ratio kept lean to minimize pollution depends on selecting an optimum instant of ignition with respect to sensed variables. Microprocessor sends these timing signals to the distributor.

When the combined instrument is put under microprocessor control, not only is the task of calibration reduced from hours to minutes and the task of monitoring for drift greatly simplified but also a large number of samples can be analyzed automatically.

The mass spectrometer can be directed to generate spectra for every chromatogram peak or to analyze only those spectra that meet predetermined criteria. The tandem machine can work overnight, producing successive chromatograms and checking their mass spectra with the signatures stored in its memory. If so directed, it will even print out or store in memory the top 10 "hits" in each comparison for review by the operator in the morning. This is almost completely automated analytical chemistry of a very sophisticated kind that can be set to detect trace amounts of anything from narcotics in urine to contaminants in lemon oil. All that is required for such automation is the provision of appropriate microprocessors and memories.

For decades the automatic factory has been forecast as imminent, and yet it never arrives. The basic reason is that automation involves far more than the development of sequential machines. It calls for feedback mechanisms that sense anomalies in the system, analyze them and take the appropriate corrective action. In the motion picture *Modern Times* Charlie Chaplin is fed lunch while he is at work by an automaton that keeps putting the food everywhere but where his mouth is. Charlie was a little short for the machine, and it could not adapt. In a large manufacturing process thousands of things can go wrong and send a nonadaptive system into a spin.

It will be a long time before the majority of production processes can be truly automated. Nevertheless, as individual machines are made smarter we can expect them to cope successfully with an ever widening range of problems. A good example is the evolution of traffic-signal systems. The first automatic traffic signals were purely sequential; they were fixed in their timing and were independent of one another. The flow of traffic was greatly aided many years ago by synchronizing the system and also by changing its timing to accommodate such variations in traffic flow as occur during rush hours. Even so, anomalous local traffic flows often called for manual intervention.

Traffic signals today can be made to sense and count the number of cars in the main stream of traffic and even the accumulation of cars in access lanes. They can be programmed to adjust their own timing to keep traffic flow at an optimum; adaptive control of this kind is a trivial task for a microprocessor. Moreover, once the microprocessor has been

NUMERICALLY CONTROLLED MACHINE TOOL (*foreground*) is computer-programmable. In the background is the keyboard and memory component; programs for various tasks are entered and called up as needed. Seen at work at the Wilsey Tool Company, Inc., of Quakertown, Pa., the tool does electric-discharge machining; a moving wire is the electrode. It is made by Andrew Engineering of Hopkins, Minn. Its cutting face appears in the photograph below.

TYPICAL TASK for a traveling-wire EDM is cutting gear teeth; in this instance the teeth are on the inside face of the gear wheel. For most kinds of inside cuts the traveling wire first has to be manually threaded through a starter hole. Thereafter the EDM needs no supervision. When a slow and complex series of cuts is programmed, it can run unattended for 60 hours.

incorporated into the traffic signal, it simplifies communication with a central controller. When smart traffic signals become ubiquitous and are linked to a control center, the traffic cop at the intersection will become obsolete.

An evolution of much the same kind can be seen in machine tools. Such devices as turret-head lathes and screw machines have performed complex sequential operations for a long time. The "program" was part of the hardware: cams and ratchets sequenced the machine through a series of steps to manufacture a particular part. The variety of sequences was necessarily limited, however, and the cost of new hardware programs was high.

In the early 1960's numerically controlled machine tools made their appearance: machines coupled to what came to be called a minicomputer. Such machines could mill, drill, bore and tap holes anywhere and at any angle on a workpiece of almost any shape. The appropriate toolhead was selected automatically by a punched-paper-tape program that was read by an electronic computer-controller. The same program positioned the tool in three coordinates and dictated the speed of the tool spindle and the depth and rate of the cut. To revise the final shape of the workpiece or to make an entirely different one it was necessary only to amend the old punched tape or to provide a new one.

Numerically controlled machines of this kind filled an important need in intermediate-volume production, where the output of workpieces is substantial but not high enough to warrant a mass-production line (where each of a series of machines performs only one operation). The computer-coupled machines were quite expensive, however, and a substantial fraction of the cost was in the computer-controller. Today microprocessor-based controllers no larger than desktop calculators are taking the place of the larger first-generation controllers, and tape cassettes are replacing the rolls of punched paper tape. Microelectronic controllers add relatively little to the cost of the machine tool.

Controllers based on microprocessors also allow keyboard access to the machine tool. When a future shop is equipped with machines of this kind, each with stored programs that can be keyboarded remotely, it will be possible for a central computer to preside over truly automatic production. About the only human intervention required will be the loading of stock at one end, the hauling away of finished workpieces at the other end and the periodic replacement of dulled cutters.

We can expect to see further progress in electronic controls for automobiles. The microprocessor is now making it economically feasible to extend electronic controls to such functions as ignition timing and carburetion. The motivation to develop such systems is now strong: they not only promote fuel economy but also aid in the meeting

CONTINUOUS CASTING OF STEEL SLABS calls for a complex automatic routine of water-cooling. This machine is the constant-radius continuous-strand caster at the Burns Harbor, Ind., complex of the Bethlehem Steel Corporation; the slab is more than six feet wide and can range from eight inches to a foot in thickness. Water temperature, valve pressure and rate of flow at successive stages are held within predetermined limits by the Foxboro Company's SPEC 200 system, a 200-loop feedback control utilizing microprocessors.

of Federal engine-emission standards.

Good ignition timing is a function of the speed of the engine and the vacuum in the intake manifold. The mechanisms used in today's cars can only approximate the optimums, whereas the application of microprocessors can ensure correct timing at all speeds and loads. Ideally these devices should play the role of monitors and not be responsible for delivering each charge of fuel and firing each spark. Such a monitor would compare actual performance with ideal performance and make the necessary adjustments to the existing basic systems. If the microprocessor were used in this way, it would have the advantage of a greatly reduced data rate, ordering only a few corrections per second. Furthermore, it would fail safe: if the microprocessor were to go out, the engine would still operate, its performance simply being reduced to the level that prevails today. The same microprocessor could continuously monitor the health of the vehicle and report in plain English on a LED display any troubles such as low fuel, overheated engine or overheated or worn brakes. The cost of such systems lies not so much in the microelectronic components as in the sensors needed to feed the data to the microprocessor and the actuators needed to carry out its instructions.

It appears that control systems and their associated sensing components are increasingly coming to resemble the human nervous system. Microprocessors in satellite controllers, by enabling "intelligence" to be distributed broadly throughout the control system, greatly reduce the amount of data that must be transmitted over data links and handled by the central computer. In the human nervous system much control is also at the local level and much processing of input data is done in ganglia. The nervous system is a network of sensors and microprocessors connected by data links to a central computer. Both the satellite processors and the central computer have a lot of firmware: subroutines that enable us to do 99 percent of what we do without "thinking," that is, without invoking our highest control centers. Here too microelectronic control systems resemble the human control system: the central computer merely has to order that an operation be done, not tell every smaller machine every step of how to do it.

The examples I have cited are only a few of literally hundreds of ways in which the microprocessor will improve and simplify the operation of the machines that serve us daily. Integrated-circuit technology has so reduced the cost of computation and of logical systems that most of the devices we use can now be much more cooperative and intelligent than they have been in the past. These decades will be recognized in the future as the beginning of the robotic revolution.

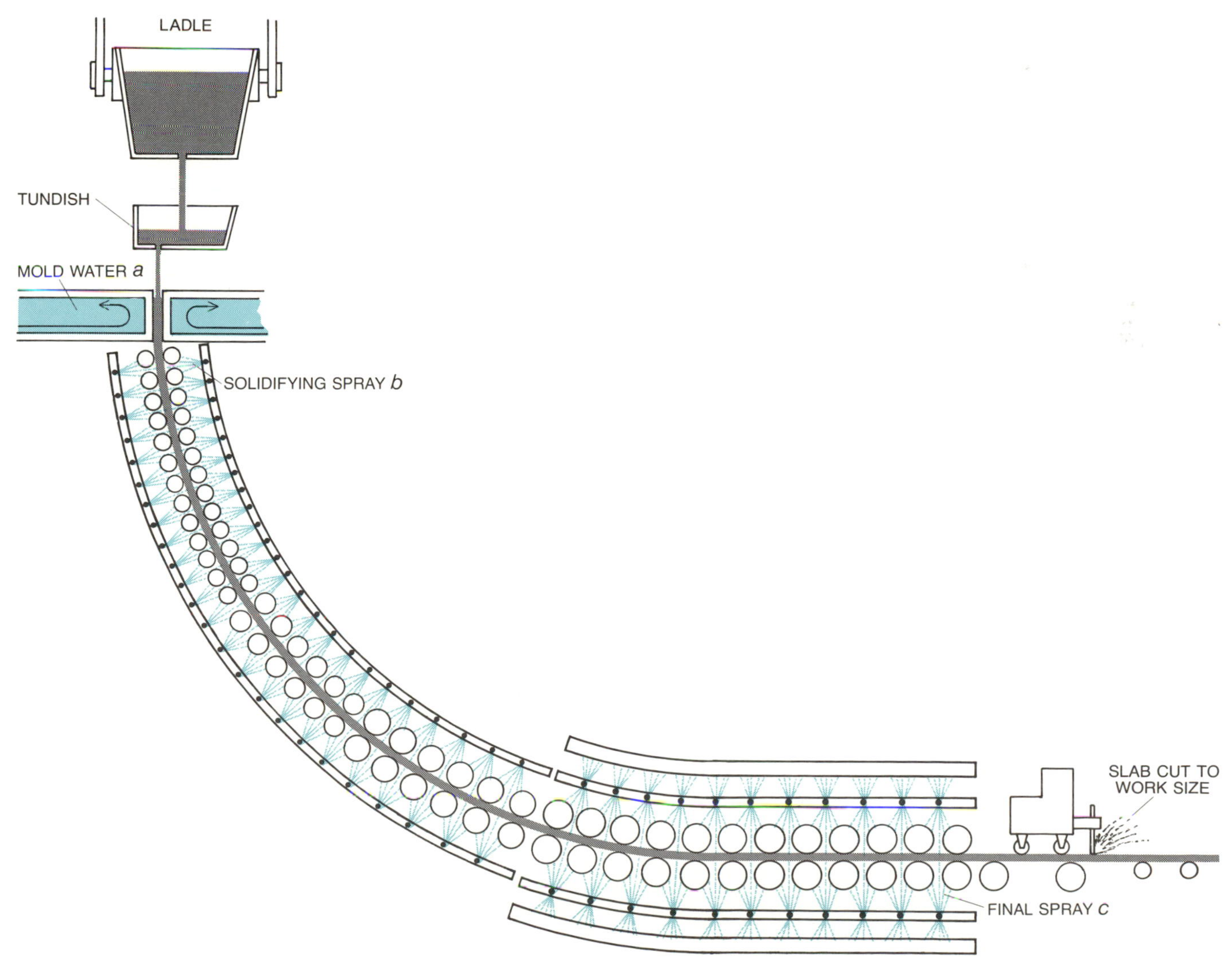

STEEL SLAB IS FIRST COOLED in the mold (*a*) where the molten steel starts to solidify. The descending slab is then further cooled by spraying water on its sides (*b*). Both the main frame and the rollers of the final flatbed component (*c*) are also water-cooled. Temperature, pressure and flow sensors supply the input data for the SPEC 200 feedback-control loops that adjust valve settings as necessary.

9

THE ROLE OF MICROELECTRONICS IN COMMUNICATION

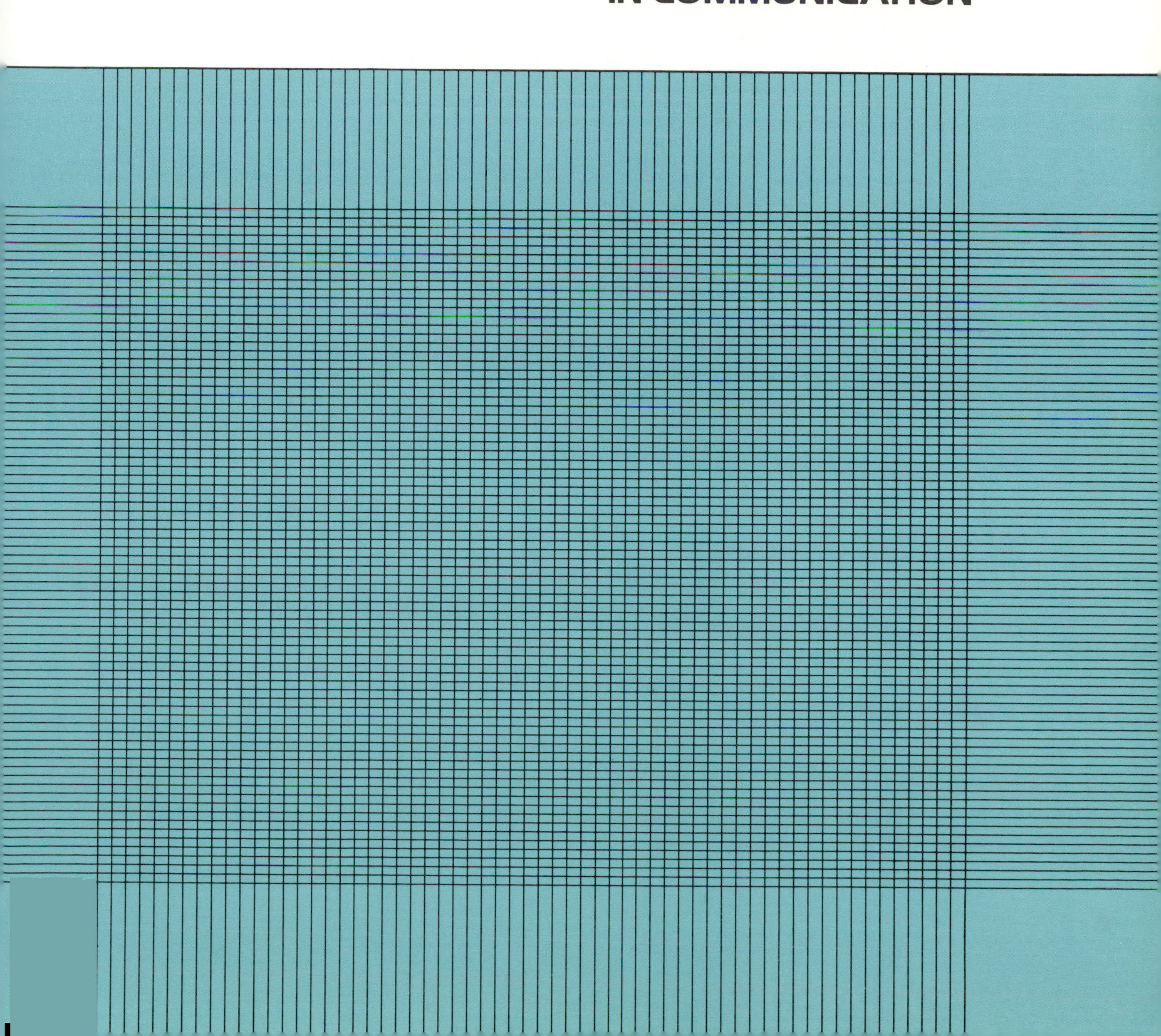

The Role of Microelectronics in Communication

by JOHN S. MAYO

The essence of systems such as the telephone, radio and television is signal processing. The large capacity, high reliability, and low cost of microelectronic devices make them ideal for such purposes

It is possible now to have a telephone that "remembers" frequently called numbers, any one of which can be reached by pushing a single button. On television one can frequently see events in Europe and other areas overseas while they are happening. Computers "talk" digitally to one another over telephone lines at remarkably high speeds. These developments and many more reflect the impact of microelectronics on communication.

Until perhaps 15 years ago the vacuum tube was the dominant active component of the electronic circuits that are fundamental to the operation of telecommunication systems. Its limitations were made evident by the rapid growth of communication by telephone, radio and television. The vacuum tube was too large, required too much power and was too unreliable to meet the need of those systems for large numbers of signal-processing devices clustered in complex circuits and required to operate with extreme reliability. The alternative came in the form of the transistor, which provided electronic gain in a semiconductor and was small and reliable. It led to the silicon integrated circuit, a revolution in electronics and a vast improvement in telecommunication.

Today, as preceding articles in this issue have shown, millions of circuit elements are simultaneously fabricated on a thin "wafer" of silicon. Typically the wafer holds several hundred copies of the same circuit and is divided into small chips holding one circuit each. The chips are usually packaged individually as integrated-circuit components. The designer of communication equipment employs these components to build complex systems, usually by placing a number of integrated circuits on a circuit board containing the printed wiring necessary to interconnect them.

A typical communication system requires a number of circuit technologies. For example, signals transmitted with high power or at high frequency often call for currents and voltages that cannot be handled by conventional integrated circuits. Some circuits require highly stable components; often they are best provided by thin-film techniques, which involve depositing conductors, insulators and resistive films on a ceramic substrate. To properly describe the role of microelectronics in communication I shall discuss several of the more important circuit technologies, with primary emphasis on silicon integrated circuits because they have had by far the greatest impact on modern telecommunication.

Microelectronics in the form of integrated circuits is an important factor in telecommunication largely because of the combined effects of low cost, high reliability and wide applicability. As increasing numbers of circuit elements are fabricated on a silicon chip, the cost of a basic circuit function decreases markedly. A circuit function of outstanding importance in communication systems (and also in computing systems) is the digital logic "gate." It controls the flow of information, providing an output signal only when the input signals are in prescribed states. From this basic element large digital signal-processing systems can be built. Hence the cost of a logic gate has a considerable influence on the cost of communication equipment: on terminals, which provide the interface between people or machines and communication channels; on switching machines, which establish communication paths, and on the equipment that processes signals so that they can be transmitted over wires and cables, by radio and by light waves.

Digital technology has progressed rapidly from the logic gate consisting of vacuum tubes to the logic gate consisting of transistors to the integrated logic gate and now to thousands of logic gates within a single integrated circuit. During this evolution the cost has gone from about $10 for the vacuum-tube gate to about one cent per gate in an integrated circuit incorporating many gates. The cost will soon reach .1 cent. With the cost of such a crucial circuit declining more than a thousandfold equipment involving complicated signal processing, as the telephone with a memory does, has become economically feasible.

Low cost, however, is not enough. A communication link may call for thousands of circuit elements all working together. The failure of any one element may break the link. Reliability is therefore as important as cost.

Again the logic gate serves as an example. The vacuum-tube logic gate was distinctly unreliable. A logic gate consisting of discrete transistors proved to be about 1,000 times more reliable than the equivalent vacuum-tube gate. A modern integrated-circuit gate is at least 100 times more reliable than the gate consisting of discrete transistors. In the progression from vacuum tubes to integrated circuits, then, the reliability of the logic gate has improved by a factor of 100,000.

The switching machines that control the routing of telephone calls provide an example of what this enhanced reliability means to a communication system. Such a machine handles up to 100,000 telephone lines simultaneously. It scans the incoming lines and detects when a customer is calling for service (having lifted his telephone off the hook). Then it collects the dialing information, connects the caller with the person he is calling, records the information needed for billing and disconnects when the caller hangs up.

The "brain" of the machine is an electronic processor consisting mostly of a central control and components that store data. The processor also handles numerous other chores, including the diagnosis of faults and failed circuits. Since the processor is crucial to the operation of as many as 100,000 telephones, it must have a high degree of reliability. The objective set for it is that it have no more than two hours of "down" time in 40 years. (Most of that time is expected to be the result of inadvertent mistakes by operating and maintenance workers rather than of the failure of electronic components.)

The unit of measurement of the failure rate of an electronic component is the FIT: one failure in 10^9 operating hours. Consider the processor of a large switching machine. Its central control may have 40,000 gates. If it can be repaired in half an hour, it could fail once in 10 years and still meet the objective. Circuit failures are random events, not easily predicted, and there are other sources of failure in any large system. Allowing a reasonable margin for these factors, the acceptable mean time to failure per gate in the processor is about 10^{11} hours: a failure rate of approximately .01 FIT per gate.

Such a rate cannot be achieved easily. Large-scale integrated circuits almost achieve it, however, and they have the added advantage of low cost. The cost makes it possible to provide redundancy that further reduces the possibility of outage and also protects against failure in other elements, such as wiring and the power supply. Redundancy keeps the system functioning, and the low failure rate of the gates makes redundancy work and keeps the cost of maintenance at a reasonable level.

The size of equipment is almost always important in telecommunication, particularly with satellites and spacecraft. Even on the ground electronic equipment that is small is easier to handle and ship and takes up less space in buildings. All these benefits lead to lower costs. The memory portion of a processor for a large electronic switching system provides an example of what developments in microelectronics have done to size.

With the technology of the early 1960's the memory for a local electronic switching system required a lineup of equipment racks 104 feet wide. Data storage was on sheets of magnetic material. By the 1970's the same amount of memory could be loaded into small, tightly packaged toroidal ferrite "cores" in a lineup of equipment racks about eight feet wide. Then came integrated circuits and semiconductor memories that could provide first 1,000 bits and then 4,000 bits of storage on a single silicon chip. In a circuit consisting of 4,000-bit chips the memory for the same switching system was packaged in a single rack 2.2 feet wide. Now integrated circuits with 16,000 bits per chip are available, making it possible to put the memory in about a quarter of a rack.

Communication systems of the future

CIRCUIT PACK for the memory unit of a telephone-switching system illustrates how microelectronics has made it possible to greatly diminish the size of communication equipment. The pack is approximately 13 inches long and eight inches wide; 161 such packs in a rack 2.2 feet wide handle the switching for up to 100,000 telephones. As recently as 15 years ago a comparable memory unit for a local electronic switching system would have required a lineup of equipment racks 104 feet wide. Here the gold rectangles are *n*-channel metal-oxide-semiconductor (*n*-MOS) memory devices. Most of the black rectangles near the top of the pack are decoders. The four white rectangles at the lower center are hybrid integrated circuits embodying the program for the memory unit. The memory unit for a local switching center stores not only permanent information needed for the computer-controlled operation of the switching machine but also transient information such as numbers dialed, duration of calls and billing data.

TELEPHONE WITH MEMORY is now in service in parts of the U.S. as a result of advances in microelectronics. With the three small buttons at the top one can record a number that is called frequently. Each such number can then be reached by pushing one of the buttons below the top three. The button at the bottom right, next to the label that reads "Last number dialed," is for times when one gets a busy signal. The telephone "remembers" the number, which the caller can try again (without the need of redialing) by pushing only that button.

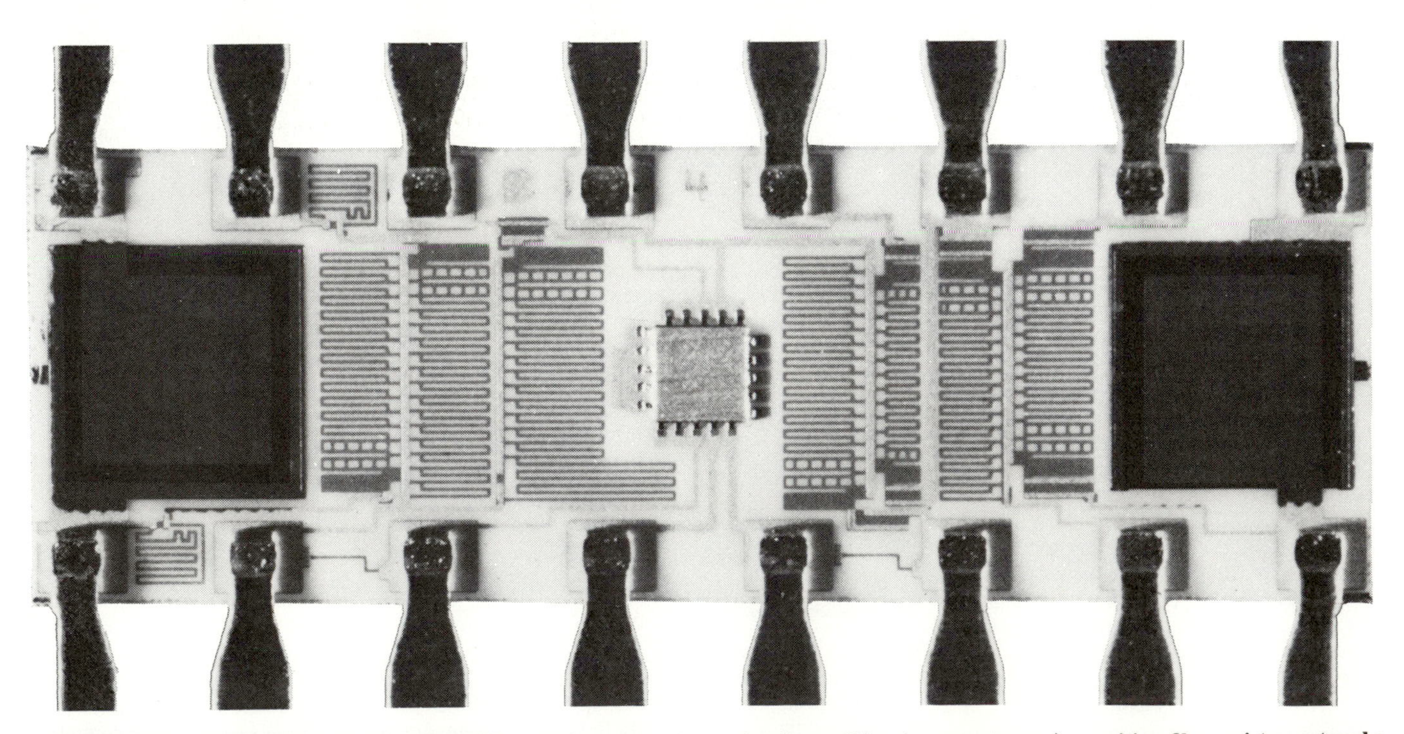

VOICE-FREQUENCY ELECTRICAL FILTER, which is one of the most basic circuits in telecommunication, demonstrates the effect of microelectronics in making components of telecommunication systems smaller and more reliable. The role of the filter is to select one audio communication band and exclude interference from it, which is particularly important when several voice signals are being multiplexed onto a single wire. The filter consists of a silicon integrated-circuit amplifier (*square at center*), precision-film resistor networks (*meandering lines*) and precision-film capacitors (*large squares at each end*). The filter is fabricated by thin-film techniques on a ceramic substrate 1.93 centimeters long and .66 centimeter wide. The integrated-circuit filter replaces the bulky and expensive devices, consisting of many large inductors and capacitors, that were required to build a comparable filter before the advent of microelectronics.

are sure to use more memory and logic. The prospect would have been forbidding when the memory required a lineup of equipment racks 104 feet wide. Now it is easy to conceive of systems employing tens of times more memory than the systems of today have. Memory is only one example of ways in which microelectronics is making new concepts in communication systems possible.

Microelectronics has also had a significant impact on analogue circuits, although the results are not as dramatic as they have been with digital logic and memory. Consider the circuit perhaps most basic to telecommunication: the voice-frequency electrical filter, which can select one audio band and reject all interference from outside the band. Such filters are particularly important in systems that multiplex a number of voice channels onto a single wire; the filter confines each voice channel to a certain frequency band and thereby prevents the channels from interfering with one another.

For almost 50 years the filters were made from large inductors and capacitors. Over the period from 1920 to 1970 designers succeeded in making the inductors and capacitors progressively smaller, but the filter was still bulky and expensive. By 1969 progress in stabilizing resistance and capacitance elements against the effects of time and temperature and in making amplifiers inexpensively with integrated circuits led to an equivalent filter that was somewhat smaller and cheaper. The use of integrated amplifiers (built on a single silicon chip) enabled the designer to build the equivalent of the large inductors and capacitors with small capacitors and resistors. Success in this endeavor required the small resistors and capacitors to be highly stable against time and temperature.

This requirement happened to mesh neatly with the development of microcircuits on a different front, which involved work on thin conducting and insulating films on ceramic substrates. From this work came the thin-film integrated circuit I have mentioned. Film circuits made of tantalum are both cheap and highly stable. For this reason they have been employed for some time as audio filters in the tone-generating circuit of the push-button telephone, to mention just one application. The tones generated by the circuit must be extremely stable over long periods of time, since the frequencies of the tones identify the digits of the number being called.

By 1973 this technology had advanced to the point where the equivalent of the old voice-band filter made of large inductors and capacitors could be fabricated on a small ceramic substrate. By 1975 the size of the substrate had been reduced enough to make the filter appear externally to be one small com-

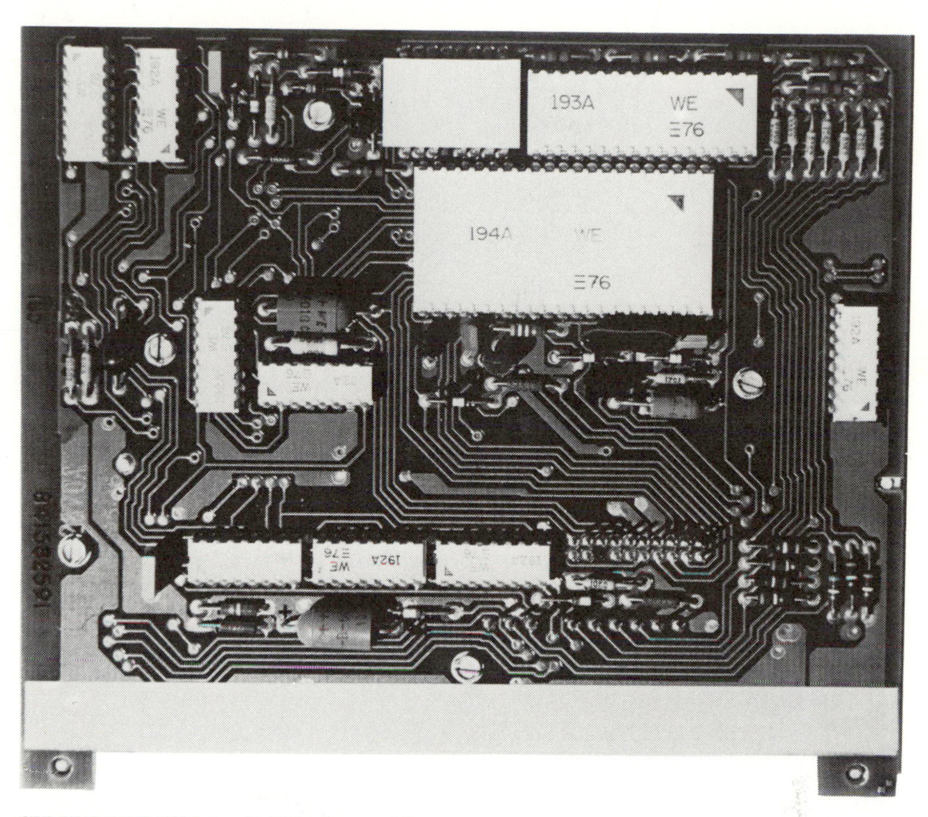

CIRCUIT BOARD for the telephone with a memory carries out the recording and "memorizing" functions of the telephone. The board is mounted on the back side of the memory section and is oriented here as one would see it by turning over the white board on the telephone so that the three top buttons are at the left. The two white rectangles at the top left are memory units, as are the six similar rectangles elsewhere on the board. Each unit is an integrated circuit that can retain four telephone numbers. Larger white rectangles at the right center are also integrated circuits that contain the program for the system and handle the dialing operation.

MEMORY CHIP for the "Touchamatic" telephone is shown as one would see it by opening up one of the memory units in the photograph at the top of this page. The chip is normally mounted in the center of a white ceramic substrate and contains the circuitry for storing and retrieving four telephone numbers. It is linked with four of the buttons on the front of the telephone.

SOLID-STATE TELEVISION CAMERA manufactured by the Fairchild Camera and Instrument Corporation is representative of the effect that microelectronics has had on television cameras. The key unit of such a camera is an imaging device of the type shown in the photograph below. Because of its solid-state components this camera (6.5 inches long including the lens) is much smaller than a television camera made with a standard vidicon camera tube.

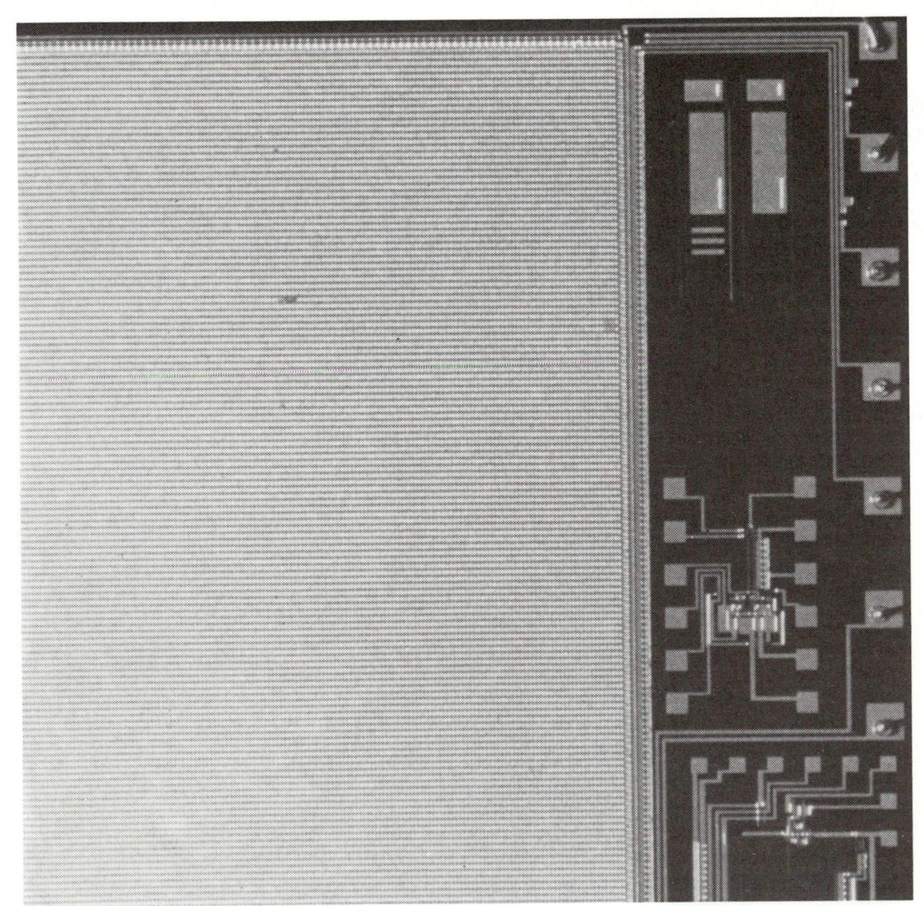

PORTION OF IMAGING CHIP for the solid-state television camera is characterized by a fine array of sensors consisting of charge-coupled devices, which release charge carriers in proportion to the light energy reaching them. The chip contains 185,440 such sensors, about a fourth of them visible in the lighter portion of the photomicrograph, and other circuitry.

ponent, quite like a standard integrated-circuit package. It is expected that soon the tantalum film will be placed directly on the surface of the silicon amplifier chip. The entire filter will be microscopic. This is not the limit, however, of the capability of microcircuits in filtering analogue signals. Time-shared digital integrated circuits and charge-coupled devices (CCD's) both show great promise in extending the improvement of electronic filters, and they may be even smaller than the thin-film filter.

Size is particularly important in communication terminals. Telephones of all kinds, private branch exchanges (the interior switching system that many large organizations have for handling telephone traffic in and out), terminals that couple computers and sensors to communication channels, and radios of all types are handled by people and occupy space in offices, homes, automobiles and many other places. The small size and other desirable features of microcircuits open new vistas in terminal equipment. Such a terminal can be amazingly "smart" while being small and compact. It is not unusual for a terminal of this kind to contain a microprocessor, which, as is indicated elsewhere in this issue, is a small computer on a silicon chip. The microprocessor in a communication terminal costs tens of dollars and yet can do computations that a decade ago would have required a large computer costing hundreds of thousands of dollars.

Size and weight are closely related. They are both extremely important in space communication. It is hard to see how the space program could have been accomplished without the smallness, lightness and reliability provided by solid-state circuits.

From the space program came the ability to launch communication satellites. When a satellite handles commercial traffic, it must be competitive with ground-based systems. Cost is important and is strongly influenced by the size and weight of the electronic gear on the satellite. The requirements of frequency and power do not allow such gear to be made entirely of integrated circuits, but circuits of that type play an important role in the economics of communication-satellite systems. Moreover, without the reliability of the solid-state components it would have been difficult to make satellite systems viable at all.

What I have said so far indicates that modern communication equipment requires a good deal of built-in intelligence. Many examples can be cited, including telephones that handle cash transactions, data terminals that send information at prodigious rates, electronic switching machines with built-in diagnostic capability, private branch exchanges with tailor-made capabilities, radios with clever frequency control, television sets that tune themselves, high-fidelity audio equipment with highly specialized features and so on.

A feature of integrated circuits that offers both benefits and problems in these applications is their limited power-handling capability. A benefit is that signal-conditioning circuits can be built at quite low power levels. A problem is that integrated circuits do not interact well with the hostile environment of the real world. Circuits connected to cables and antennas are not easily protected from lightning strokes, and integrated circuits do not readily ring telephones or provide strong radio signals for communication through the atmosphere.

Intelligence does not require a specific level of power. A circuit performing an intelligence function can often operate at a power level just sufficient to drive circuit nodes to voltages a few times greater than the voltage of any inadvertent noise on the circuit. In integrated circuits much of the circuit wiring is on the silicon chip and the circuit nodes are small, so that little power is required. As the scale of integration of circuits has increased it has been both necessary and possible to decrease power levels in order that complex circuits on a single chip can be operated at acceptable temperatures. Only a few input and output leads that connect to other components must be driven with sufficient power to deliver strong signals.

It is often possible to reduce power levels still further by choosing a technology that does not require much power in its active circuit elements. One such technology is represented by complementary-metal-oxide-semiconductor (CMOS) circuits. A microprocessor employing about 8,000 transistors has been designed in this technology on a chip that is less than a quarter-inch on a side. It is specifically designed as the control element for communication equipment. The chip can execute 434 different instructions, operates at speeds of up to two megahertz (two million cycles per second) and yet consumes less than .1 watt of power. A system that provides such a large amount of intelligence with such low power consumption enables communication equipment to be operated in places that have either no commercial power supply or an unreliable one. It allows complex remote equipment connected to long wires or cables to be supplied with power over the same small conductors that carry the signals.

Progress in microelectronics has been much slower in the areas of communication that call for high power. A circuit's ability to handle power is limited by the maximum temperature at which the semiconductor chip can be operated. A chip containing many transistors will become hotter than a chip containing only one transistor.

Solid-state devices can be employed in radio-frequency circuits of modest power, but usually they must be in the form of discrete transistors. At audio frequencies integrated power amplifiers have helped to reduce the cost and size of radios and recording equipment. Solid-state switches of modest power are beginning to replace electromechanical switches in large communication systems. In power-conversion equipment solid-state components are of increasing help in improving efficiency and lowering costs.

The pressure to expand the capability of microelectronics to serve communication circuits of higher power is considerable. Circuits able to withstand hundreds of volts would be needed to hold off surges of lightning on exposed cable systems and to transmit even the moderate amount of power required to actuate the bell on a telephone. It is unlikely that integrated circuits will ever be available for these purposes at the scale of integration now found in circuits handling intelligence. High voltages can break down any material of sufficiently small dimensions and can impair the reliability of the device in other ways. A high-voltage circuit must therefore have relatively wide spacing between its elements.

Many communication needs can best be met with digital technology. Signals transmitted as digits are not degraded as long as the digits are correctly received. A large amount of noise or "cross talk," which would be objectionable in an analogue signal of the type that usually carries the human voice, may have no degrading effect whatever on digital signals. As long as the noise or cross talk is somewhat weaker than the digital signal the presence of each binary digit can be detected and the digit can be regenerated fully. The signal is therefore stripped of noise and cross talk. This is a powerful advantage when the signal is being transmitted through a noisy medium.

A digital communication system is likely to operate at a high pulse rate and to employ a large number of digital logic gates. Such circuits are small and well matched to modern microelectronics technology. Voice signals have been

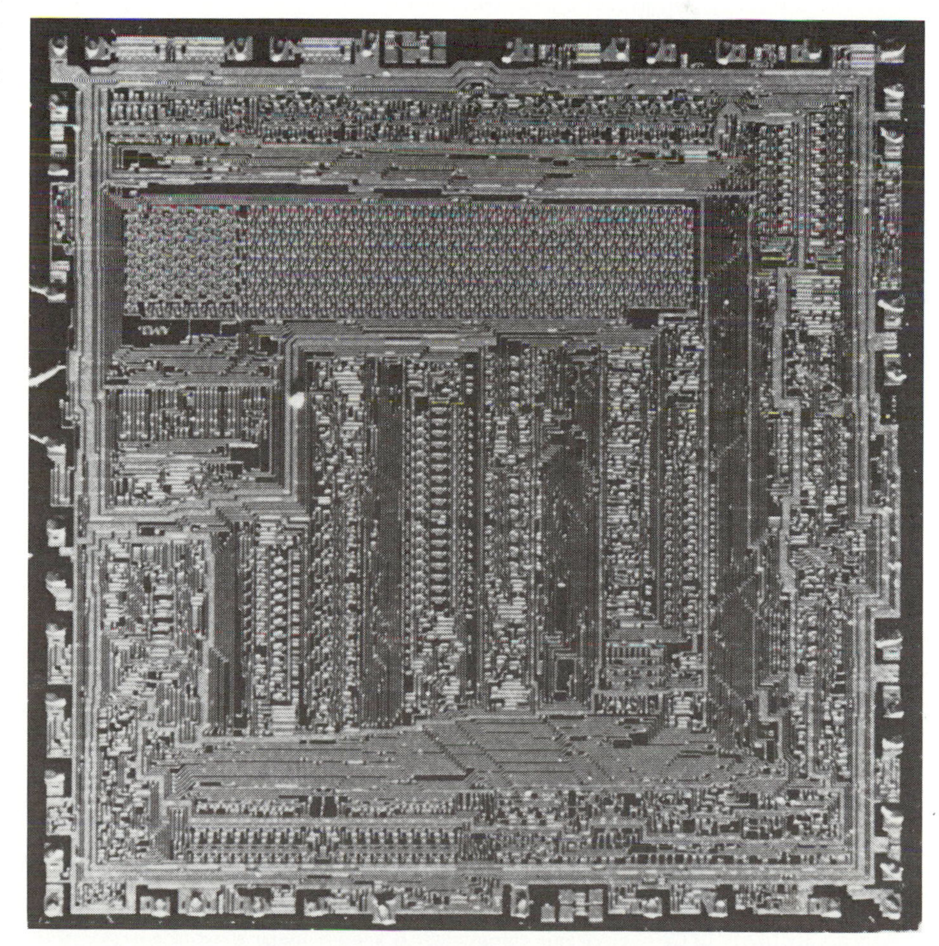

MODEM CHIP is central to the operation of the modulator-demodulator shown in the upper illustration on page 107. The chip is a large-scale integrated circuit that turns the digital, or square-wave, signals from a computer or some other data-processing machine into the modulated-wave format required on most of the telephone lines in the U.S. It is likely that over a number of years the telephone lines will be modified to carry even voices as digital signals.

transmitted as digits in parts of the U.S. telephone system since 1962. The transmissions employ pulse-code modulation, wherein the voice signals are repeatedly sampled at a high rate and the samples are coded into a stream of binary numbers. The binary numbers are transmitted as digital signals, which are periodically regenerated along the route to remove noise, cross talk and distortion. At the receiving end the binary numbers are converted back to signal samples from which the voice sound can be reconstituted.

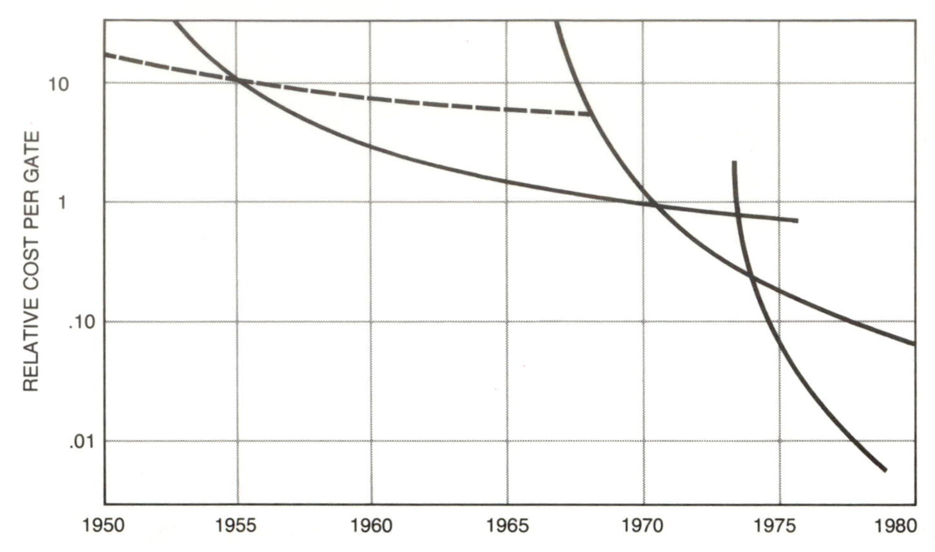

REDUCTION IN COST of components of communication systems is one of the major effects of microelectronics. It is illustrated here for the logic gate, which is a key element of communication and computing systems. The curves, read from the left, represent respectively vacuum-tube logic gates, gates made with discrete transistors, gates embodying small-scale integration (a few gates on each chip) and gates embodying large-scale integration (thousands of gates per chip). In each instance the device represented is a digital logic gate of the best quality.

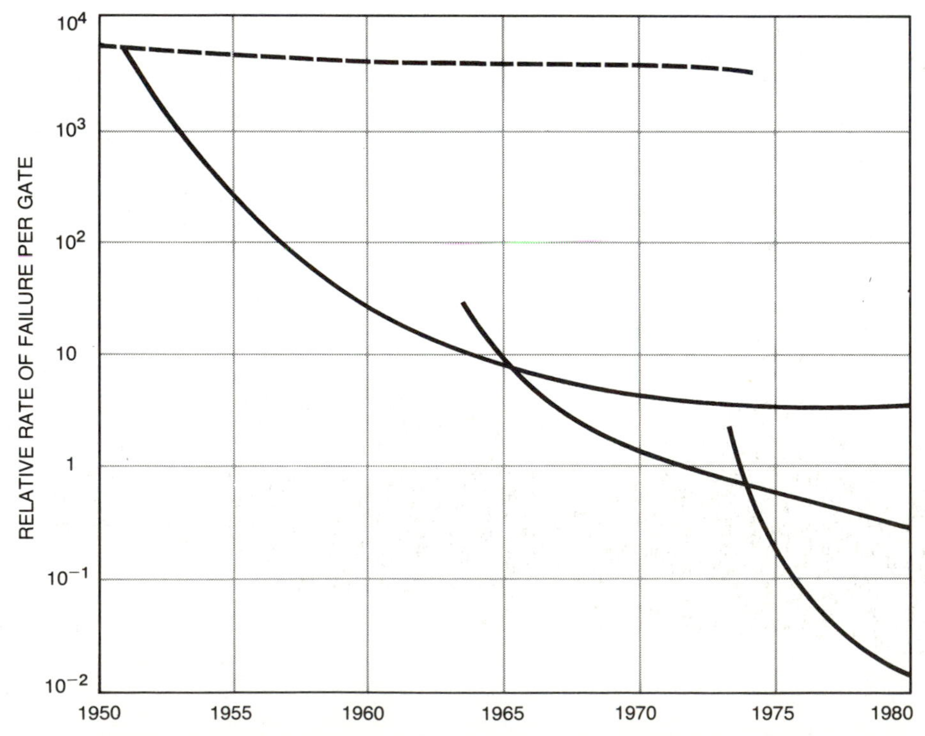

GAIN IN RELIABILITY of electronic components is another major effect of microelectronics on communication. The two developments (reduced cost and increased reliability) together have had a profound impact on communication. In this diagram as in the upper one on this page the curves represent high-quality logic gates based respectively on vacuum tubes, discrete transistors, small-scale integration and large-scale integration. A logic gate is an organizer, providing an output signal only when the input signals it receives are in prescribed states.

The advantages of digital communication have been known for some time. The extent of its application has been controlled mostly by the relative economics of digital and analogue circuitry. The development of integrated circuits has greatly changed the economics, tipping the scale in favor of digital operation in many domains.

A disadvantage of digital signals is that they require more bandwidth for transmission than analogue signals. Therefore analogue techniques remain attractive in well-shielded mediums of limited bandwidth, such as coaxial cables, and in radio propagation where the bandwidth is limited by the available spectrum. Much work, embodying a wide variety of approaches, has been directed at reducing the bit rate required for transmitting voice communication digitally. The low cost of microelectronics is essential to almost all of them. The channel rate per voice ranges from about 2,400 bits per second in systems that are sophisticated but of low quality to some 64,000 bits per second in pulse-code-modulation systems of high quality. The lower rates inevitably introduce degradation of the signal, which may or may not be acceptable in a given application.

An approach that is of particular interest at present is the time-assignment-speech-interpolation (TASI) concept originally developed by Bell Laboratories for cable circuits carrying voice signals overseas by analogue techniques. The concept exploits the brief silences in ordinary conversations to take a channel away from a momentarily silent speaker and assign it to someone who is speaking. The juggling is so swift and accurate that the speakers are unaware of it. The concept can now be executed with microcircuits and digital techniques, converting a large number of voice signals into digits in such a way as to considerably reduce the effective bit rate per voice channel.

Another concept that has been enhanced by microelectronics is stored-program control. First in the form of powerful processors associated with large equipment such as switching machines and more recently in the form of microprocessors embedded in all types of equipment, microelectronics has brought to communication a rapidly expanding array of programmable devices. The service and operational features of many large systems can now be altered by rewriting or adding to the program stored in the memory.

Until recently a communication system had to be either rewired or replaced to change the type of service offered. With stored-program control the old hardware can be programmed to perform new functions, which relate just as much to more efficient use of the people who operate and maintain the system as to service features seen by the user. The systems are almost always digital, employing a great deal of logic and memory. Microelectronics has vastly expanded the horizons of stored-program control by offering logic and memory circuits that are inexpensive and highly reliable.

Although stored-program control is a simple concept, the development of a program that is free of "bugs" is difficult. It is not unusual for the development of a program for a small microprocessor of the type that might be used in communication terminals to cost $100,000. The cost of developing a program for large equipment, such as an electronic switching system, runs to millions of dollars.

Silicon is to the electronics revolution what steel was to the Industrial Revolution. Seldom, however, can a communication system be built entirely of silicon circuits. Several other microelectronic technologies are important.

For example, a number of semiconductor materials (gallium arsenide and gallium phosphide among them) emit light when they conduct current. Light-emitting diodes serve widely as indicators and illuminators. They also display in numerals and letters the readout of digital signals. Such devices are extremely important in communication because they provide the translations re-

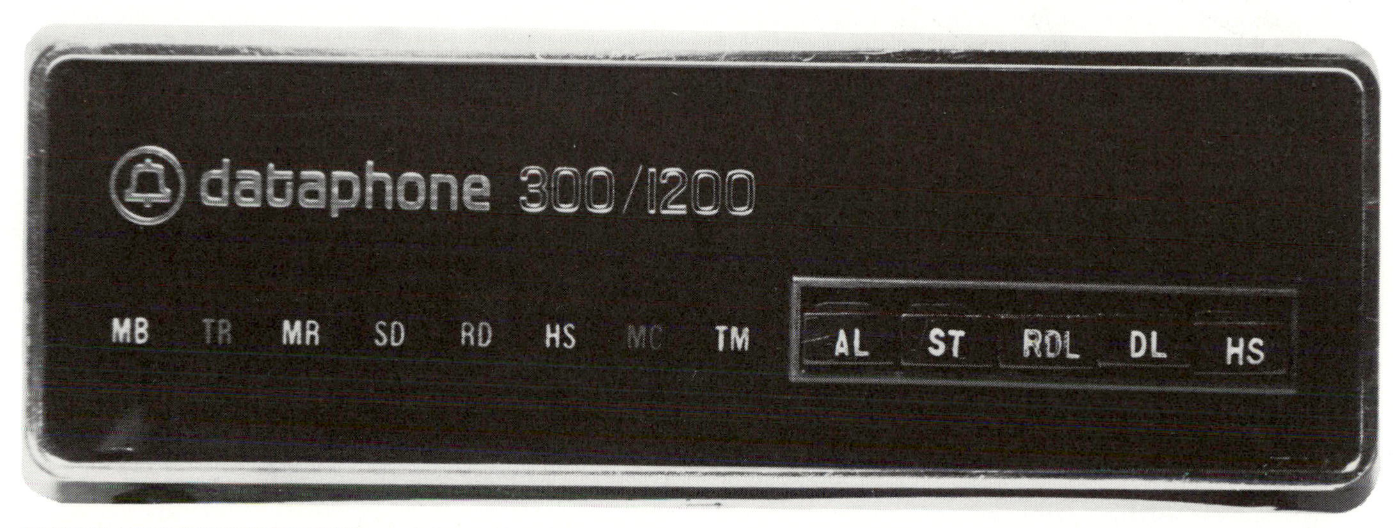

MODULATOR-DEMODULATOR made by the Bell System serves to transmit digital data, as from a computer or a business machine, over an analogue circuit of the type that carries most of the voice traffic over the telephone network in the U.S. The device is known as a modem (a contraction of modulator-demodulator) or a data set. At the receiving end a similar instrument makes the signals digital again.

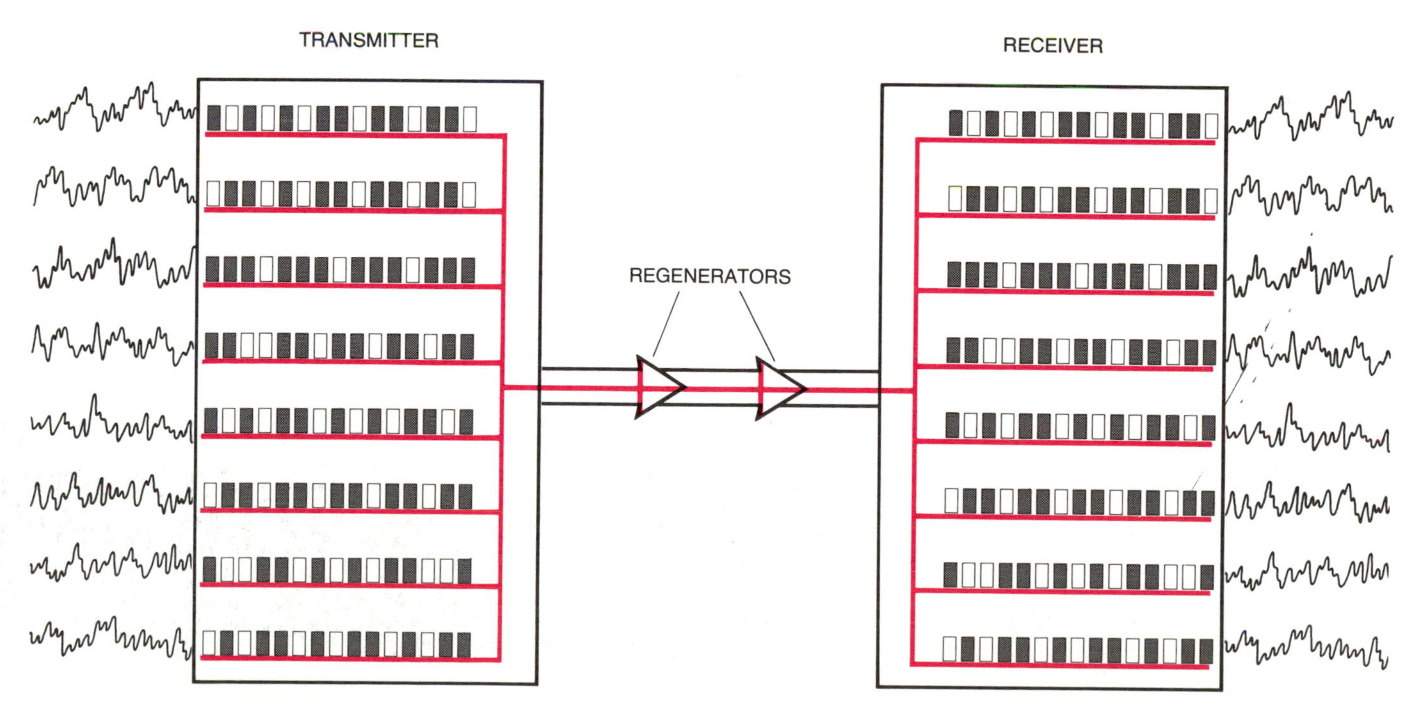

DIGITAL TRANSMISSION of voice signals is done increasingly in the U.S. telephone system as a result of developments in microelectronics. Sound waves (*left*) generated by voices are sampled at a high rate and the samples are converted into digital signals, which are transmitted over telephone lines as either "on" pulses (*white*) or "off" pulses (*black*). Periodically the signals are regenerated during transmission. At the receiving end the digital signals are reconverted into voice signals. Microcircuits play an important role not only in converting speech into digits but also in multiplexing a large number of conversations onto a cable and in regenerating signals during transmission.

quired to couple electrical signals to the human brain without the necessity of printing the results on paper.

Of even greater potential importance to communication is the solid-state laser. It is a sophisticated cousin of the light-emitting diode, being basically a complex structured diode that emits a highly collimated and monochromatic beam of light. It will be particularly useful in light-wave communication systems. The problem has been with the laser's reliability. Just five years ago the best solid-state lasers failed after only a few hours of operation. Now such lasers can be made to last a million hours, which is enough to make the device an attractive element in the telecommunication network.

Magnetic-bubble circuits are another promising product of communication research. Magnetic bubbles are small, cylindrical magnetic domains that can persist and move about in certain materials. A garnet of special composition has been fabricated to support magnetic bubbles that are a few micrometers in diameter. Indeed, the highest scale of circuit integration yet achieved has been by means of magnetic-bubble memory devices. Garnet chips with up to 64,000 bits of memory per chip can be fabricated routinely, and 256,000-bit chips have been made on an exploratory basis. Magnetic-bubble devices with four 64,000-bit chips per package are now providing service in the telephone network in the form of a machine with no moving parts that delivers recorded announcements. Magnetic bubbles have the highly desirable characteristic of not disappearing when power to the circuit is interrupted.

A functionally similar semiconductor arrangement is the charge-coupled device. It can transfer a chain of electric charges across a semiconductor in a way much like the motion of a train of magnetic bubbles in a garnet. Unfortunately the charge in a CCD does disappear when power is interrupted. The CCD has the advantage of using processing and lithographic techniques essentially the same as those employed for the fabrication of integrated circuits. Indeed, it is an integrated circuit of a special configuration. It offers the additional advantage of high mobility of charge carriers, leading to devices that are many times faster than the magnetic-bubble device. The technology of the charge-coupled device has been applied in the solid-state television camera and in serial memory devices. It also shows promise as a vehicle for integrated electronic filters in communication systems.

MICROELECTRONICS AND COMPUTER SCIENCE

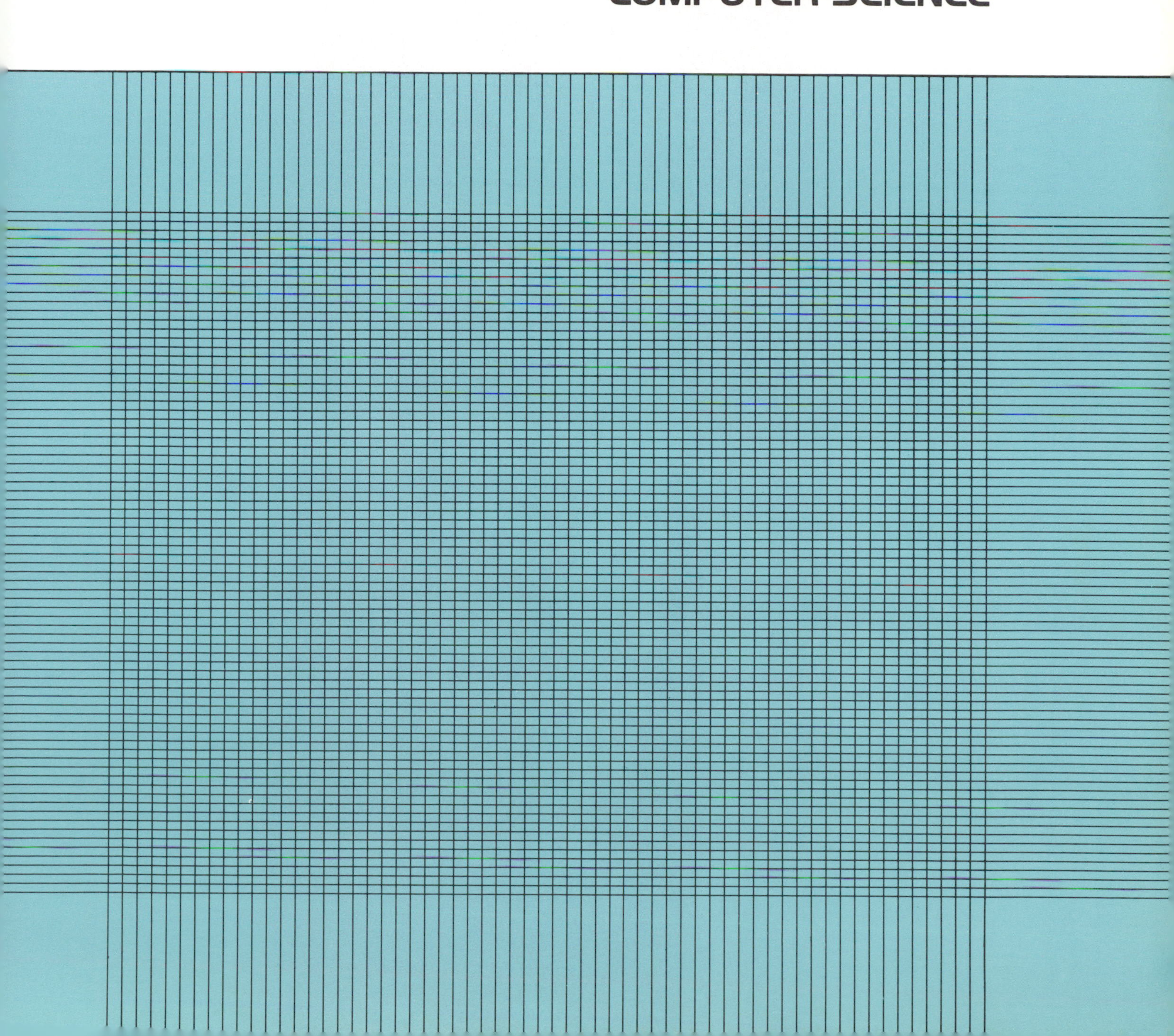

Microelectronics and Computer Science

by IVAN E. SUTHERLAND and CARVER A. MEAD

Large-scale integration makes logic elements fast and cheap, leaving movement of data still slow and expensive. New theories and designs are required, based on parallel processing and geometric regularity.

Computer science has grown up in an era of computer technologies in which wires were cheap and switching elements were expensive. Integrated-circuit technology reverses the cost situation, making switching elements essentially free and leaving wires as the only expensive component. In an integrated circuit the "wires," actually conducting paths, are expensive because they occupy most of the space and consume most of the time. Between integrated circuits the wires, which may be flat conducting paths on a printed circuit board, are expensive because of their size and delaying effect. Computer theory is just beginning to take the cost reversal into consideration. As a result computer design has not yet begun to take advantage of the full range of capabilities implicit in microelectronics. As we learn to understand the changed relative costs of logic and wiring and to take advantage of the possibilities inherent in large-scale integration we can expect a real revolution in computation, not only in the forms of computing machines but also in the theories on which their design and use are founded.

Why is it that computation theory needs to be revised? Suppose one sets out to develop some theories of computation, hoping to put them to work toward two ends: to establish upper bounds on what is computable and to serve as a guide to the design and use of computing machines. Such theories would presumably also advance understanding of computation processes and perhaps shed light on the nature of knowledge and thought. The theories might be based purely on mathematical reasoning or might also be based on fundamental physical principles. By mathematical reasoning alone one can prove many things about computers without resorting to physical principles. Only by attending to physical principles, however, can one make more quantitative statements about how long a computer of given physical dimensions must take to accomplish a given process, based on the fact that information cannot move around in the computer faster than the speed of light and that it takes a certain amount of matter, energy and space to represent one bit, or binary digit, of information with a given reliability.

Computer science as it is practiced today is based almost entirely on mathematical reasoning. It is concerned with the logical operations that take place in computing devices. It touches only lightly on the necessity to distribute logic devices in space, a necessity that forces one to provide communication paths between them. Computer science as it is practiced today has little to say about how the physical limitations to such communications bound the complexity of the computing tasks a physically realizable computer can accomplish.

That is so in part because anyone who thinks of a computer as a logical machine that performs logical, numerical or algebraic operations on data will naturally think of the machine in terms of the mathematical notation relevant to those fields. In such notations the symbol x written in one place on the page is identical in meaning with the symbol x written in another place on the page. The idea that communication in space is required if such values are to be identical as represented in a computer storage device has no place in the notation. The notation itself focuses attention on the logical operations, reflecting the fact that human beings think most effectively about only one thing at a time. A mathematical proof is a sequence of steps we absorb over a period of time, and it is easiest to think of computing devices that also do only one thing at a time. The sequential approach to mathematics is not required inside a computer, but the mathematical approach we normally take to problems does not encourage us to think of approaches other than sequential ones for the solution of problems. Nearly all computers in operation today perform individual steps on individual items of data one after another in time sequence.

It was appropriate to ignore the costs of communication when logic elements were slow and expensive and wires were relatively fast and cheap. Sequential machines are appropriate to such technologies because they can be built with a minimum number of switching elements. We have been led—by natural inclination, by our accustomed notations for mathematics and by technology—to develop a style of computing machines and a body of computing theory both of which are rendered obsolete by integrated-circuit technology. We have been able to ignore the limitations placed by physical principles on communications inside computers because those communications did not slow down our operations appreciably and were only a small part of the cost of the machines we built. By making logic elements essentially free and leaving communication cost the dominant factor, integrated-circuit technology forces us into a revolution not only in the kinds of machines we build but also in their theoretical basis.

Developing a new theoretical basis

"OM" CIRCUIT, an experimental microprocessor designed by the authors at the California Institute of Technology, is notable for its high degree of regularity, which makes it possible to pack more logic and memory functions on the chip. The main body of the chip (omitting the communication interfaces at top and bottom) is made up of 16 nearly identical columns in four groups of four; each column represents one bit of a 16-bit computer. About 40 percent of the chip (*lower portion*) is memory; middle 20 percent is the "shifter" section and top 20 percent is the arithmetic section.

for computer science will not be easy; indeed, the task has been put off in part because it is very difficult to combine notions of logic with notions of topology, time, space and distance, as a new theory will require. In this article we shall outline some of the elements such a theory must include, first by examining the inadequacies of a simple existing theory applicable to small logic networks. Then we shall see how the changes in the relative costs of wiring and logic must change the nature of the computers built in the future. Finally we hope to outline some elements we feel belong in a theory of computation appropriate to the new structures. Such a theory will be quite unlike the present basis of computer science, and so we feel justified in describing as revolutionary the effect of integrated-circuit technology both on the design of computing machines and on the intellectual framework within which such machines are exploited.

Most computer-science curriculums include a course in switching theory, even though it is largely irrelevant to the present-day practice of computer design. Switching theory, which was developed to help design the relay-operated switching networks of automatic telephone systems, provided guides that enabled a designer to formulate a network with the minimum number of relays for accomplishing some given logical operation. It has been extended to the design of networks of newer kinds of logic elements, for example a logic network with

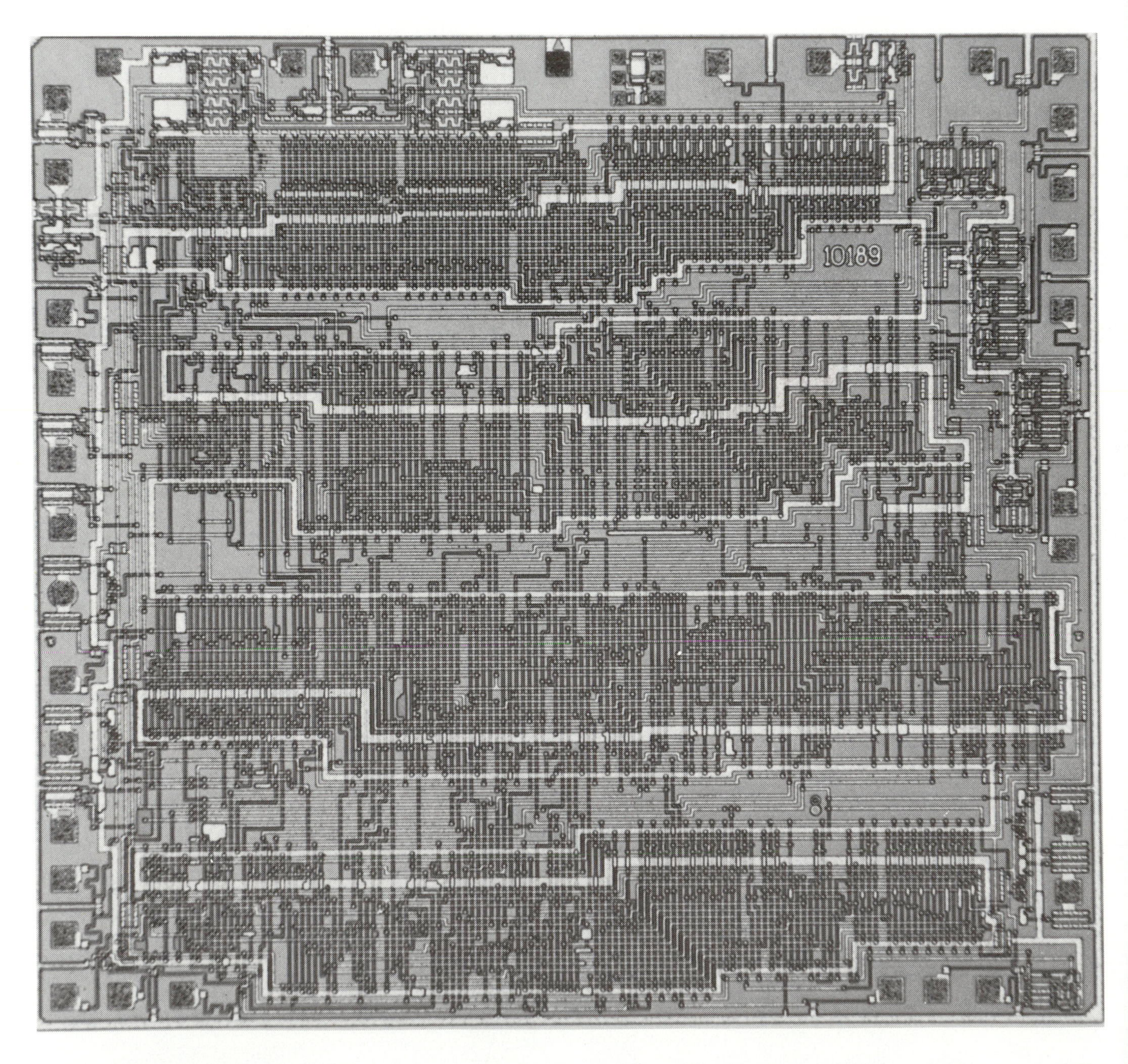

INTERCONNECTIONS among the logic elements of an integrated circuit have become more expensive than the elements themselves. Moreover, as the complexity of a randomly wired array of elements increases, the interconnections become longer and more numerous. Even at modest levels of complexity "wiring" occupies most of the available space on an integrated-circuit chip. This is a comparatively simple integrated circuit dating from about 1971. Note that the linear connectors running between active elements occupy most of the space.

the minimum number of conventional logic gates.

There is no guarantee, however, that such a minimum-number network will occupy the minimum space in an integrated circuit or perform its task in the minimum time. Integrated-circuit designers find they can often add transistors to a design and thereby save space or time, because adding to the minimum number may simplify the pattern of conductors in the design and may speed up its operation. Switching theory does minimize the number of switching components, but it ignores the cost and delay of the communication paths. In today's technology the area of a circuit devoted to communication between elements usually far exceeds the area devoted to switching elements, and communication delays are much longer than logic delays. What is needed, therefore, is a theory that minimizes the cost of computational tasks, considering not only the cost in area and time of the switching elements but also the much larger area and time costs of transporting data from one place to another. Because switching theory as it is known today is based on an obsolete cost function it is largely useless for the design of integrated circuits.

Switching theory is even less useful at the level of design where one is combining integrated circuits into a larger system. In most cases it costs much more to test, package and interconnect integrated circuits than to manufacture the circuits themselves. These costs are largely independent of the particular function of the circuit involved. Even if the cost of the circuits is ignored, communication from one integrated-circuit chip to another is much slower than communication on a single chip. Given a catalogue of standard circuits, there is great motivation to introduce more complex integrated circuits because fewer of them are required, so that the large cost of mounting and interconnecting them is reduced. In fact, designers often specify integrated circuits containing superfluous elements because there is no cost advantage to eliminating the unneeded switching elements. Switching theory has nothing to say about these important issues of cost and speed.

Although the cost of communication has so far found no real place in the theoretical results of computer science, it does play a role in the thinking of practical designers. Seymour Cray, the designer of many of the most powerful computers, cites the "thickness of the mat" and "getting rid of the heat" as the two major problems of machine design. It is obvious that controlling the geometry of the interconnections is essential. If connections can be made to follow regular patterns, they can be produced by less expensive methods and can also be

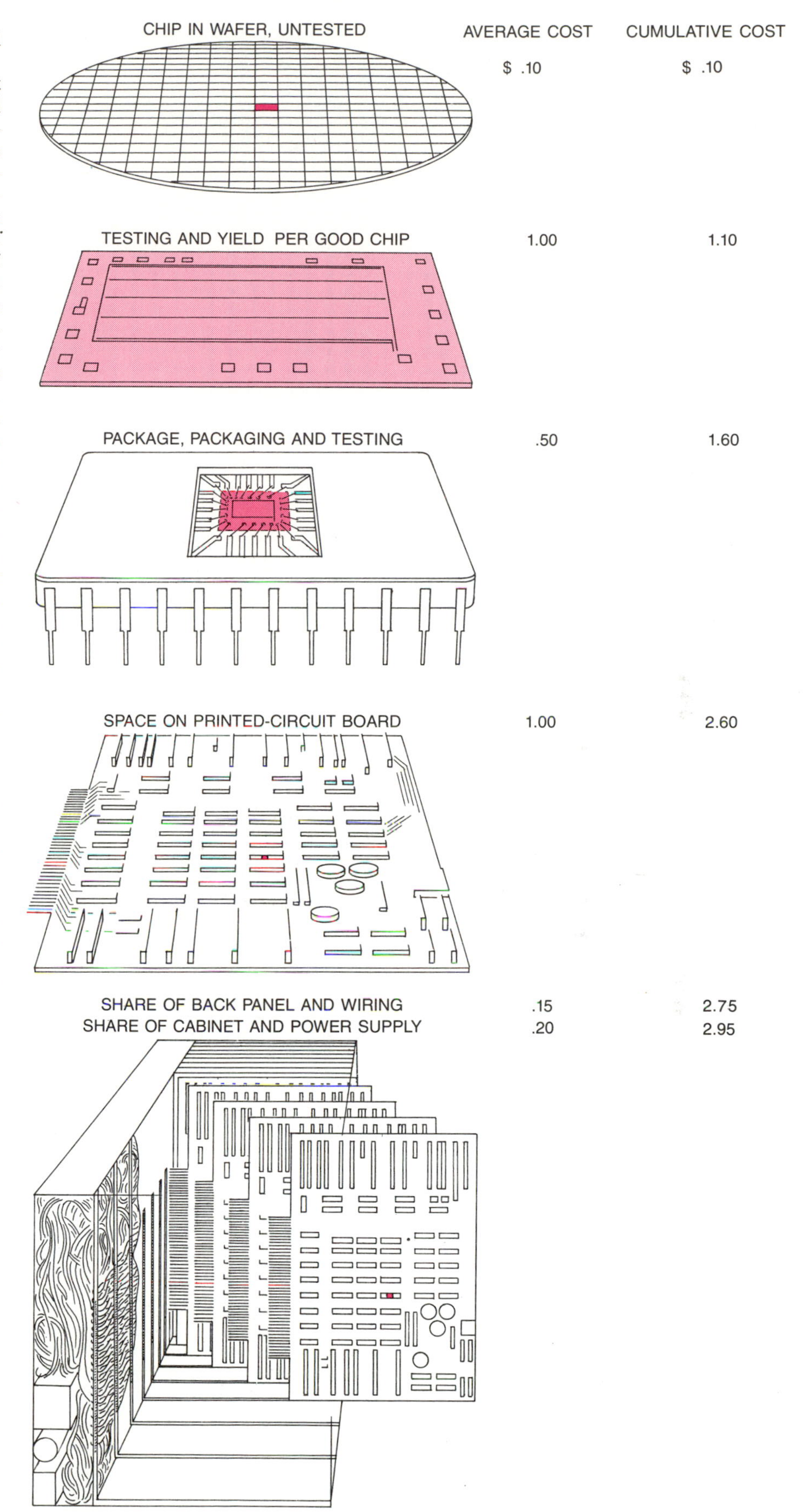

COST OF AN INTEGRATED CIRCUIT is a small part of the cost of a complete system. As is shown here, the cost of a single typical integrated-circuit die in a wafer is only 10 cents. Given about a 20 percent yield of good chips, after packaging and testing each good chip costs $1.60. Assuming that 100 chips are assembled on each of 20 printed-circuit boards, the cost per chip is almost doubled by each chip's share of board, back panel, cabinet and power supply.

THICK MAT OF WIRES covers the back panel of a large general-purpose computer, in this case the Cray Research, Inc., CRAY-1. Moving data over the wires takes time and costs money, and the thickness of the mat makes repairs difficult. Arranging elements of a computer so that wires are all parallel would greatly reduce the complexity and thickness of the mat.

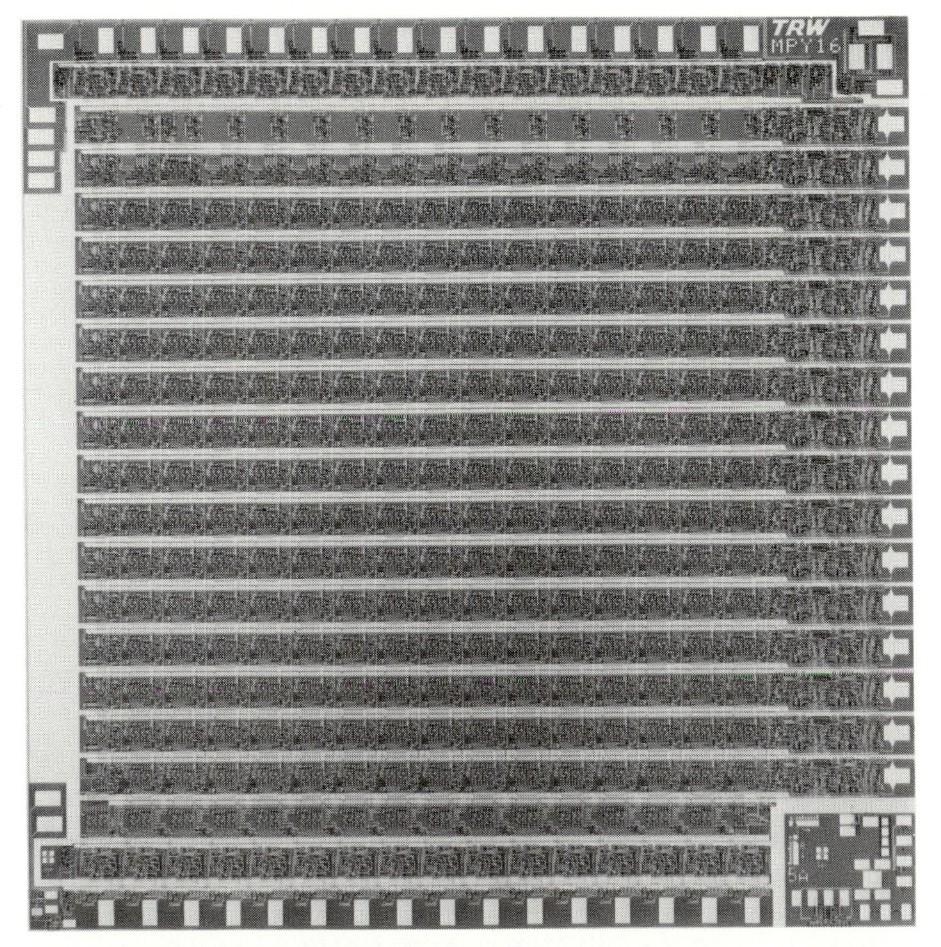

REGULARITY is a characteristic of memory circuits and of certain arithmetic circuits, such as this 16-bit multiplier array made by TRW Inc. The chip, about .28 inch square, contains more than 18,000 transistors and resistors. Regularity makes for high logic-element density.

made to occupy less space and so be faster.

If the geometry of interconnection paths is not carefully controlled, the space required for them grows more than linearly as the number of logic elements to be connected is increased. This nonlinear growth comes about because bigger systems require more wires, which are on the average also longer. Because the interconnection paths grow both in number and in length the total area or volume devoted to communication becomes disproportionately larger: to interconnect twice as many randomly placed devices requires four times as much communication space. To accommodate greater wiring space larger printed circuit boards must have wider spacing between components than small boards have; Los Angeles suffers more from freeway congestion than Plains, Ga., does.

Not only do longer communication paths occupy a disproportionate amount of space but also they function more slowly than short ones. That is because signals traveling even at the speed of light take some time to travel down a path and also because longer paths store more energy. (Inside integrated circuits the speed limit set by the speed of light is not yet an important issue because the distances are short compared with the switching times of the logic elements; the energy-storage delays, however, are important.) Before a signal path can be switched from one electrical state to another, the energy stored in the path must be removed and converted into heat. One must either design a larger driving circuit to provide for the larger power required to switch long wires quickly or suffer the delays of passing the larger amounts of energy through a less powerful driver. More powerful drivers must themselves be driven, and they are therefore not only larger in area but also inherently slower than small drivers.

Moreover, the heat generated by the more powerful drivers must be dissipated in some structure, which itself occupies space. It is quite possible that the signaling energies required in a given technology and the size of the structures provided to dissipate heat may set an upper limit to the complexity of the systems that can be built in that technology. Above such a limit the increase in wire length required to provide the space to house what is required to drive longer wires may exceed the original increase in length of wires that was made possible by the larger drivers! There is so far no theory addressing the limits to speed and complexity that may be imposed by this possibility.

The disproportionate growth of interconnections can be avoided by building

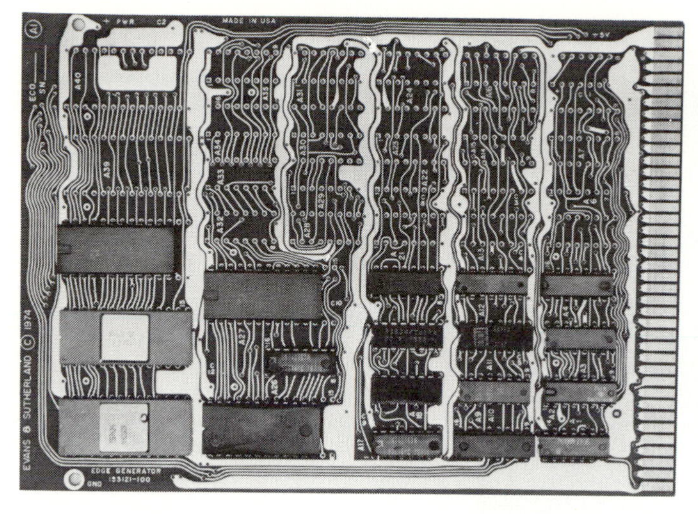

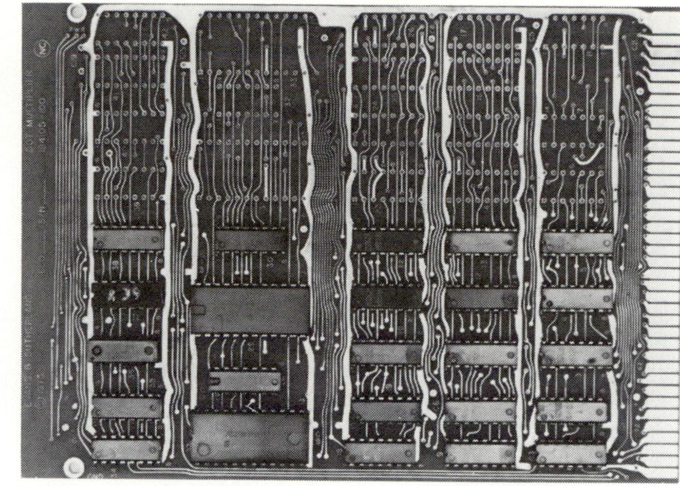

PRINTED-CIRCUIT BOARDS can also be designed to minimize the preponderance of communication paths. Regularity decreases the amount of wiring (*top*). A five-by-seven-inch board of irregular logic made by the Evans & Sutherland Computer Corporation (*top left*) is compared with a more regular memory board of the same size (*top right*). The larger the board, the greater the preponderance of wiring. **An 8½-by-10-inch board, the Digital Equipment Corporation's LSI-11 microcomputer, shows how much area is occupied on a conventional large board by communication paths (*bottom*). If the close packing characteristic of the regularly wired memory circuits that can be seen in the top right portion of board could have been attained throughout the board, the board would have been only half as big.**

very regular patterns of interconnection. There is already a trend toward very regular wiring patterns for integrated circuits and the interconnections among circuits. Read-only memories, for example, implement complex and irregular logic functions with a simple and very regular integrated circuit pattern. This regularity is desirable not only because it makes the specification of such functions simple but also because it may be the most efficient layout from an interconnection point of view. We believe regular patterns of wiring will play an increasing role in future designs. In part, computer science will become the study of the regularity of these structures.

COMMUNICATION "BUS" connects a computer's central processing unit to memory modules and other peripheral units (*top*). It is typically a flat cable of between 20 and 100 long wires that are tapped as they pass through each of the connected units. Photograph shows Digital Equipment Corporation's UNIBUS (*wide, light-colored flat cable*) connecting two disk memory units (*upper left*) with central processing unit (*lower right*) of DEC's PDP-11/40 computer.

The architecture of a typical computer includes a single logical processing element that communicates with a random-access memory through 20 to 100 long wires combined into a "bus," which, like its namesake, provides public transportation for data but is actually more like a telephone party line. The communication bus is often a flexible cable 50 to 100 feet long. A signaling protocol is specified for the bus so that all the units to which it is connected communicate in a common way and avoid interfering with one another. The great advantage of a bus structure in a computer is that any unit connected to the bus can communicate directly with all other units. Moreover, the protocol and the bus structure may survive several generations of hardware development, so that a line of computing equipment can adopt new storage devices, new input-output units and even new processing elements. In addition, the number of switching elements devoted to communication in each unit on the bus is minimized because each unit needs to communicate with only the one bus to send messages anywhere.

The drawback of the bus structure is that it provides a communication bottleneck. Consider a typical computer with, say, one million words (32 million bits) of integrated-circuit storage built out of 2,048 circuits that store some 16,000 bits each. Any one reference to memory can potentially sense the values of 128 bits on each of the 2,048 integrated circuits constituting the memory. Of these quarter-million bits to which access is obtained on the integrated circuits only 2,048 (one from each integrated circuit) are delivered outside the integrated-circuit package, and of these 2,048 only 32 are delivered over the communication bus to the logical processing unit of the computer. It is assumed that the communication bus connects the memory and the logical processing unit; we assert that in fact it separates them. Each memory access in a large computer wastes access to many thousands of bits

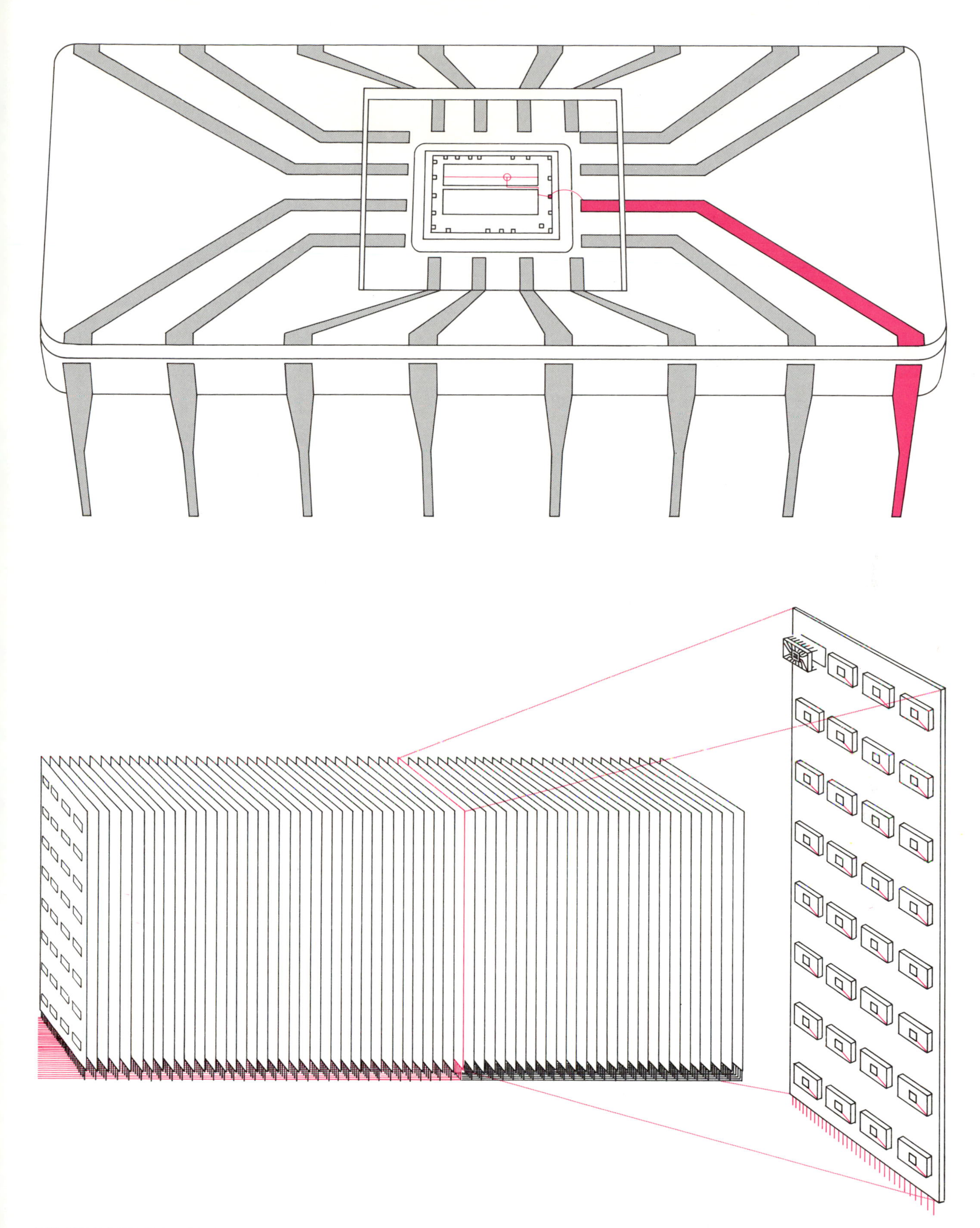

TYPICAL MEMORY CHIP has 16,384 bits arranged in a 128 × 128 array (*top*). An entire row of 128 bits can be accessed at one time, but a selector enables only a single bit to pass to an output pin (*dark color*). A typical memory system is made up of 2,048 such chips, say 64 groups of 32 (*bottom*). Only 32 chips can place their outputs on the 32 wires that join the bus to the central processor. Of the 262,144 (128 × 2,048) bits that moved less than a millimeter on each chip, only 2,048 moved three millimeters to get off their chip and only 32 moved a meter to the processor. In other words, the bus utilizes only about an eight-thousandth of the memory chips' available "bandwidth."

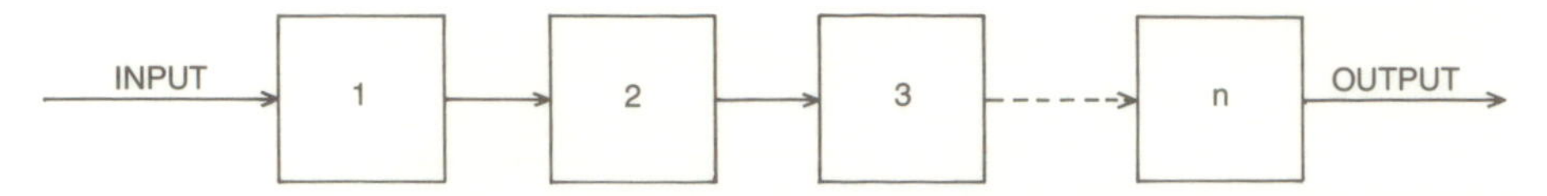

"PIPELINE" PROCESSOR is one of three kinds of parallel processor, illustrated on this page, that have been effective. In a pipeline processor data are passed along from one specialized processing element to the next, with each element performing a successive operation on the data. The pipeline is analogous to an assembly line: all operations are conducted simultaneously but not on the same material. The pipeline configuration is optimum as long as the same basic type of operation is to be performed; it is less effective when the operations are variable.

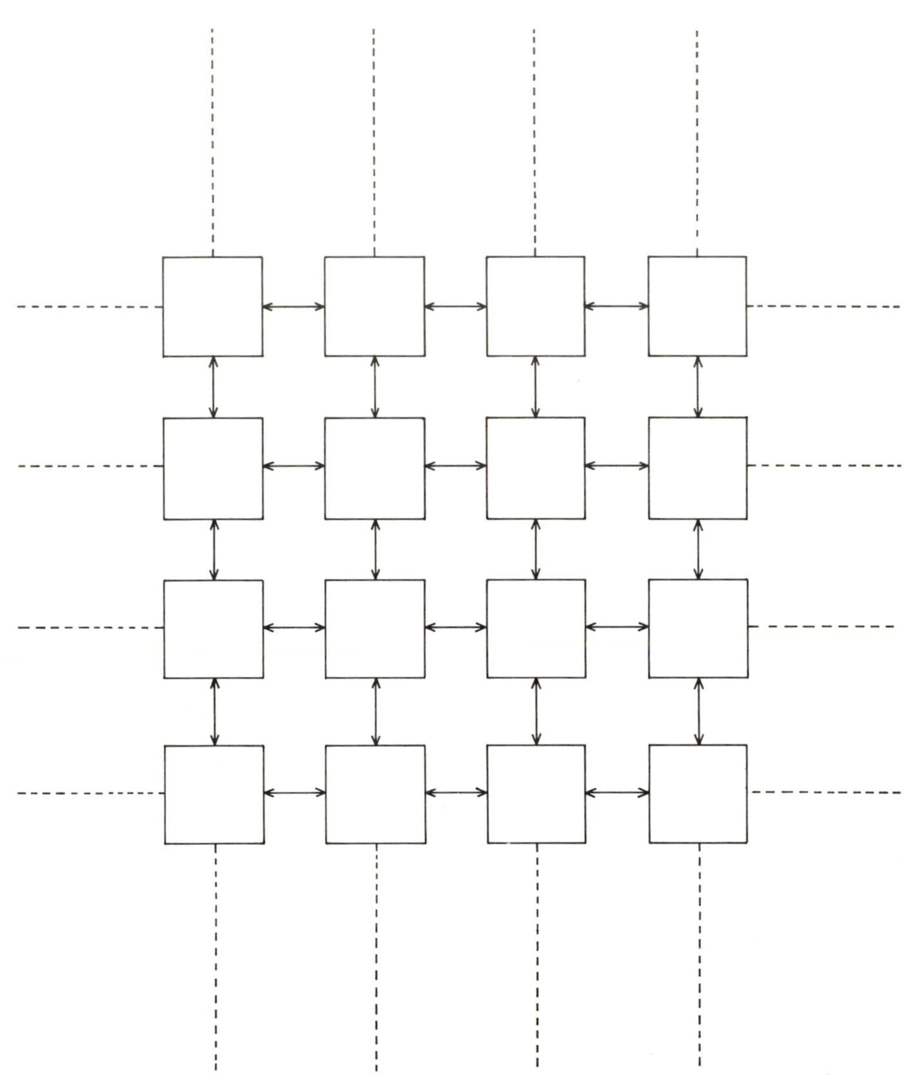

ARRAY PROCESSOR is effective when much identical processing is to be done on many items of data. All the processors receive the same instructions, like a company of soldiers drilling "by the numbers." The limitation here is that individual computations must depend only on the data in a particular element and its immediate neighbors. This can be effective, however, in operations such as weather simulation, where local atmospheric interactions are significant.

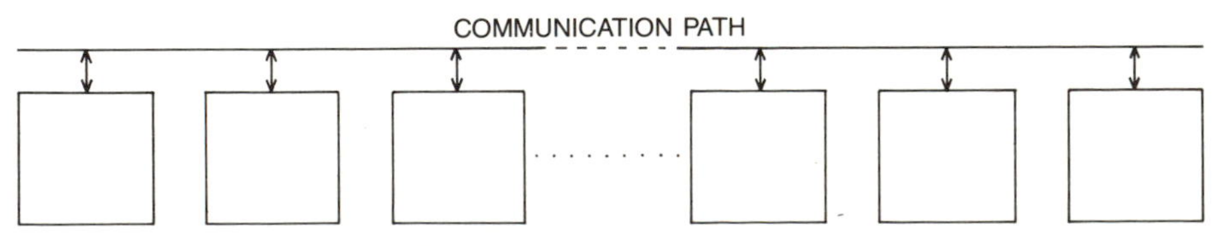

INDEPENDENT PROCESSORS connected by a communication path constitute the most flexible arrangement for parallel execution of different operations. Tasks are given to each processor as is required, as they are to the individual workers in a cottage industry. The system works best when each element can do much processing and need not communicate much with other elements; bottlenecks develop when tasks require elements to wait for the party line.

by selecting only a few bits to send over the memory bus to the central processing unit. This waste is tolerated for two reasons. First, it simplifies our conception of the machine and matches it to our natural inclination to do one thing at a time. Second, it provides a single, simple interface between various parts of the machine.

We pay a high price for this convenience. In an age when memories and logical processing elements were made by different technologies, we had little choice. Now, however, with the silicon integrated circuit dominating both the memory and the logical processing tasks in computers, there is little justification for continuing to accept such waste. Now it is possible to distribute the memory bus over many thousands of integrated circuits, in effect giving each logic element the memory it needs by moving information less than a millimeter from memory to processing facilities located together in the same integrated circuits on each of many thousands of chips. We are just beginning to explore systems with this unconventional architecture. To employ them effectively we must learn how to match the complexities of given problems to the simple fixed patterns of communication provided in the systems we can build.

Machines in which large numbers of logic elements operate simultaneously are called parallel processors. (To some extent, to be sure, every computing machine is a parallel processor. The separate bits that together represent a number are moved simultaneously on parallel communication paths; binary addition is performed by an adder circuit that operates on all the bits of the number at once; multiplication is performed either by sequential addition or, in faster models, by including a number of separate adder circuits and operating them in parallel. Levels of parallelism above basic arithmetic, however, are rare in today's computers.) The simplest form of real parallel processing now available has a few independent processors operating on a common memory; a typical large computing system has from two to a few dozen processors at work. Typically, however, these processors serve quite independent functions and their very existence may be hidden from the user. For example, a separate processor may be involved in communication with the user's keyboard, in operating magnetic-tape or magnetic-disk input-output units or in scheduling the resources of the central processor. Such "multiprocessing" systems have little impact on the user's algorithms.

Three kinds of systems that can truly be classed as parallel processors have been built. In one of them, the "pipeline" processor, several processing elements,

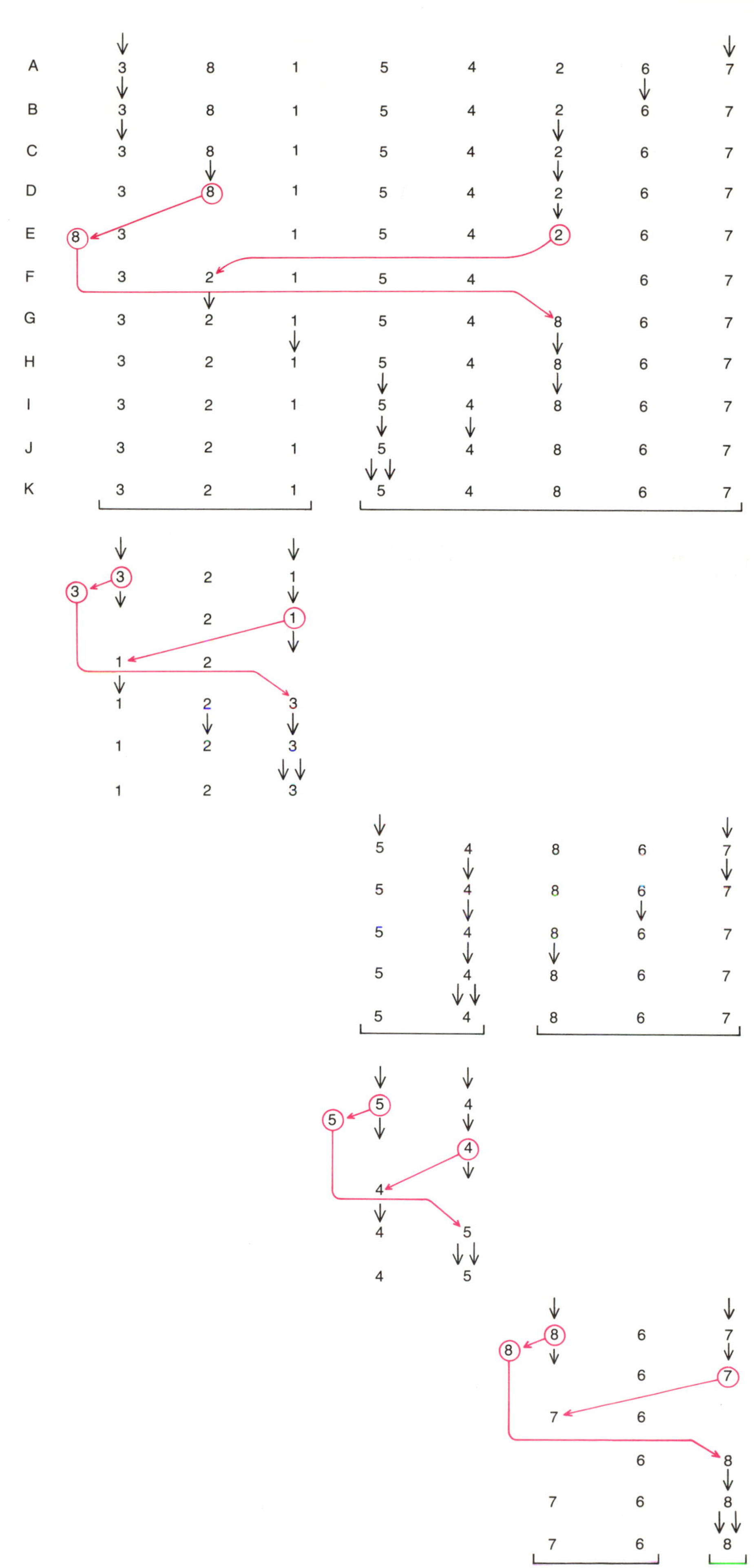

each of which is specialized for some particular task, are connected in sequence. The work to be processed flows through these processors much as workpieces move along an assembly line. Communication is simple because information flows along a fixed pathway and has only a short distance to move between processing steps. A pipeline processor gains efficiency for the same reasons an assembly line does: functions are specialized and communications are minimized. Pipelining enables the arithmetic sections of very fast computers to process sequences of numbers with a high overall speed. Pipeline processors are less effective where the tasks to be performed are highly variable.

In a second form of parallel processor many identical processing elements are brought to bear on separate parts of a problem under the control of a single instruction sequence. Several such machines have been built, of which the largest and best known is ILLIAC IV. A modern parallel processor of this type was proposed at a recent Rand Corporation workshop on hydrodynamic simulations. In this hypothetical machine there would be 10,000 processors, each with arithmetic capability and memory, each built on a single integrated circuit and all under the command of a common instruction device. All the processors would execute commands in rigid lock-step. The processors would be arranged in a square array, 100 × 100, and each would communicate data only with adjacent processors to its north, south, east and west in the array, with relatively slow bit-serial communication on a single wire in each direction. We estimate that such a machine would take about five microseconds to communicate a single 64-bit number from one processor to its neighbor, which is very slow by today's standards. Of course, it could communicate 10,000

"QUICKSORT" is a typical sequential algorithm for arranging numbers in ascending order. Numbers pointed to by arrows are compared with the number farthest to the left. If the pointer farthest to the right indicates a number greater than the reference value (*row A*), it is advanced to the left until it rests on a number less than the reference value (*C*). Then the left pointer is advanced to the right until it rests on a number greater than the reference value (*D*). At that stage the numbers pointed to are interchanged (*E–G*). The process is continued until the pointers rest on the same number (*K*); at that stage all numbers to right of pointers are greater than the reference value, and all numbers to left are less than or equal to it. The same algorithm is then applied to each subset; to complete the sorting illustrated here requires five more steps than are shown. To sort n numbers requires $n(\log_2 n)$ comparisons of numbers that may be stored in distant locations. The communication cost is the dominant cost of executing the algorithm.

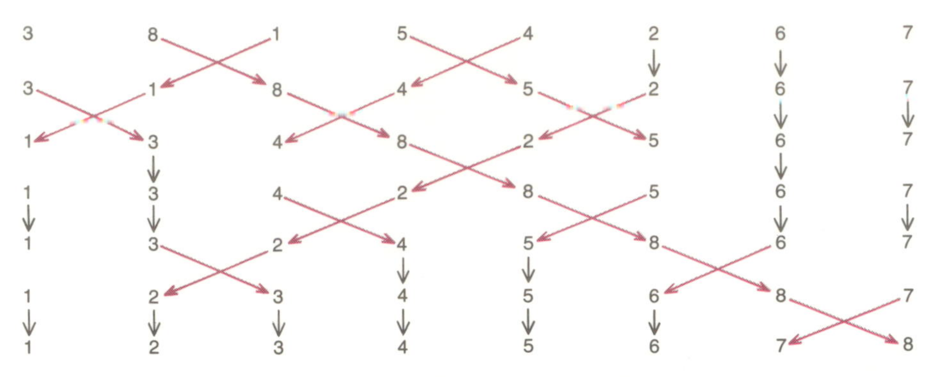

PARALLEL EXECUTION speeds the sorting task. In this algorithm adjacent members of number pairs are compared and are interchanged if the left-hand member is larger than the right-hand one. (In the first row a "pair" is defined as two numbers the left one of which is in an even column; in the second row the left member of each pair is in an odd column, and so on alternatingly.) Sorting the entire set requires $n^2/2$ comparisons, always of nearby numbers. Interchange sorting has been considered slow, but if comparison and interchange elements are attached to each memory element in integrated circuits, comparisons required for one sweep can be accomplished in one memory cycle and the sorting can be completed in n cycles at most.

such numbers in any one five-microsecond period. It would also take about five microseconds to perform a multiplication, but again it could perform 10,000 multiplications in that time, for an average rate of two billion multiplications per second. Machines such as this one are called array processors or single-instruction-stream, multiple-data-stream machines. They are most suitable for highly regular tasks such as hydrodynamic computations, numerical simulation of the weather and the inversion of large matrixes.

A third type of parallel processor is one where separate, independent processors under separate, self-contained control structures perform independent parts of the task, communicating data and instructions as is required. The advent of the microprocessor has, of course, suggested to many people the possibility of making systems consisting of thousands of separate microprocessors and having them work in concert on large tasks. Few such multiple-instruction-stream, multiple-data-stream machines have been built, and their properties are poorly understood.

The challenge in designing or using a parallel processor of any of these three types lies in discovering ways in which simple patterns of communication within the processor can be made to match the communication tasks inherent in the problem being solved. As integrated-circuit technology progresses there will be individual circuits of increasing speed and complexity. No relief is in sight, however, for the costs and delays inherent in communicating information from one circuit to another. To provide better communication will require more connections to the integrated circuit, a bigger housing for it, more or larger communication-driving circuits and consequently more heat dissipation. To obtain maximum performance from large computing systems programmers will have to face up to the limitations on communication that are imposed by physical reality. High-performance communication cannot be provided from every element to every other element; the programmer will have to match his formulation of the problem to the available communication paths. Although this is a difficult task, success in accomplishing it will provide unprecedented processing power.

We believe that just as an important part of today's computer science concerns itself with sequences of instructions distributed in time, so an important aspect of computer science in the future will be the study of sets of communications distributed in space. If processors can communicate only with their nearest neighbors, what kinds of arrangements are possible? Obviously one can wire processors in a linear string. Such processors can operate in the pipeline fashion described above, with each one passing data along to the next, or by performing common operations under command from a central instruction device. Such near-neighbor connections are highly effective for tasks such as sorting, in which the local communication of data suffices. Alternatively, one can connect processors in an array, with each processor having more than two neighbors. Such arrays have a basic structure much like the structure of a crystal, and various forms of local communication are possible. In our laboratory at the California Institute of Technology we are considering the properties of the different communication paths that might be included in such structures.

Some years ago R. S. Gaines and C. Y. Lee, who were then working at Bell Laboratories, described three types of interconnection path. One kind of interconnection has connections that are common to all processors; it is effective for sending commands to the processors and for "broadcasting" to all processors certain values that may be important in the course of a computation. A second kind of communication path enables each processor to "talk" simultaneously to its neighbors. Such a path can handle the communications required in pipelining or, if each processor has its own storage function, open up a space anywhere in the store by moving information simultaneously away from the location of the desired gap. A third kind of communication path enables the processing elements to say something collectively about their results. Such a path can indicate whether no processor, one processor or more than one contains a given condition, which of the processors contains the smallest value or which are between the beginning and end of a particular string.

In our laboratory we have taken on the task of building and using some simple parallel processors involving one-, two- and three-dimensional interconnection patterns. We hope to learn more about the relation of communication paths to the performance of such processors. We have become convinced that the performance of parallel processors can depend critically on the design of communication paths that enable processing elements to make collective statements about their actions. Without such paths how does one find the smallest value stored in the array? How can one identify the set of processors that lie between two processors with designated properties? How does one obtain answers from a number of processors in sequence when more than one of them has something to report? Such paths are electrically complex. Either they involve each local processor as the driving element in a "global" communication task or they require intermediate circuitry specialized for collecting such information, and such intermediate circuitry inevitably introduces time delays and cost. The best structures for this kind of communication appear to be similar to those in the carry circuits of fast parallel adders, but the communication costs of such circuits have not yet been adequately analyzed.

A cornerstone of computer science today is the theoretical analysis of sequential algorithms. There is a large and growing body of theory for selecting efficient algorithms for sequential machines. This body of theory, as one might expect, focuses on algorithms that minimize the number of logical operations required to accomplish some task. For example, it has been shown that for putting numbers into sequence, "quicksort" algorithms are the best to use because they require only $n(\log_2 n)$ comparisons. This body of theory assumes that all data elements in storage are equally accessible and that the movement of data is free.

Data elements in storage are never really equally accessible, although they can arbitrarily be made equally inaccessible in a random-access device by making the access time for all elements as slow as the time required for the most inaccessible element. Because information in a random-access storage device must be moved over long distances the data rate in a random-access storage device of a given technology is inevitably lower than what can be achieved with a more orderly sequential-access mechanism. Moreover, transporting data from a memory cell to a comparison circuit is never really free; in most machines the transportation time far exceeds the comparison time. And so, for example, an analysis of algorithms seeking to show that a particular sorting algorithm is best is based on giving what may be a less than optimum machine the task of performing that algorithm; given a different structure, sorting might be done much faster.

Work on the theory of algorithms has not yet focused on the true relation of computing costs to communication costs. Given that one starts with a blank piece of silicon and is free to place wires, logic gates and so on anywhere one chooses, what choices should one make to accomplish a given computation task in the least time or on the smallest possible area of silicon? We have no basis at all for making sensible choices as to the computing structures we should build. We do have a large body of experience with a particular structure: sequential machines with random-access storage. It may be that such machines are effective because overall they best match a wide variety of computing tasks. There is mounting evidence, however, that a parallel structure can outperform the standard computer by many orders of magnitude on tasks that are suitable for parallel execution. As such unconventional structures have appeared, a wider range of tasks is being discovered for which they are suitable. Certainly one would expect to obtain better performance by paying attention to the real costs of systems, which is to say to communication, than by simply considering the cost—now an almost vanishing cost—of logical processing.

We believe adequate theories that account properly for the costs of communication will be an important guide for designing the machines that have been made possible by the integrated-circuit revolution. We believe such theories will have their basis in the study of regularity, so that computer science will come to include a body of theory akin to that of topology or crystallography. Although such a development is revolutionary in some ways, it is essentially a continuation of the search for regularity in all programming tasks. Computer scientists will simply add geometric regularity to the logical regularity they have already come to know and value.

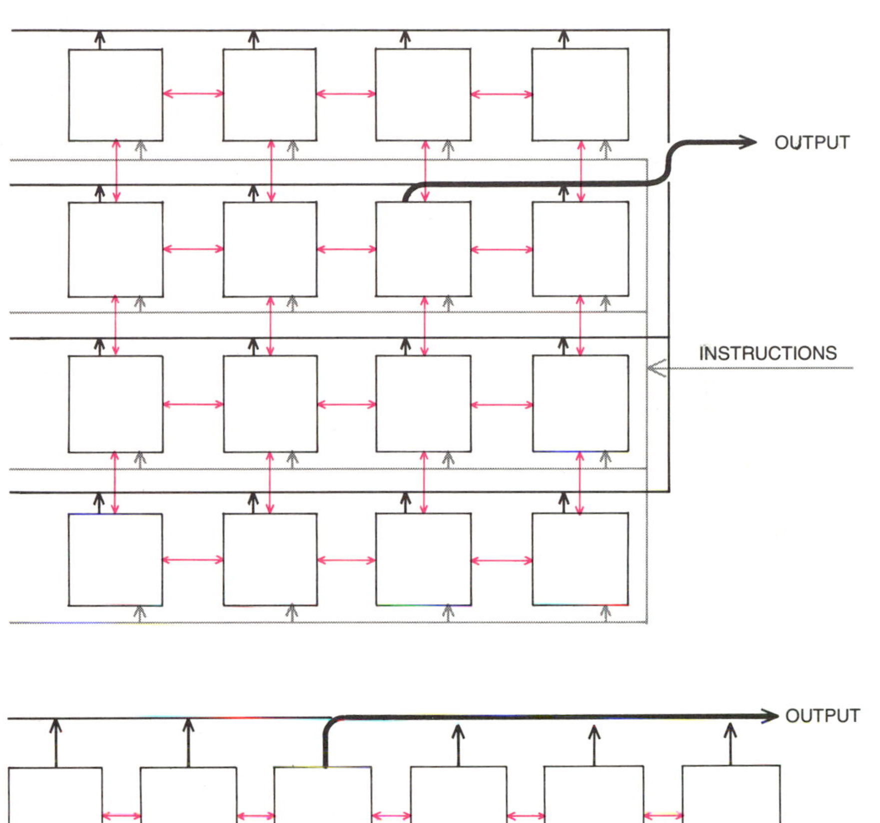

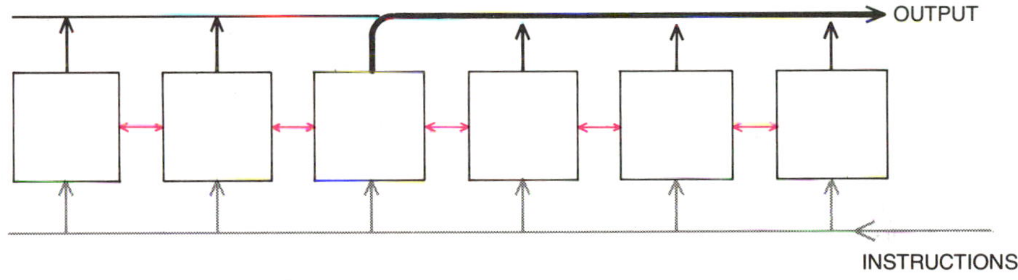

WIRES THAT INTERCONNECT MODULES of an array processor can be exploited in three distinct ways. One type of connection "broadcasts" information from a control center to all modules (*gray*). A second type (*black*) moves information from a selected module to a control destination, one module at a time (*heavy black line*). A third type passes data from each module to its nearest neighbor (*color*); in this case all the modules can "talk" at the same time.

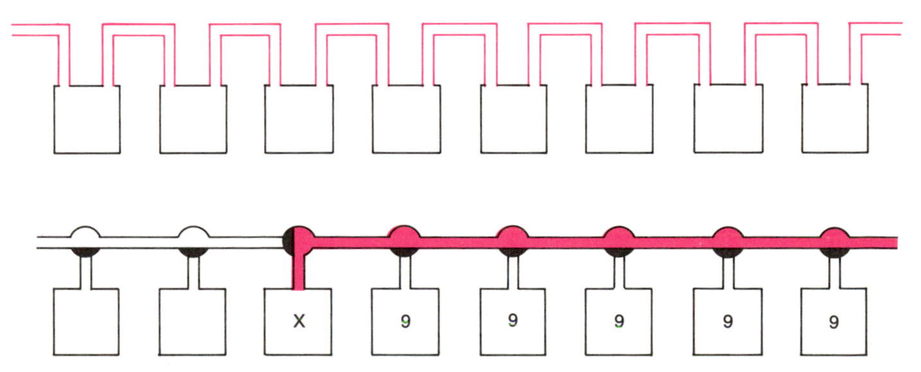

TWO SUBTYPES of the third type of wiring are in common use. In one subtype (*top*) information passes into a module during one step, is processed and then passed on to the next module during the next step. The other subtype (*bottom*) is designed merely to "discover" something about a module (or a number of modules collectively) and to do so quickly: information is moved past a module unchanged during a single processing step, provided that the module is in a specified state. Such wiring might discover, for example, where the 9's are in a parallel adder.

MICROELECTRONICS AND THE PERSONAL COMPUTER

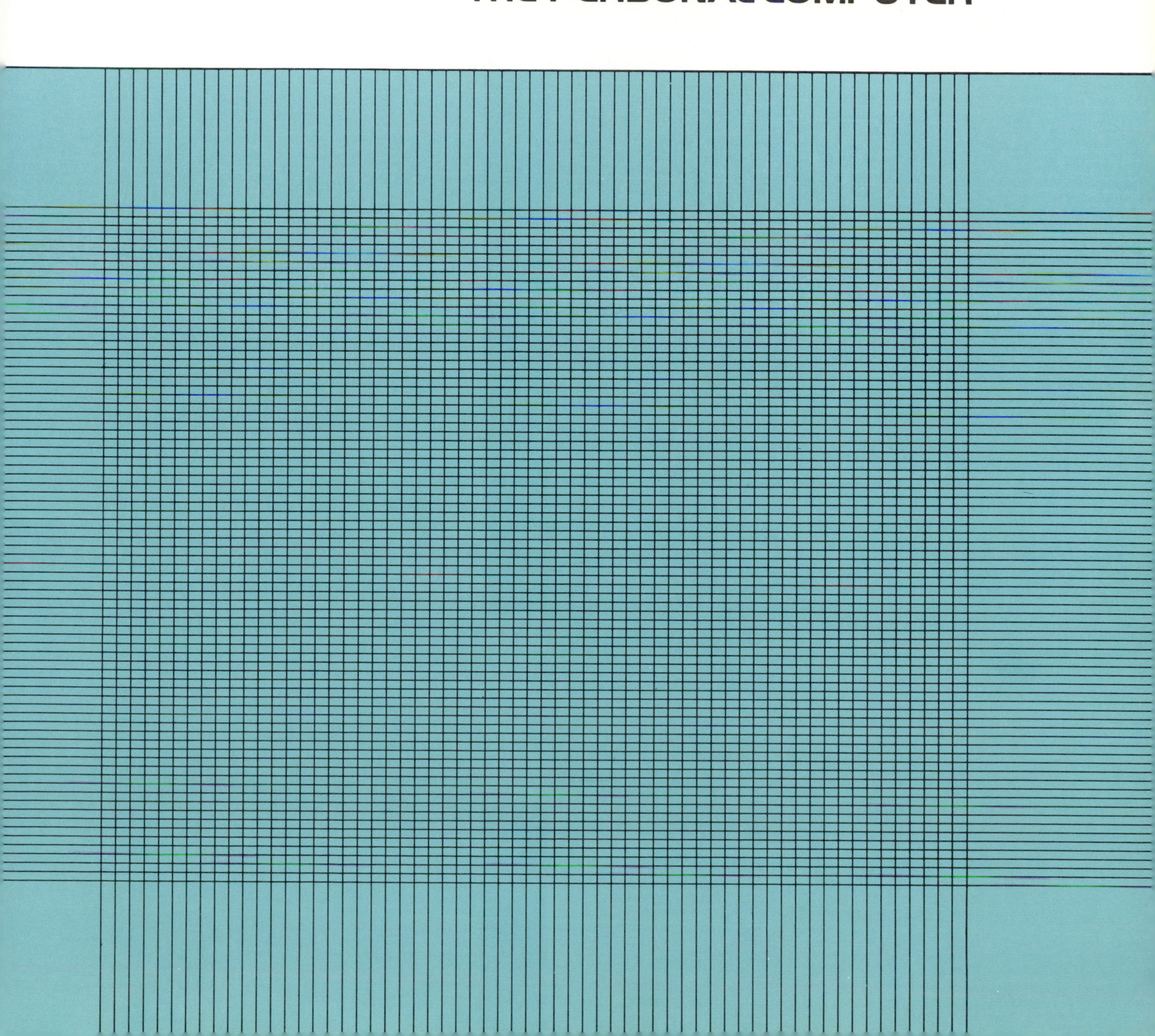

Microelectronics and the Personal Computer

by ALAN C. KAY

Rates of progress in microelectronics suggest that in about a decade many people will possess a notebook-size computer with the capacity of a large computer of today. What might such a system do for them?

The future increase in capacity and decrease in cost of microelectronic devices will not only give rise to compact and powerful hardware but also bring qualitative changes in the way human beings and computers interact. In the 1980's both adults and children will be able to have as a personal possession a computer about the size of a large notebook with the power to handle virtually all their information-related needs. Computing and storage capacity will be many times that of current microcomputers: tens of millions of basic operations per second will manipulate the equivalent of several thousand printed pages of information.

The personal computer can be regarded as the newest example of human mediums of communication. Various means of storing, retrieving and manipulating information have been in existence since human beings began to talk. External mediums serve to capture internal thoughts for communication and, through feedback processes, to form the paths that thinking follows. Although digital computers were originally designed to do arithmetic operations, their ability to simulate the details of any descriptive model means that the computer, viewed as a medium, can simulate any other medium if the methods of simulation are sufficiently well described. Moreover, unlike conventional mediums, which are passive in the sense that marks on paper, paint on canvas and television images do not change in response to the viewer's wishes, the computer medium is active: it can respond to queries and experiments and can even engage the user in a two-way conversation.

The evolution of the personal computer has followed a path similar to that of the printed book, but in 40 years rather than 600. Like the handmade books of the Middle Ages, the massive computers built in the two decades before 1960 were scarce, expensive and available to only a few. Just as the invention of printing led to the community use of books chained in a library, the introduction of computer time-sharing in the 1960's partitioned the capacity of expensive computers in order to lower their access cost and allow community use. And just as the Industrial Revolution made possible the personal book by providing inexpensive paper and mechanized printing and binding, the microelectronic revolution of the 1970's will bring about the personal computer of the 1980's, with sufficient storage and speed to support high-level computer languages and interactive graphic displays.

Ideally the personal computer will be designed in such a way that people of all ages and walks of life can mold and channel its power to their own needs. Architects should be able to simulate three-dimensional space in order to reflect on and modify their current designs. Physicians should be able to store and organize a large quantity of information about their patients, enabling them to perceive significant relations that would otherwise be imperceptible. Composers should be able to hear a composition as they are composing it, notably if it is too complex for them to play. Businessmen should have an active briefcase that contains a working simulation of their company. Educators should be able to implement their own version of a Socratic dialogue with dynamic simulation and graphic animation. Homemakers should be able to store and manipulate records, accounts, budgets, recipes and reminders. Children should have an active learning tool that gives them ready access to large stores of knowledge in ways that are not possible with mediums such as books.

How can communication with computers be enriched to meet the diverse needs of individuals? If the computer is to be truly "personal," adult and child users must be able to get it to perform useful activities without resorting to the services of an expert. Simple tasks must be simple, and complex ones must be possible. Although a personal computer will be supplied with already created simulations, such as a general text editor, the wide range of backgrounds and ages of its potential users will make any direct anticipation of their needs very difficult. Thus the central problem of personal computing is that nonexperts will almost certainly have to do some programming if their personal computer is to be of more than transitory help.

To gain some understanding of the problems and potential benefits of personal computing my colleagues and I at the Xerox Palo Alto Research Center have designed an experimental personal computing system. We have had a number of these systems built and have studied how both adults and children make use of them. The hardware is faithful in capacity to the envisioned notebook-

COMPUTER SIMULATIONS generated on a high-resolution television display at the Evans & Sutherland Computer Corporation show the quality of the images it should eventually be possible to present on a compact personal computer. The pictures are frames from two dynamic-simulation programs that revise an image 30 times per second to represent the continuous motion of objects in projected three-dimensional space. The sequence at the top, made for the National Aeronautics and Space Administration, shows a space laboratory being lifted out of the interior of the space shuttle. The sequence at the bottom, made for the U.S. Maritime Administration, shows the movement of tankers in New York harbor. Ability of the personal computer to simulate real or imagined phenomena will make it a new medium of communication.

size computer of the 1980's, although it is necessarily larger. The software is a new interactive computer-language system called SMALLTALK.

In the design of our personal computing system we were influenced by research done in the late 1960's. At that time Edward Cheadle and I, working at the University of Utah, designed FLEX, the first personal computer to directly support a graphics- and simulation-oriented language. Although the FLEX design was encouraging, it was not comprehensive enough to be useful to a wide variety of nonexpert users. We then became interested in the efforts of Seymour A. Papert, Wallace Feurzeig and others working at the Massachusetts Institute of Technology and at Bolt, Beranek and Newman, Inc., to develop a computer-based learning environment in which children would find learning both fun and rewarding. Working with a large time-shared computer, Papert and Feurzeig devised a simple but powerful computer language called LOGO. With this language children (ranging in age from eight to 12) could write programs to control a simple music generator, a robot turtle that could crawl around the floor and draw lines, and a television image of the turtle that could do the same things.

After observing this project we came to realize that many of the problems involved in the design of the personal computer, particularly those having to do with expressive communication, were brought strongly into focus when children down to the age of six were seriously considered as users. We also realized that children require more computer power than an adult is willing to settle for in a time-sharing system. The best outputs that time-sharing can provide are crude green-tinted line drawings and square-wave musical tones. Children, however, are used to finger paints, color television and stereophonic records, and they usually find the things that can be accomplished with a low-capacity time-sharing system insufficiently stimulating to maintain their interest.

Since LOGO was not designed with all the people and uses we had in mind, we decided not to copy it but to devise a new kind of programming system that would attempt to combine simplicity and ease of access with a qualitative improvement in expert-level adult programming. In this effort we were guided, as we had been with the FLEX system, by the central ideas of the programming language SIMULA, which was developed in the mid-1960's by Ole-Johan Dahl and Kristen Nygaard at the Norwegian Computing Center in Oslo.

Our experimental personal computer

EXPERIMENTAL PERSONAL COMPUTER was built at the Xerox Palo Alto Research Center in part to develop a high-level programming language that would enable nonexperts to write sophisticated programs. The author and his colleagues were also interested in using the experimental computer to study the effects of personal computing on learning. The machine is completely self-contained, consisting of a keyboard, a pointing device, a high-resolution picture display and a sound system, all connected to a small processing unit and a removable disk-file memory. Display can present thousands of characters approaching the quality of those in printed material.

is self-contained and fits comfortably into a desk. Long-term storage is provided by removable disk memories that can hold the equivalent of 1,500 printed pages of information (about three million characters). Although image displays in the 1980's will probably be flat-screened mosaics that reflect light as liquid-crystal watch displays do, visual output is best supplied today by a high-resolution black-and-white or color television picture tube. High-fidelity sound output is produced by a built-in conversion from discrete digital signals to continuous waveforms, which are then sent to a conventional audio amplifier and speakers. The user makes his primary input through a typewriterlike keyboard and a pointing device called a mouse, which controls the position of an arrow on the screen as it is pushed about on the table beside the display. Other input systems include an organlike keyboard for playing music, a pencillike pointer, a joystick, a microphone and a television camera.

The commonest activity on our personal computer is the manipulation of simulations already supplied by the SMALLTALK system or created by the user. The dynamic state of a simulation is shown on the display, and its general course is modified as the user changes the displayed images by typing commands on the keyboard or pointing with the mouse. For example, formatted textual documents with multiple typefaces are simulated so that an image of the finished document is shown on the screen. The document is edited by pointing at characters and paragraphs with the mouse and then deleting, adding and restructuring the document's parts. Each change is instantly reflected in the document's image.

In many instances the display screen is too small to hold all the information a user may wish to consult at one time, and so we have developed "windows," or simulated display frames within the larger physical display. Windows organize simulations for editing and display, allowing a document composed of text, pictures, musical notation, dynamic animations and so on to be created and viewed at several levels of refinement. Once the windows have been created they overlap on the screen like sheets of paper; when the mouse is pointed at a partially covered window, the window is redisplayed to overlap the other windows. Those windows containing useful but not immediately needed information are collapsed to small rectangles that are labeled with a name showing what information they contain. A "touch" of the mouse causes them to instantly open up and display their contents.

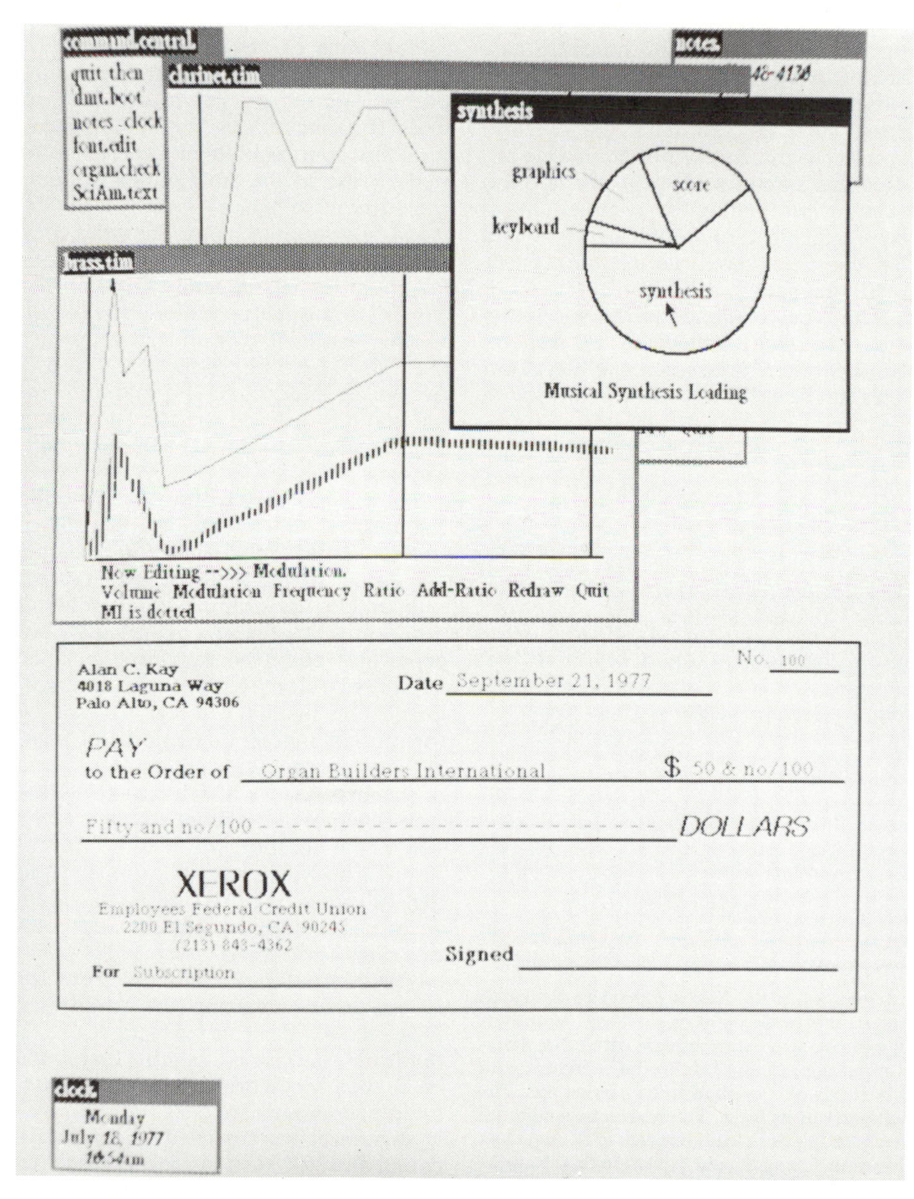

"WINDOWS," display frames within the larger display screen, enable the user to organize and edit information at several levels of refinement. Once the windows are created they overlap on the screen like sheets of paper. When a partially covered window is selected with the pointing device, the window is redisplayed to overlap the other windows. Images with various degrees of symbolic content can be displayed simultaneously. Such images include detailed halftone drawings, analogical images such as graphs and symbolic images such as numbers or words.

In the present state of the art software development is much more difficult and time-consuming than hardware development. The personal computer will eventually be put together from more or less standard microelectronic components, but the software that will give life to the user's ideas must go through a long and arduous process of refinement if it is to aid and not hinder the goals of a personal dynamic medium.

For this reason we have over the past four years invited some 250 children (aged six to 15) and 50 adults to try versions of SMALLTALK and to suggest ways of improving it. Their creations, as imaginative and diverse as they themselves, include programs for home accounts, information storage and retrieval, teaching, drawing, painting, music synthesis, writing and games. Subsequent designs of SMALLTALK have been greatly influenced and improved by our visitors' projects.

When children or adults first encounter a personal computer, most of them are already involved in pursuits of their own choosing. Their initial impulse is to exploit the system to do things they are already doing: a home or office manager will automate paperwork and accounts, a teacher will portray dynamic and pictorial aspects of a curriculum, a child will work on ways to create pictures and games. The fact is that people naturally start to conceive and build personal tools. Although man has been characterized as the toolmaking species, toolmaking itself has historically been the province of technological specialists. One reason is that technologies frequently require special techniques, materials, tools and physical conditions. An important property of computers, however, is that very general tools for using them can be built by anyone. These tools are made from the same materials and with the same effort as more specific creations.

Initially the children interact with our computer by "painting" pictures and drawing straight lines on the display screen with the pencillike pointer. The children then discover that programs can create structures more complex than any they can create by hand. They learn that a picture has several representations, of which only the most obvious—the image—appears on the screen. The most important representation is the editable symbolic model of the picture stored in the memory of the computer. For example, in the computer an image of a truck can be built up from models of wheels, a cab and a bed, each a different color. As the parts of the symbolic model are edited its image on the screen will change accordingly.

Adults also learn to exploit the properties of the computer medium. A professional artist who visited us spent several months building various tools that resembled those he had worked with to create images on paper. Eventually he discovered that the mosaic screen—the indelible but instantly erasable storage of the medium—and his new ability to program could be combined to create rich textures of a kind that could not be created with ink or paint. From the use of the computer for the impoverished simulation of an already existing medium he had progressed to the discovery of the computer's unique properties for human expression.

One of the best ways to teach nonexperts to communicate with computers is to have them explore the levels of abstraction at which images can be manipulated. The manipulation of images follows the general stages of intellectual growth. For a young child an image is something to make: a free mixture of forms and colors unconnected with the real world. Older children create images that directly represent concepts such as people, pets and houses. Later analogical images appear whose form is closely related to their meaning and purpose, such as geometric figures and graphs. In the end symbolic images are used that stand for concepts that are too abstract to analogize, such as numbers, algebraic and logical terms and the characters and words that constitute language.

The types of image in this hierarchy are increasingly difficult to represent on the computer. Free-form and literal images can be easily drawn or painted with lines and halftones in the dot matrix of the display screen with the aid of the mouse or in conjunction with programs that draw curves, fill in areas with tone or show perspectives of three-dimensional models. Analogical images can also be generated, such as a model of a simulated musical intrument: a time-sequenced graph representing the dynamic evolution of amplitude, pitch variation and tonal range.

Symbolic representations are particularly useful because they provide a means of handling concepts that are difficult to portray directly, such as generalizations and abstract relations. Moreover, as an image gets increasingly complex its most important property, the property of making local relations instantly clear, becomes less useful. Communication with computers based on symbols as they routinely occur in natural language, however, has proved to be far more difficult than many had supposed. The reason lies in our lack of understanding of how human beings exploit the context of their experience to make sense of the ambiguities of common discourse. Since it is not yet understood how human beings do what they do, getting computers to engage in similar activities is still many years in the future. It is quite possible, however, to invent artificial computer languages that can represent concepts and activities we do understand and that are simple enough in basic structure for them to be easily learned and utilized by nonexperts.

The particular structure of a symbolic language is important because it provides a context in which some concepts are easier to think about and express than others. For example, mathematical notation first arose to abbreviate concepts that could be expressed only as ungainly circumlocutions in natural language. Gradually it was realized that the form of an expression could be of great help in the conception and manipulation of the meaning for which the expression stood. A more important advance came when new notation was created to represent concepts that did not fit into the culture's linguistic heritage at all, such as functional mappings, continuous rates and limits.

The computer created new needs for language by inverting the traditional process of scientific investigation. It made new universes available that could be shaped by theories to produce simulated phenomena. Accordingly symbolic structures were needed to communicate concepts such as imperative descriptions and control structures.

Most of the programming languages in service today were developed as symbolic ways to deal with the hardware-level concepts of the 1950's. This approach led to two kinds of passive building blocks: data structures, or inert con-

trait name	description
name	box; picture; activity
location	□
angle	□
size	□
new	location ← center. angle ← 0. size ← 100.
show	♂ paint black. shape
erase	♂ paint background. shape
shape	♂ up; goto location; turn angle; down. 1 to 4 do (♂ go size; turn 90)
grow □	erase. size ← size + □. show

SMALLTALK is a new programming language developed at the Xerox Palo Alto Research Center for use on the experimental personal computer. It is made up of "activities," computerlike entities that can perform a specific set of tasks and can also communicate with other activities in the system. New activities are created by enriching existing families of activities with additional "traits," or abilities, which are defined in terms of a method to be carried out. The description of the family "box" shown here is a dictionary of its traits. To create a new member of the family box, a message is sent to the trait "new" stating the characteristics of the new box in terms of specific values for the general traits "location," "angle" and "size." In this example "new" has been filled in to specify a box located in the center of the screen with an angle of zero degrees and a side 100 screen dots long. To "show" the new box, a member of the curve-drawing family "brush" is given directions by the open trait "shape." First the brush travels to the specified location, turns in the proper direction and appears on the screen. Then it draws a square by traveling the distance given by "size," turning 90 degrees and repeating these actions three more times. The last trait on the list is open, indicating that a numerical value is to be supplied by the user when the trait is invoked by a message. A box is "grown" by first erasing it, increasing (or decreasing) its size by the value supplied in the message and redisplaying it.

Message Interaction	Pictorial Effect	Commentary
box new named "joe"! box:joe		An offspring of the family "box" is created and is named "joe."
joe turn 30! ok		The box joe receives the message and turns 30 degrees.
joe grow −15! ok		Joe becomes smaller by 15 units.
joe erase! ok		Joe disappears from the screen.
joe show! ok		Joe reappears.
box new named "jill"! box:jill	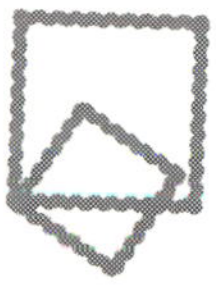	A new box appears.
jill turn −10! ok	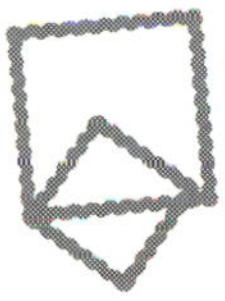	Only jill turns. Joe and jill are independent activities.
1 to 10! interval:1 2 3 4 5 6 7 8 9 10		An interval stands for a sequence of numbers.
forever! interval:1 2 3 4 5 6 7 8 9 10 11...		Forever is the infinite interval. It must be terminated by hitting an escape key.
1 to 10 do (joe turn 20)! ok	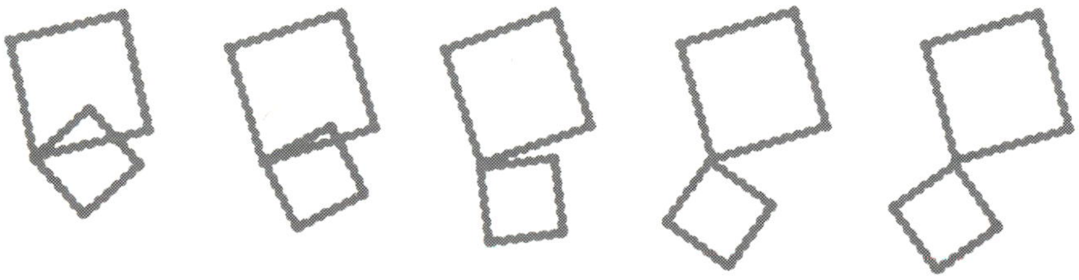	Joe spins.
forever do (joe turn 11. jill turn −13)! ok	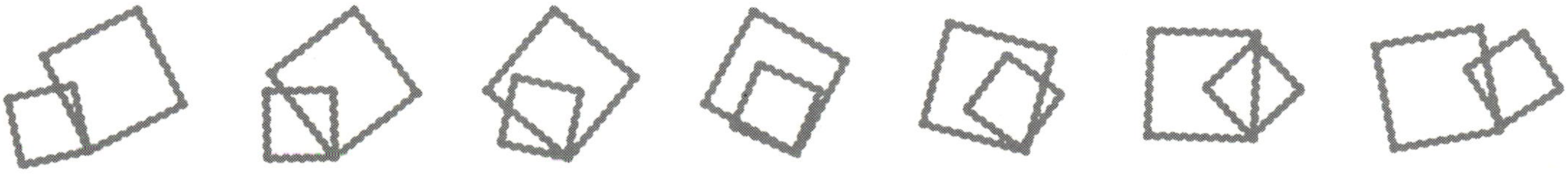	A simple parallel movie of joe and jill spinning in opposite directions is created by combining forever with a turn request to both joe and jill.

SMALLTALK LEARNING SEQUENCE teaches students the basic concepts of the language by having them interact with an already defined family of activities. First, offspring of the family box are created, named and manipulated, and a second family of activities called "interval" is introduced. Offspring of the interval and box families are then combined to generate an animation of two spinning boxes.

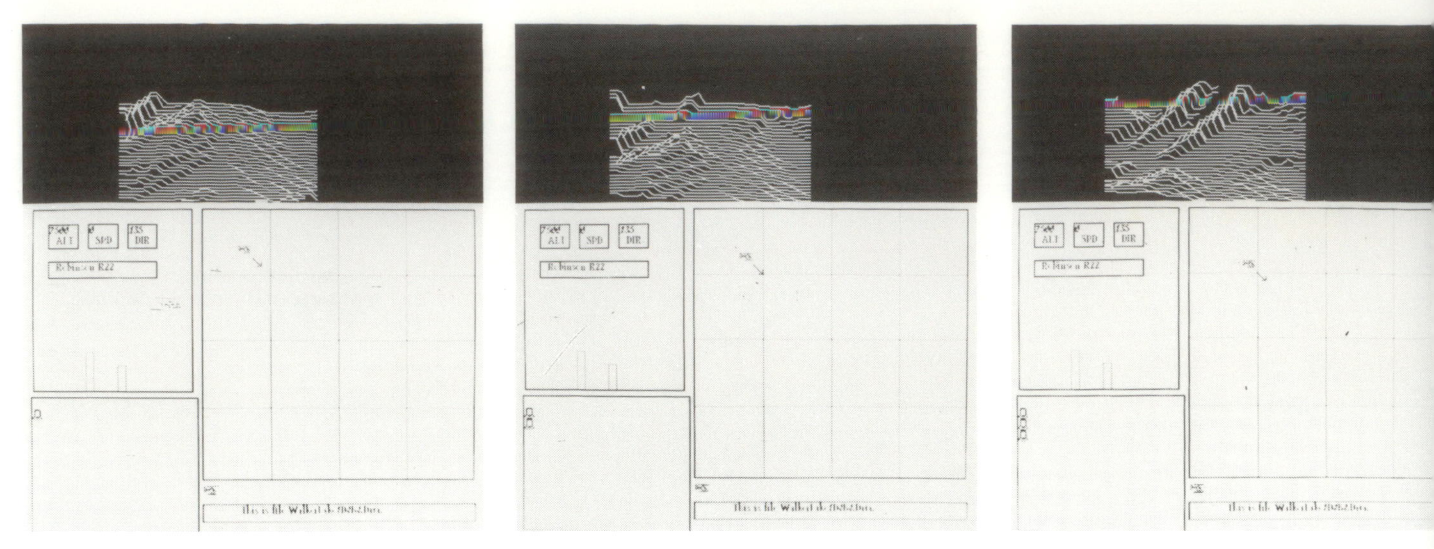

HELICOPTER SIMULATION was developed by a 15-year-old student. The user directs the helicopter where to go with the pointing device, which controls the position of the black arrow on the screen. The window at the top shows the changing topography of the terrain

struction materials, and procedures, or step-by-step recipes for manipulating data. The languages based on these concepts (such as BASIC, FORTRAN, ALGOL and APL) follow their descriptions in a strictly sequential manner. Because a piece of data may be changed by any procedure that can find it the programmer must be very careful to choose only those procedures that are appropriate. As ever more complex systems are attempted, requiring elaborate combinations of procedures, the difficulty of getting the entire system to work increases geometrically. Although most programmers are still taught data-procedure languages, there is now a widespread recognition of their inadequacy.

A more promising approach is to devise building blocks of greater generality. Both data and procedures can be replaced by the single idea of "activities," computerlike entities that exhibit behavior when they are sent an appropriate message. There are no nouns and verbs in such a language, only dynamically communicating activities. Every transaction, description and control process is thought of as sending messages to and receiving messages from activities in the system. Moreover, each activity belongs to a family of similar activities, all of which have the ability to recognize and reply to messages directed to them and to perform specific acts such as drawing pictures, making sounds or adding numbers. New families are created by combining and enriching "traits," or properties inherited from existing families.

A message-activity system is inherently parallel: every activity is constantly ready to send and receive messages, so that the host computer is in effect divided into thousands of computers, each with the capabilities of the whole. The message-activity approach therefore enables one to dynamically represent a system at many levels of organization from the atomic to the macroscopic, but with a "skin" of protection at each qualitative level of detail through which negotiative messages must be sent and checked. This level of complexity can be safely handled because the language severely limits the kinds of interactions between activities, allowing only those that are appropriate, much as a hormone is allowed to interact with only a few specifically responsive target cells. SMALLTALK, the programming system of our personal computer, was the first computer language to be based entirely on the structural concepts of messages and activities.

The third and newest framework for high-level communication is the observer language. Although message-activity languages are an advance over the data-procedure framework, the relations among the various activities are somewhat independent and analytic. Many

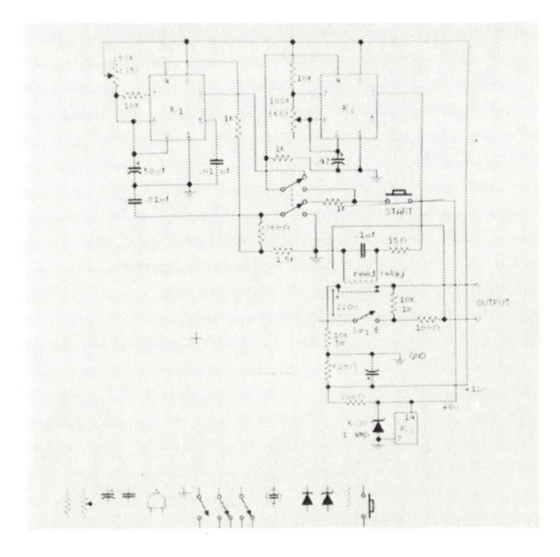

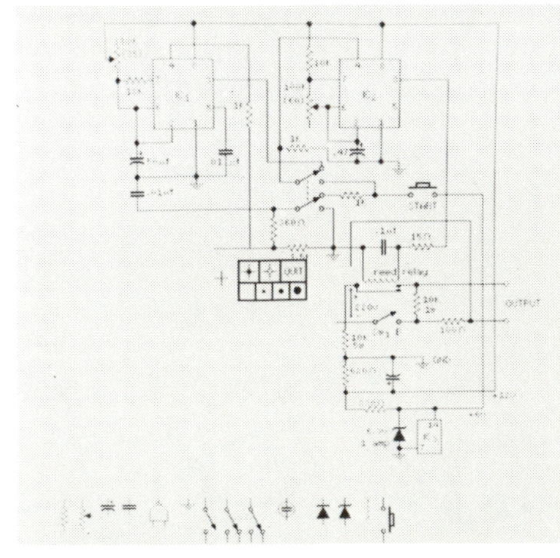

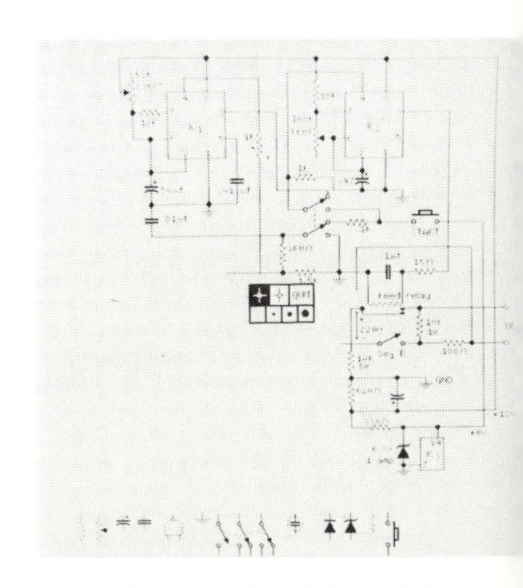

CIRCUIT-DRAWING PROGRAM that was developed by a 15-year-old boy enables a user to construct a complex circuit diagram by selecting components from a "menu" displayed at the bottom of the screen. The components are then positioned and connected with the

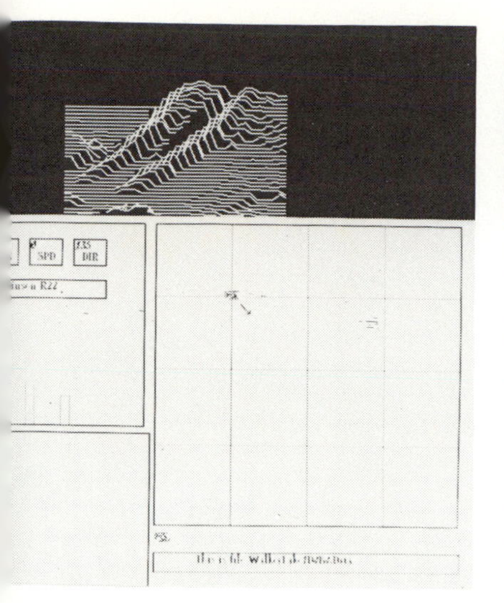

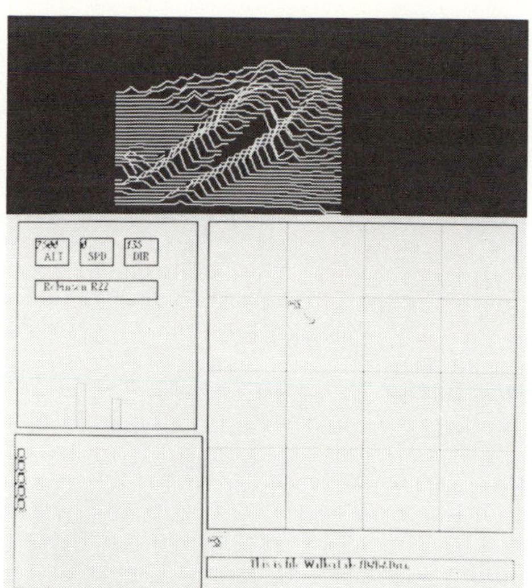

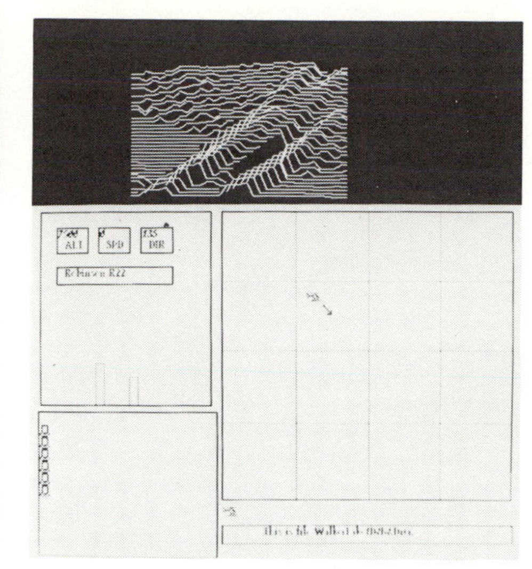

below as the helicopter flies over it. (Actual terrains were obtained from *Landsat* maps.) A third window keeps track of the helicopter's altitude, direction and speed. The variety of events that can be simulated at the same time demonstrates the power of parallel processing.

concepts, however, are so richly interwoven that analysis causes them virtually to disappear. For example, 20th-century physics assigns equal importance to a phenomenon and its context, since observers with different vantage points perceive the world differently. In an observer language, activities are replaced by "viewpoints" that become attached to one another to form correspondences between concepts. For example, a dog can be viewed abstractly (as an animal), analytically (as being composed of organs, cells and molecules), pragmatically (as a vehicle by a child), allegorically (as a human being in a fairy tale) and contextually (as a bone's way to fertilize a lawn). Observer languages are just now being formulated. They and their successors will be the communication vehicles of the 1980's.

Our experience, and that of others who teach programming, is that a first computer language's particular style and its main concepts not only have a strong influence on what a new programmer can accomplish but also leave an impression about programming and computers that can last for years. The process of learning to program a computer can impose such a particular point of view that alternative ways of perceiving and solving problems can become extremely frustrating for new programmers to learn.

At the beginning of our study we first timidly considered simulating features of data-procedure languages that children had been able to learn, such as BASIC and LOGO. Then, worried that the imprinting process would prevent stronger ideas from being absorbed, we decided to find a way to present the message-activity ideas of SMALLTALK in concrete terms without dilution. We did so by starting with simple situations that embodied a concept and then gradually increasing the complexity of the examples to flesh out the concept to its full generality. Although the communicationlike model of SMALLTALK is a rather abstract way to represent descriptions, to our surprise the first group and succeeding groups of children who tried it appeared to find the ideas as easy to learn as those of more concrete languages.

For example, most programming languages can deal with only one thing at a time, so that it is difficult to represent with them even such simple situations as children in a school, spacecraft in the sky or bouncing balls in free space. In SMALLTALK parallel models are dealt with from the start, and the children seem to have little difficulty in handling them. Actually parallel processing is remarkably similar to the way people think. When you are walking along a street, one part of your brain may be thinking about the route you are taking, another part may be thinking about the dinner you are going to eat, a third

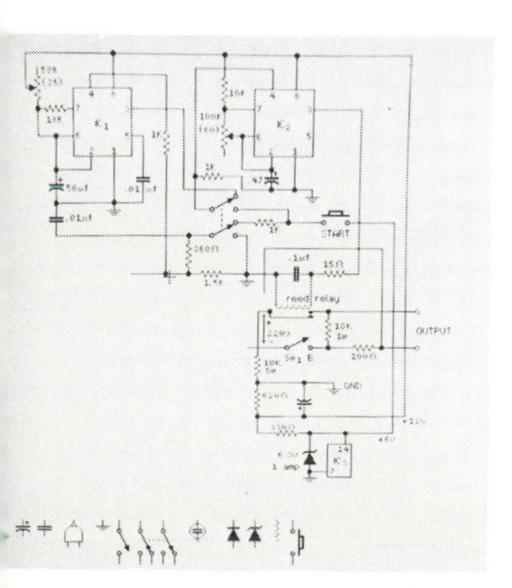

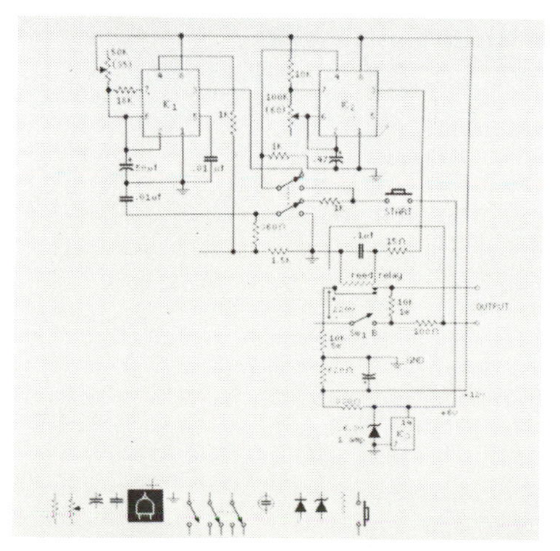

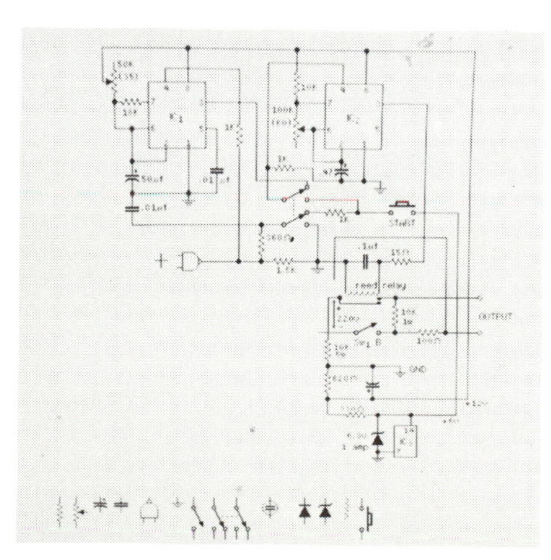

pointing device. An additional menu can be generated on the screen by pushing a button on the pointing device; this menu supplies solid and open dots and lines of various widths. In the sequence shown here two components are selected and added to a circuit diagram.

HORSE-RACE ANIMATION shows the capabilities of the experimental personal computer for creating dynamic halftone images. The possible range of such simulations is limited only by the versatility of the programming language and the imagination of the child or adult user. In this sequence, images of horses, riders and background are called up independently from the storage files and arranged for the racing simulation with the pointing device. A single typed command then causes the two horses and riders to race each other across screen.

part may be admiring the sunset, and so forth.

Another important characteristic of SMALLTALK is the classification of objects into families that are generalizations of their properties. Children readily see themselves as members of the family "kids," since they have common traits such as language, interests and physical appearance. Each individual is both a member of the family kids and has his or her own meaning for the shared traits. For example, all kids have the trait eye color, but Sam's eyes are blue and Bertha's are brown. SMALLTALK is built out of such families. Number symbols, such as 2 or 17, are instances of the family "number." The members of this family differ only in their numerical value (which is their sole property) and share a common definition of the different messages they can receive and send. The symbol of a "brush" in SMALLTALK is also a family. All the brush symbols have the ability to draw lines, but each symbol has its own knowledge of its orientation and where it is located in the drawing area.

The description of a programming language is generally given in terms of its grammar: the meaning each grammatical construction is supposed to convey and the method used to obtain the meaning. For example, various programming languages employ grammatical constructions such as (PLUS 3 4) or 3 ENTER 4 + to specify the intent to add the number 3 to the number 4. The meaning of these phrases is the same. In the computer each should give rise to the number 7, although the actual methods followed in obtaining the answer can differ considerably from one type of computer to the next.

The grammar of SMALLTALK is simple and fixed. Each phrase is a message to an activity. A description of the desired activity is followed by a message that selects a trait of the activity to be performed. The designated activity will decide whether it wants to accept the message (it usually does) and at some later time will act on the message. There may be many concurrent messages pending for an activity, even for the same trait. The sender of the message may decide to wait for a reply or not to wait. Usually it waits, but it may decide to go about other business if the message has invoked a method that requires considerable computation.

The integration of programming-language concepts with concepts of editing, graphics and information retrieval makes available a wide range of useful activities that the user can invoke with little or no knowledge of programming. Learners are introduced to SMALLTALK by getting them to send messages to already existing families of activities, such

MUSIC CAN BE REPRESENTED on the personal computer in the form of analogical images. Notes played on the keyboard are "captured" as a time-sequenced score on the display.

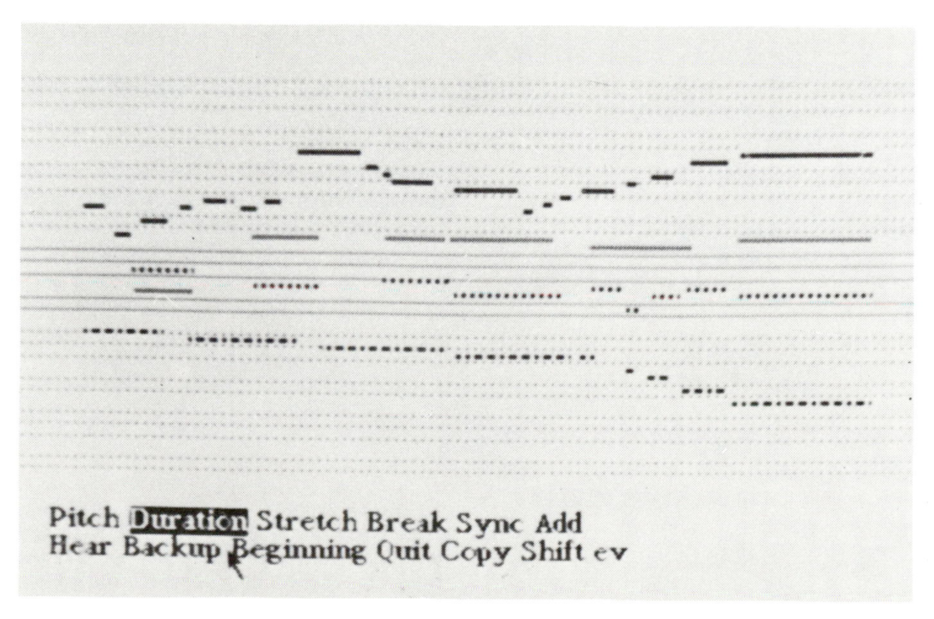

MUSICAL SCORE shown here was generated as music was played on the keyboard. The simplified notation represents pitch by vertical placement and duration by horizontal length. Notes can be shortened, lengthened or changed and the modified piece then played back as music.

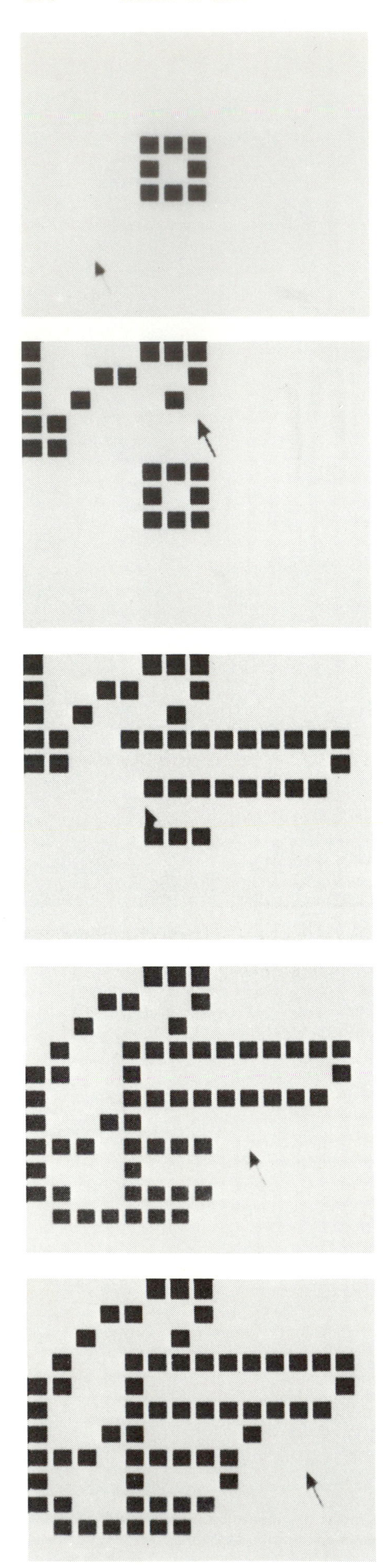

DISPLAY FONTS can be designed on personal computer by constructing them from a matrix of black-and-white squares. When the fonts are reduced, they approach the quality of those in printed material. The image of a pointing hand shown here is a symbol in SMALLTALK representing the concept of a literal word, such as the name associated with an activity.

as the family "box," whose members show themselves on the screen as squares. A box can individually change its size, location, rotation and shape. After some experience with sending messages to cause effects on the display screen the learner may take a look at the definition of the box family. Each family in SMALLTALK is described with a dictionary of traits, which are defined in terms of a method to be carried out. For example, the message phrase "joe grow 50" says: Find the activity named "joe," find its general trait called "grow ___" and fill in its open part with the specific value 50. A new trait analogous to those already present in the family definition (such as "grow" or "turn") can easily be added by the learner. The next phase of learning involves elaboration of this basic theme by creating games such as space war and tools for drawing and painting.

There are two basic approaches to personal computing. The first one, which is analogous to musical improvisation, is exploratory: effects are caused in order to see what they are like and errors are tracked down, understood and fixed. The second, which resembles musical composition, calls for a great deal more planning, generality and structure. The same language is used for both methods but the framework is quite different.

From our study we have learned the importance of a balance between free exploration and a developed curriculum. The personal computing experience is similar to the introduction of a piano into a third-grade classroom. The children will make noise and even music by experimentation, but eventually they will need help in dealing with the instrument in nonobvious ways. We have also found that for children the various levels of abstraction supplied by SMALLTALK are not equally accessible. The central idea of symbolization is to give a simple name to a complex collection of ideas, and then later to be able to invoke the ideas through the name. We have observed a number of children between the ages of six and seven who have been able to take this step in their computer programs, but their ability to look ahead, to visualize the consequences of actions they might take, is limited.

Children aged eight to 10 have a gradually developing ability to visualize and plan and are able to use the concept of families and a subtler form of naming: the use of traits such as size, which can stand for different numerical values at different times. For most children, however, the real implications of further symbolic generality are not at all obvious. By age 11 or 12 we see a considerable improvement in a child's ability to plan general structures and to devise comprehensive computer tools. Adults advance through the stages more quickly than children, and usually they create tools after a few weeks of practice. It is not known whether the stages of intellectual development observed in children are absolutely or only relatively correlated with age, but it is possible that exposure to a realm in which symbolic creation is rewarded by wonderful effects could shorten the time required for children to mature from one stage to the next.

The most important limitation on personal computing for nonexperts appears when they conceive of a project that, although it is easy to do in the language, calls for design concepts they have not yet absorbed. For example, it is easy to build a span with bricks if one knows the concept of the arch, but otherwise it is difficult or impossible. Clearly as complexity increases "architecture" dominates "material." The need for ways to characterize and communicate architectural concepts in developing programs has been a long-standing problem in the design of computing systems. A programming language provides a context for developing strategies, and it must supply both the ability to make tools and a style suggesting useful approaches that will bring concepts to life.

We are sure from our experience that personal computers will become an integral part of peoples' lives in the 1980's. The editing, saving and sifting of all manner of information will be of value to virtually everyone. More sophisticated forms of computing may be like music in that most people will come to know of them and enjoy them but only a few will actually become directly involved.

How will personal computers affect society? The interaction of society and a new medium of communication and self-expression can be disturbing even when most of the society's members learn to use the medium routinely. The social and personal effects of the new medium are subtle and not easy for the society and the individual to perceive. To use writing as a metaphor, there are three reactions to the introduction of a new medium: illiteracy, literacy and artistic creation. After reading material became available the illiterate were those who were left behind by the new medium. It was inevitable that a few creative individuals would use the written word to express inner thoughts and ideas. The most profound changes were brought about in the literate. They did not necessarily become better people or better members of society, but they came to view the world in a way quite different from the way they had viewed it before, with consequences that were difficult to predict or control.

We may expect that the changes resulting from computer literacy will be as far-reaching as those that came from literacy in reading and writing, but for most people the changes will be subtle and not necessarily in the direction of their idealized expectations. For example, we should not predict or expect that the personal computer will foster a new revolution in education just because it could. Every new communication medium of this century—the telephone, the motion picture, radio and television—has elicited similar predictions that did not come to pass. Millions of uneducated people in the world have ready access to the accumulated culture of the centuries in public libraries, but they do not avail themselves of it. Once an individual or a society decides that education is essential, however, the book, and now the personal computer, can be among the society's main vehicles for the transmission of knowledge.

The social impact of simulation—the central property of computing—must also be considered. First, as with language, the computer user has a strong motivation to emphasize the similarity between simulation and experience and to ignore the great distances that symbols interpose between models and the real world. Feelings of power and a narcissistic fascination with the image of oneself reflected back from the machine are common. Additional tendencies are to employ the computer trivially (simulating what paper, paints and file cabinets can do), as a crutch (using the computer to remember things that we can perfectly well remember ourselves) or as an excuse (blaming the computer for human failings). More serious is the human propensity to place faith in and assign higher powers to an agency that is not completely understood. The fact that many organizations actually base their decisions on—worse, take their decisions from—computer models is profoundly disturbing given the current state of the computer art. Similar feelings about the written word persist to this day: if something is "in black and white," it must somehow be true.

Children who have not yet lost much of their sense of wonder and fun have helped us to find an ethic about computing: Do not automate the work you are engaged in, only the materials. If you like to draw, do not automate drawing; rather, program your personal computer to give you a new set of paints. If you like to play music, do not build a "player piano"; instead program yourself a new kind of instrument.

A popular misconception about computers is that they are logical. Forthright is a better term. Since computers can contain arbitrary descriptions, any conceivable collection of rules, consistent or not, can be carried out. Moreover, computers' use of symbols, like the use of symbols in language and mathematics, is sufficiently disconnected from the real world to enable them to create splendid nonsense. Although the hardware of the computer is subject to natural laws (electrons can move through the circuits only in certain physically defined ways), the range of simulations the computer can perform is bounded only by the limits of human imagination. In a computer, spacecraft can be made to travel faster than the speed of light, time to travel in reverse.

It may seem almost sinful to discuss the simulation of nonsense, but only if we want to believe that what we know is correct and complete. History has not been kind to those who subscribe to this view. It is just this realm of apparent nonsense that must be kept open for the developing minds of the future. Although the personal computer can be guided in any direction we choose, the real sin would be to make it act like a machine!

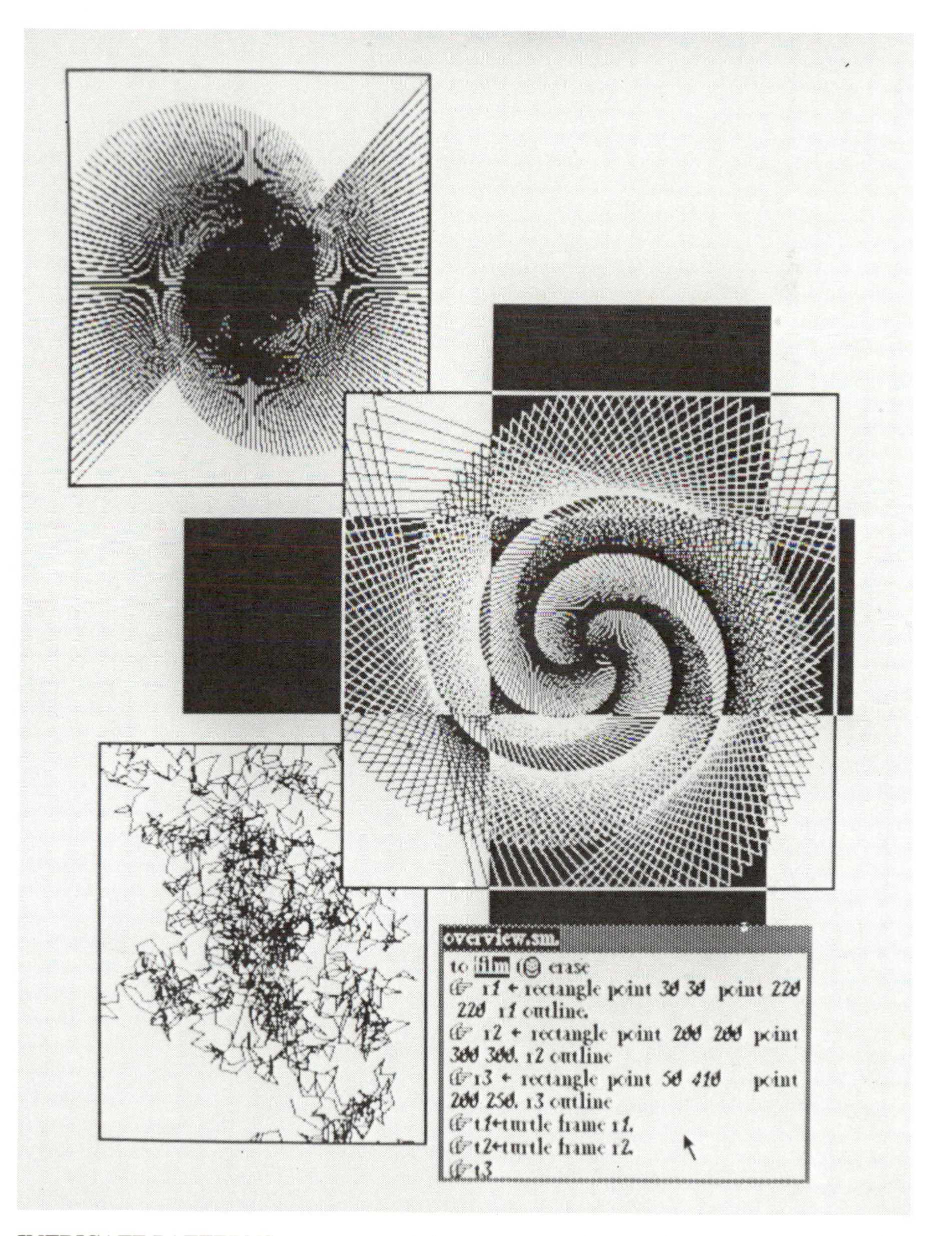

INTRICATE PATTERNS can be generated on the personal computer with very compact descriptions in SMALLTALK. They are made by repeating, rotating, scaling, superposing and combining drawings of simple geometric shapes. Students who are learning to program first create interesting free-form or literal images by drawing them directly in the dot matrix of the display screen. Eventually they learn to employ the symbolic images in the programming language to direct the computer to generate more complex imagery than they could easily create by hand.

THE AUTHORS

BIBLIOGRAPHIES

INDEX

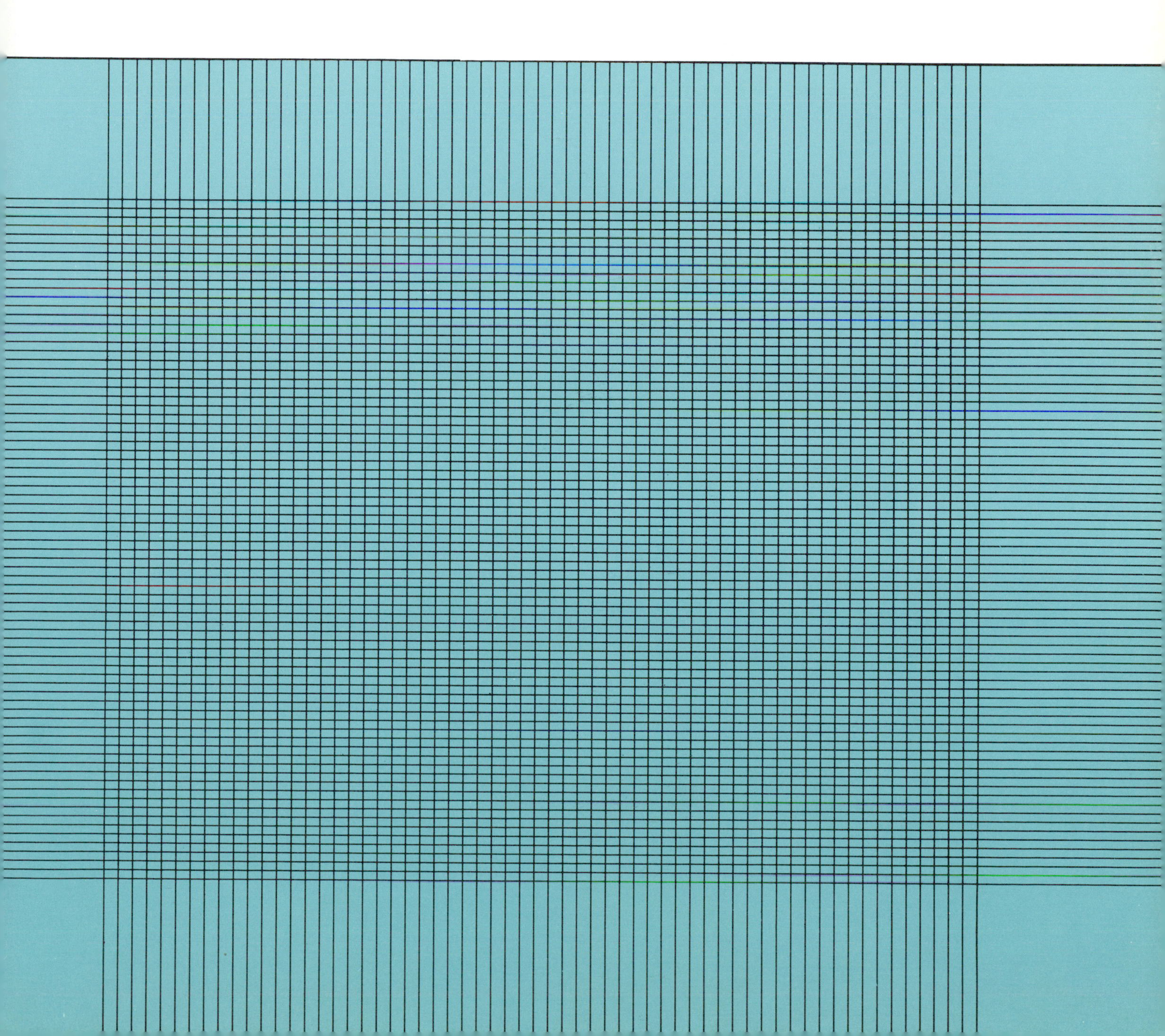

THE AUTHORS

ROBERT N. NOYCE ("Microelectronics") is chairman of the Intel Corporation. He attended Grinnell College and the Massachusetts Institute of Technology, receiving his Ph.D. in physical electronics from M.I.T. in 1953. Thereafter he worked first on the research staff of the Philco Corporation and then at the Shockley Semiconductor Laboratory, where he took part in the design and development of silicon transistors. In 1957 he helped to found the Fairchild Semiconductor Corporation, where during the next 10 years he served as research director, vice-president and general manager. He left Fairchild in 1968 to cofound Intel, a company "devoted to making the large-scale integration of microelectronic circuits a reality."

JAMES D. MEINDL ("Microelectronic Circuit Elements") is professor of electrical engineering at Stanford University and director of the Stanford Electronics Laboratories. He was educated at Carnegie-Mellon University, receiving his Ph.D. in electrical engineering in 1958. After two years with the Westinghouse Electric Corporation he joined the Army Electronics Laboratory, serving from 1959 to 1967 in several capacities, including as director of the Integrated Electronics Division. In 1967 he went to Stanford as associate professor of electrical engineering. Meindl's current research is focused on the medical applications of integrated-circuit technology, specifically the development of ultrasonic medical diagnostic instruments.

WILLIAM C. HOLTON ("The Large-Scale Integration of Microelectronic Circuits") is director of the Advanced Components Laboratory of Texas Instruments Incorporated. He attended the University of North Carolina, where he was graduated in 1952 with a B.S. in physics. After two and a half years in the Navy he took up graduate work in physics at the University of Illinois, obtaining his Ph.D. in 1960. That year he joined Texas Instruments as a member of the technical staff. In 1967 he was named manager of the Quantum Electronics branch, where he conducted research and development on lasers and integrated optics. He was appointed to his present position in 1973. Holton is currently responsible for directing research and development work on large-scale integrated circuits, electron-beam and X-ray lithography, microwave devices and displays. He still finds time, however, to crew on the racing yacht *Business Machine*.

WILLIAM G. OLDHAM ("The Fabrication of Microelectronic Circuits") is professor of electrical engineering at the University of California at Berkeley. He was educated at Carnegie-Mellon University, receiving his Ph.D. in 1963. After a year on the research staff of the Siemens Corporation in Germany he joined the Berkeley faculty. In 1969 he spent a sabbatical year at the Technische Universität in Munich, and in 1974 he went on leave to the Intel Corporation, where he managed the development of a microelectronic computer memory with a storage capacity of 16,000 bits. Oldham's current research interests focus on techniques for fabricating microelectronic devices of submicrometer dimensions. Outside the laboratory he spends much of his time adding to his extensive collection of working mechanical antiques, such as music boxes and player pianos.

DAVID A. HODGES ("Microelectronic Memories") is professor of electrical engineering and computer sciences at the University of California at Berkeley. He received his bachelor's degree in electrical engineering at Cornell University and his Ph.D. in electrical engineering from Berkeley in 1966. After graduation he worked at the Bell Laboratories, first in the components area at Murray Hill, N.J., and then as head of the Systems Elements Research Department at Holmdel, N.J. He joined the Berkeley faculty in 1970 and has since served as a consultant in the semiconductor and communications-equipment industries.

HOO-MIN D. TOONG ("Microprocessors") is assistant professor of computer science and engineering at the Massachusetts Institute of Technology and also assistant professor of management science at M.I.T.'s Sloan School of Management. He was educated at M.I.T., where he received his electrical engineering and master's degrees in 1969 and his Ph.D. in 1974. Thereafter he joined the M.I.T. faculty. In addition to teaching undergraduate courses in computer science he has been giving lectures on microprocessors throughout the U.S., Europe and the Far East. Toong is in charge of the Digital Systems Laboratory at M.I.T., where his research has been concentrated on software and hardware problems of multiprocessor and distributed-computing systems. He is also applying microprocessor technology to methods for conserving energy and controlling its utilization.

LEWIS M. TERMAN ("The Role of Microelectronics in Data Processing") is a member of the research staff at the Thomas J. Watson Research Center of the International Business Machines Corporation. He was educated at Stanford University, receiving his Ph.D. in 1961. He then joined the Watson Research Center, where he worked on computer logic design and in 1963 was made manager of a group working on magnetic-memory technology. Two years later he became interested in the use of microelectronic devices for computer memories and was one of its earliest advocates. He is currently interested in the application of charge-coupled devices to memories and in the implications of large-scale integration for systems design. An amateur trumpet player, Terman is a devotee of jazz and has a collection of more than 2,000 jazz records. He is the grandson of Lewis M. Terman, the well-known Stanford psychologist.

BERNARD M. OLIVER ("The Role of Microelectronics in Instrumentation and Control") is vice-president for research and development of the Hewlett-Packard Company and a member of the company's board of directors. He was educated at Stanford University and the California Institute of Technology, where he received his M.S. in 1936. Following a year of study in Germany under an exchange scholarship, he returned to Cal Tech and obtained his Ph.D., magna cum laude, in 1940. He then went to work at Bell Laboratories and over the next 12 years participated in the development of automatic tracking radar, television transmission, information theory and coding systems. In 1952 he joined Hewlett-Packard as director of research; he was appointed to his present position in 1957. Oliver holds more than 50 patents in the field of electronics, and he has served as a member of the President's Commission on the Patent System. In 1966 he was elected to the National Academy of Engineering and in 1973 to the National Academy of Sciences.

JOHN S. MAYO ("The Role of Microelectronics in Communication") is vice-president of electronics technology at Bell Laboratories. Born in North Carolina, he was educated at North Carolina State University, where he received his Ph.D. in electrical engineering in 1955. He then went to Bell Laboratories, initially doing research on transistorized digital computers and on the use of computers in defense systems. He also developed the method of transmitting information in digital form known as pulse-code modulation. Later he applied the technique to the design of the command decoder and switching unit of the Telstar communications satellite and to the development of the Picturephone system. Mayo went to his present job in 1975. He is responsible for directing the development of electronic components, energy sources, power supplies and building-environment designs for the operating telephone companies in the Bell System.

IVAN E. SUTHERLAND and CARVER A. MEAD ("Microelectronics and Computer Science") are respectively director of the computer-science department and professor of computer science at the California Institute of Technology. Sutherland was educated at the Carnegie Institute of Technology, Cal Tech and M.I.T., receiving his Ph.D. in electrical engineering from M.I.T. in 1963. He then entered the Army Signal Corps and was stationed at the National Security Agency, where he designed a new line of computer-display equipment. In 1964 Sutherland was made director for information-processing techniques at the Advanced Research Projects Agency (ARPA) of the Department of Defense, staying in that position after his discharge from the Army. He left ARPA in 1966 to become associate professor of electrical engineering at Harvard University. Two years later he moved to Salt Lake City to become president of the Evans & Sutherland Computer Corporation, of which he was cofounder. At the same time he continued his research at the University of Utah, where he helped to design a computer-graphics system that could create color images with lifelike shading and perspective. He joined the Cal Tech faculty in 1976, but he still is a consultant for Evans & Sutherland and serves on the Defense Science Board. Mead was educated at Cal Tech, where he received his Ph.D. in electrical engineering in 1959. His research began in the area of solid-state-device physics, with a brief foray into the biophysics of nerve membranes. It was during his calculation of the physical limitations on how small a transistor could be and still operate that he became absorbed in the problems of managing the complexity of large-scale integrated systems. Mead escapes occasionally from the world of high technology to raise hazelnuts on his 70-acre ranch in Oregon.

ALAN C. KAY ("Microelectronics and the Personal Computer") is a principal scientist and head of the Learning Research Group at the Xerox Palo Alto Research Center. He received his B.A. in mathematics from the University of Colorado at Boulder and, after a short career as a professional jazz guitarist, studied computer science at the University of Utah, obtaining his Ph.D. in 1969. He then became a research associate and lecturer at the Stanford University Artificial Intelligence Project. He moved to Xerox in 1971. "I have always been equally attracted to the arts and the sciences," he writes. "Eventually I discovered that the world of computers provides a satisfying environment for my blend of interests."

BIBLIOGRAPHIES

Readers interested in further explanation of the subjects covered by the articles in this issue may find the following lists of publications helpful.

MICROELECTRONICS

MICROCOMPUTERS/MICROPROCESSORS: HARDWARE, SOFTWARE, AND APPLICATIONS. John L. Hilburn and Paul M. Julich. Prentice-Hall, Inc., 1976.

SPECIAL ISSUE ON MICROPROCESSOR TECHNOLOGY AND APPLICATIONS. *Proceedings of the IEEE,* Vol. 64, No. 6; June, 1976.

MICROELECTRONIC CIRCUIT ELEMENTS

SEMICONDUCTOR ELECTRONICS. James F. Gibbons. McGraw-Hill Book Company, 1966.

PHYSICS AND TECHNOLOGY OF SEMICONDUCTOR DEVICES. A. S. Grove. John Wiley & Sons, Inc., 1967.

THE THEORY AND PRACTICE OF MICROELECTRONICS. Sorab K. Ghandhi. John Wiley & Sons, Inc., 1968.

MICROPOWER CIRCUITS. James D. Meindl. John Wiley & Sons, Inc., 1969.

BASIC INTEGRATED CIRCUIT ENGINEERING. Douglas Hamilton and William Howard. McGraw-Hill Book Company, 1975.

THE LARGE-SCALE INTEGRATION OF MICROELECTRONIC CIRCUITS

INTEGRATED CIRCUITS: A BASIC COURSE. Robert G. Hibberd. McGraw-Hill Book Company, 1969.

METAL-OXIDE-SEMICONDUCTOR TECHNOLOGY. William C. Hittinger in *Scientific American,* Vol. 229, No. 2, pages 48–57; August, 1973.

DESIGN OF DIGITAL COMPUTERS. Hans W. Gshwind and Edward J. McCluskey. Springer-Verlag, 1975.

MICROCOMPUTERS. André G. Vacroux in *Scientific American,* Vol. 232, No. 5, pages 32–40; May, 1975.

PHYSICS OF COMPUTER MEMORY DEVICES. S. Middelhoek and P. Dekker. Academic Press, Inc., 1976.

THE SMALL ELECTRONIC CALCULATOR. Eugene W. McWhorter in *Scientific American,* Vol. 234, No. 3, pages 88–98; March, 1976.

THE FABRICATION OF MICROELECTRONIC CIRCUITS

INTEGRATED CIRCUIT FABRICATION PROCESSES. Alan B. Grebene in *Analog Integrated Circuit Design.* Van Nostrand Reinhold Company, 1972.

MOS INTEGRATED CIRCUITS: THEORY, FABRICATION, DESIGN AND SYSTEMS APPLICATIONS OF MOS LSI. Engineering staff of American Micro-systems, Inc., edited by William M. Penny and Lillian Lau. Van Nostrand Reinhold Company, 1972.

MOS/LSI DESIGN AND APPLICATION. William N. Carr and Jack P. Mize, edited by Robert E. Sawyer and John R. Miller. McGraw-Hill Book Company, 1972.

METAL-OXIDE-SEMICONDUCTOR TECHNOLOGY. William C. Hittinger in *Scientific American,* Vol. 229, No. 2, pages 48–57; August, 1973.

MICROELECTRONIC MEMORIES

SEMICONDUCTOR MEMORIES. Edited by David A. Hodges. IEEE Press, 1972.

MASS MEMORIES. Rein Turn in *Computers in the 1980s.* Columbia University Press, 1974.

RANDOM-ACCESS MEMORIES. Rein Turn in *Computers in the 1980s.* Columbia University Press, 1974.

DIGITAL MEMORIES. Carlo H. Séquin and Michael F. Tompsett in *Charge Transfer Devices.* Academic Press, Inc., 1975.

MAGNETIC BUBBLE TECHNOLOGY: INTEGRATED-CIRCUIT MAGNETICS FOR DIGITAL STORAGE AND PROCESSING. Edited by Hsu Chang. IEEE Press, 1975.

SPECIAL ISSUE ON LARGE CAPACITY DIGITAL STORAGE SYSTEMS. *Proceedings of the IEEE,* Vol. 63, No. 8; August, 1975.

NEW MEMORY TECHNOLOGIES. Jan A. Rajchman in *Science,* Vol. 195, No. 4283, pages 1223–1229; March 18, 1977.

MICROPROCESSORS

SPECIAL ISSUE ON MICROPROCESSOR TECHNOLOGY AND APPLICATIONS. *Proceedings of the IEEE,* Vol. 64, No. 6; June, 1976.

SPECIAL ISSUE ON SMALL SCALE COMPUTING. *Computer,* Vol. 10, No. 3; March, 1977.

TRENDS IN COMPUTERS AND COMPUTING: THE INFORMATION UTILITY. Stuart E. Madnick in *Science,* Vol. 195, No. 4283, pages 1191–1199; March 18, 1977.

THE ROLE OF MICROELECTRONICS IN DATA PROCESSING

DIGITAL COMPUTER FUNDAMENTALS. Thomas C. Bartee. McGraw-Hill Book Company, 1972.

MINICOMPUTER SYSTEMS: STRUCTURE, IMPLEMENTATION AND APPLICATION. Cay Weitzman. Prentice-Hall, Inc., 1974.

DATA PROCESSING IN 1980–1985: A STUDY OF POTENTIAL LIMITATIONS TO PROGRESS. T. A. Dolotta, M. I. Bernstein, R. S. Dickson, Jr., N. A. France, B. A. Rosenblatt, D. M. Smith and T. B. Steel, Jr. John Wiley & Sons, Inc., 1976.

TRENDS IN COMPUTER HARDWARE TECHNOLOGY. David A. Hodges in *Computer Design,* Vol. 15, No. 2, pages 77–85; February, 1976.

THE ROLE OF MICROELECTRONICS IN INSTRUMENTATION AND CONTROL

SERVO-MOTOR RESPONSE. Bernard M. Oliver in *Proceedings of the IEEE,* Vol. 53, No. 2, pages 201–202; February, 1965.

ELECTRONIC MEASUREMENTS AND INSTRUMENTATION. Edited by Bernard M. Oliver and John M. Cage. McGraw-Hill Book Company, 1971.

MICROPROCESSORS: DESIGN AND APPLICATIONS IN DIGITAL INSTRUMENTATION AND CONTROL. Walter Banks and Jayanti C. Majithia in *IEEE Transactions on Instrumentation and Measurement,* Vol. IM-25, No. 3, pages 245–249; September, 1976.

DIGITAL SYSTEMS SPAWN NEW TASKS IN MEASUREMENT. A. Santoni in *Electronics,* Vol. 49, No. 22, pages 100–106; October 28, 1976.

LSI CHIPS TAKING OVER MORE HOUSEHOLD CHORES. G. M. Walker in *Electronics,* Vol. 49, No. 22, pages 128–134; October 28, 1976.

THE ROLE OF MICROELECTRONICS IN COMMUNICATION

PRINCIPLES OF PULSE CODE MODULATION. K. W. Cattermole. American Elsevier Publishing Company, Inc., 1969.

DIGITAL CODING OF SPEECH WAVEFORMS: PCM, DPCM, AND DM QUANTIZERS. Nuggehally S. Jayant in *Proceedings of the IEEE,* Vol. 62, No. 5, pages 611–632; May, 1974.

OPTICAL TRANSMISSION OF VOICE AND DATA. Ira Jacobs and Stewart E. Miller in *IEEE Spectrum,* Vol. 14, No. 2, pages 32–41; February, 1977.

SPECIAL ISSUE: THE *1A* PROCESSOR. *The Bell System Technical Journal,* Vol. 56, No. 2; February, 1977.

SINGLE-SLICE SUPERHET. William Peil and Robert J. McFadyen in *IEEE Spectrum,* Vol. 14, No. 3, pages 54–57; March, 1977.

MICROELECTRONICS AND COMPUTER SCIENCE

THE SHORTEST PATH THROUGH A MAZE. Edward F. Moore in *The Annals of the Computation Laboratory of Harvard University: Vol. XXX, Proceedings of an International Symposium on the Theory of Switching, Part II.* Harvard University Press, 1959.

FUNDAMENTAL LIMITATIONS IN MICROELECTRONICS, I: MOS TECHNOLOGY. B. Hoeneisen and Carver A. Mead in *Solid-State Electronics,* Vol. 15, No. 7, pages 819–829; July, 1972.

LIMITATIONS IN MICROELECTRONICS, II: BIPOLAR TECHNOLOGY. B. Hoeneisen and Carver A. Mead in *Solid-State Electronics,* Vol. 15, No. 8, pages 891–897; August, 1972.

HOW BIG SHOULD A PRINTED CIRCUIT BOARD BE? Ivan E. Sutherland and Donald Oestreicher in *IEEE Transactions on Computers,* Vol. C-22, No. 5, pages 537–542; May, 1973.

MICROELECTRONICS AND THE PERSONAL COMPUTER

TOWARDS A THEORY OF INSTRUCTION. Jerome S. Bruner. Belknap Press of Harvard University Press, 1966.

ARTIFICIAL INTELLIGENCE. Seymour A. Papert and Marvin Minsky. Condon Lectures, Oregon State System of Higher Education, 1974.

PERSONAL DYNAMIC MEDIA. Alan C. Kay and Adele Goldberg in *Computer,* Vol. 10, No. 3, pages 31–41; March, 1977.

INDEX